Synthese Library

Studies in Epistemology, Logic, Methodology, and Philosophy of Science

Volume 508

The aim of *Synthese Library* is to provide a forum for the best current work in the methodology and philosophy of science and in epistemology, all broadly understood. A wide variety of different approaches have traditionally been represented in the Library, and every effort is made to maintain this variety, not for its own sake, but because we believe that there are many fruitful and illuminating approaches to the philosophy of science and related disciplines.

Special attention is paid to methodological studies which illustrate the interplay of empirical and philosophical viewpoints and to contributions to the formal (logical, set-theoretical, mathematical, information-theoretical, decision-theoretical, etc.) methodology of empirical sciences. Likewise, the applications of logical methods to epistemology as well as philosophically and methodologically relevant studies in logic are strongly encouraged. The emphasis on logic will be tempered by interest in the psychological, historical, and sociological aspects of science. In addition to monographs *Synthese Library* publishes thematically unified anthologies and edited volumes with a well-defined topical focus inside the aim and scope of the book series. The contributions in the volumes are expected to be focused and structurally organized in accordance with the central theme(s), and should be tied together by an extensive editorial introduction or set of introductions if the volume is divided into parts. An extensive bibliography and index are mandatory.

Carlo Cellucci

A Philosophy for the Future

Escaping from Irrelevance

Springer

Carlo Cellucci
Department of Philosophy
Sapienza University of Rome
Rome, Italy

ISSN 0166-6991 ISSN 2542-8292 (electronic)
Synthese Library
ISBN 978-3-031-98442-6 ISBN 978-3-031-98443-3 (eBook)
https://doi.org/10.1007/978-3-031-98443-3

This Springer imprint is published by the registered company Springer Nature Switzerland AG
The registered company address is: Gewerbestrasse 11, 6330 Cham, Switzerland

Preface

The present condition of philosophy is a peculiar one. On the one hand, there have never been so many professional philosophers as today, on the other hand, philosophy has never been considered so irrelevant.

This is because the dominant philosophy today, analytic philosophy, assumes that philosophy is not and should not be acquisition of knowledge, that is, an inquiry aimed at acquiring knowledge of the world, because acquiring knowledge of the world is the purpose of science, not philosophy. Philosophy aims only to understand what we already know.

The abandonment of the view that philosophy is acquisition of knowledge has had a very negative effect on philosophy. Philosophy has become more and more inward-looking and self-referential, of no interest to people working in other areas, and to cultured people at large.

To overcome the present impasse of philosophy, this book proposes a view of philosophy as acquisition of knowledge, according to which philosophy can open new ways to knowledge and lead to the birth of new sciences. This is what philosophy has repeatedly done until recently, and there is no reason to think that philosophy will not do it again.

Rome, Italy

Carlo Cellucci

Acknowledgments

The views presented in this book are a development of views introduced in my books *Perché ancora la filosofia*, Laterza, Rome 2008; *Rethinking logic: Logic in relation to mathematics, evolution, and method*, Springer, Cham 2013; *Rethinking knowledge: The heuristic view*, Springer, Cham 2017; and *The making of mathematics: Heuristic philosophy of mathematics*, Springer, Cham 2022.

Chapters 2, 3, and 4 use material from my articles "Philosophy at a crossroads: Escaping from irrelevance," *Syzetesis* 5 (2018), 13–53; "The most urgent task of philosophy today," *Borderless Philosophy* 2 (2019), 45–75; and "Philosophy, discovery, and advancement of knowledge," *Syzetesis* 8 (2021), 9–32.

Chapter 17 uses material from my article "The two paradigms of logic," in Emiliano Ippoliti and Fabio Sterpetti (eds.), *The heuristic view: Logic, mathematics, and science*, Springer, Cham 2025, 25–47.

For their help, in whatever form, I am especially indebted to Mirella Capozzi, Miriam Franchella, Marshall Gordon, Andreas Stephens, and Fabio Sterpetti.

This does not mean that they endorse any of the views expressed in this book or are responsible for any remaining inaccuracies.

Competing Interests The author has no competing interests to declare that are relevant to the content of this manuscript.

Contents

Chapter 1
Introduction

Abstract The great developments in the sciences since the nineteenth century have created the need to find a new role for philosophy. Several attempts have been made in this direction, but they have failed. This raises the question: Can philosophy be useful again, and what type of philosophy can be such? The main thesis of the book is that philosophy can be useful again if and only if it is acquisition of knowledge, that is, an inquiry aimed at acquiring knowledge of the world. The task of the book is to formulate the view that philosophy is acquisition of knowledge, to show that the arguments against it are not valid, and to develop the view by considering several questions about knowledge.

Keywords Philosophy as the mother of all sciences · The claim that philosophy is dying · Failure of the attempts to find a new role for philosophy · The question of whether philosophy has been replaced by the sciences · The view that philosophy is acquisition of knowledge

1.1 Philosophy as the Mother of All Sciences

From antiquity to the modern age, the role of philosophy in relation to science has been considered important. It has even been said that philosophy is the mother of all sciences.

Thus, Cicero says that philosophy is "the mother of all arts" (Cicero, *Tusculanae disputationes*, I.64).

Bacon says that "natural philosophy has to be regarded as the great mother of the sciences" (Bacon 2004, 64).

However, with the great developments in the sciences since the nineteenth century, this way of considering philosophy has changed, it has even been thought that philosophy is dying.

C. Cellucci, *A Philosophy for the Future*, Synthese Library 508,
https://doi.org/10.1007/978-3-031-98443-3_1

Thus, Bona Meyer says: “Philosophy as a science seems to be over, its solvable problems have now become the tasks of particular sciences. The daughters of the common mother now demand independence and no longer want to be supervised or corrected; they would prefer that the old and grumpy mother finally lay down in peace. Many also think that philosophy is already in its last stages and is preparing to die” (Bona Meyer 1874, 1).

1.2 The Claim That Philosophy Is Dying

In fact, some eminent philosophers have argued that philosophy is dying, but their arguments are not valid.

(1) Heidegger argues that “the sciences are now taking over as their own task what philosophy in the course of its history tried to present in part, and even there only inadequately, that is, the ontologies of the various regions of beings (nature, history, law, art)” (Heidegger 1972, 58). The “development of philosophy into the independent sciences” is “the legitimate completion of philosophy. Philosophy is ending in the present age. It has found its place in the scientific attitude of socially active humanity” (ibid.). The “end of philosophy means the completion” of philosophy, where “however, completion does not mean perfection as a consequence of which philosophy would have to have attained the highest perfection at its end” (ibid., 56). Instead, completion means “the end of philosophy in the sense of its complete dissolution into the sciences” (Heidegger 1998, 259, footnote).

But this argument is not valid. It assumes that all knowledge will be given by the present sciences, but this assumption is untenable. For, there are areas of experience that the present sciences cannot address. To investigate these areas, new sciences are necessary, and philosophy can lead to their birth, as it has done until recently (see Chap. 4).

(2) Quine argues that philosophy, in particular epistemology, has dissolved into the natural sciences because, while “the old epistemology aspired to contain, in a sense, natural science; it would construct it from sense data,” epistemology in its new setting “is contained in natural science, as a chapter of psychology” (Quine 1969, 83).

But this argument is not valid. Knowledge is based on faculties of the body that are implemented not only by processes internal to the body but also by processes external to it, which cooperate with those internal to the body forming an integrated cognitive system with them. The integration of internal processes with external ones is a continuing process, because new external processes are always being created, and the integration gives rise to more powerful cognitive systems. Psychology does not account for this fact, therefore epistemology cannot be contained in natural science as a chapter of psychology (see Chap. 13).

From this it seems fair to conclude that the claim that philosophy is dying is not valid.

1.3 Attempts to Find a New Role for Philosophy

Although the claim that philosophy is dying is not valid, the great developments in the sciences since the nineteenth century have created the need to find a new role for philosophy.

(1) Some philosophers have argued that traditional philosophy is obsolete, the new philosophy consists in the unified science of the present sciences.

Thus, Neurath argues that "philosophy as an independent system of definite doctrines is obsolete" (Neurath 1983, 49). There is only "one science with subdivisions—a unified science of sciences" (ibid., 48). The "unified science replaces all former philosophy" (ibid., 56). What "can not be regarded as unified science must be accepted as poetry or fiction" (ibid., 49). Some "still wish to separate the discussions of the conceptual foundations of the sciences from the body of scientific work and allow this to continue as 'philosophising'," but this "is not feasible," the discussion of the conceptual foundations of the sciences "is part and parcel of the work of unified science" (ibid., 52).

This, however, is based on the assumption that all knowledge of the world can be given by the unified science of the present sciences. This assumption is not valid because there are many questions about the world that the present sciences are unable to address, there is no reason to think that these questions admit of no definite answer, and that philosophy will not lead to answer them.

(2) Other philosophers have argued that, if philosophy is to survive, it must be built as a science.

Thus, Husserl argues that "philosophy, according to its historical purpose the loftiest and most rigorous of all sciences, representing as it does humanity's imperishable demand for pure and absolute knowledge," is still "incapable of assuming the form of rigorous science" (Husserl 1965, 72). Philosophy "does not have at its disposal a merely incomplete" and "imperfect doctrinal system; it simply has none whatever. Each and every question is herein controverted, every position is a matter of individual conviction, of the interpretation given by a school, of a 'point of view'" (ibid., 74–75). But "the highest interests of human culture demand the development of a rigorously scientific philosophy; consequently, if a philosophical revolution in our time is to be justified, it must without fail be animated by the purpose of laying a new foundation for philosophy in the sense of strict science" (ibid., 78).

This, however, contrasts with the fact that all attempts to build philosophy as a science have failed. In particular, Husserl tries to build philosophy as a "universal science of the world, universal, definitive knowledge, the universe of truths in themselves about the world, the world in itself" (Husserl 1970, 335). To this purpose, he starts from the constitutive activity of the self, with the intent of arriving at the "discovery of the necessary concrete manner of being of absolute (in the ultimate sense, transcendental) subjectivity in a transcendental life of constant 'world-constitution'; the new discovery, correlative to this, of the 'existing world', whose ontic meaning, as transcendentally constituted, results in a new meaning for what, in the earlier stages, was called world, world-truth, world-knowledge" (ibid., 340).

But, starting from the constitutive activity of the Self, Husserl's supposed universal science of the world remains confined to the realm of subjectivity, without ever arriving at the new discovery of the existing world.

(3) Other philosophers have argued that science has no privileged position over the other areas of culture, the validity of all areas of culture must not be evaluated in terms of their ability to produce knowledge, but in terms of their ability to achieve the goals we want to achieve with them.

Thus, Rorty argues that we must abandon the view of the logical positivists that "science was the paradigmatic human activity," and "what little there was to say about other areas of culture amounted to a wistful hope that some of them (e.g., philosophy) might themselves become more 'scientific'" (Rorty 1991–2007, I, 46). Science and the other areas of culture, including philosophy, are parts of the same web, which cannot be "divided into the bit which limns the true structure of reality and the part which does not" (ibid., I, 51). Their validity must not be evaluated in terms of "a relation between beliefs and objects called 'correspondence'" (ibid., I, 22). Instead, it must be evaluated in terms of their ability to achieve the goals we want to achieve with them, because "we continue to hold the beliefs which prove to be reliable guides to getting what we want" (Rorty 1999, 33).

This, however, contrasts with the fact that the description of the goals these philosophers want to achieve with philosophy is vague and indeterminate. Thus, Rorty says: "When I am asked" what "I take contemporary philosophy's 'mission' or 'task' to be," the best "I can do is to stammer that we philosophy professors are people who have a certain familiarity with a certain intellectual tradition," so "we can offer some advice about what will happen when you try to combine or to separate certain ideas, on the basis of our knowledge of the results of past experiments" (ibid., 19–20). But "we are not the people to come to if you want confirmation that the things you love with all your heart are central to the structure of the universe" (ibid., 20). That is, we philosophy professors are not the people to come to if you want to have an answer to the questions that interest you most.

(4) Other philosophers have argued that science is the denial of all knowledge of truth, only philosophy can give knowledge of truth because only philosophy is knowledge of the essence of things.

Thus Heidegger says that "science is the disavowal of all knowledge of truth," therefore "no one who knows will envy scientists—the most miserable slaves of modern times" (Heidegger 1994, 5–6). Science aims at exactness and security, but its "insistence on what is demonstrable" blocks "the way to what-is" (Heidegger 1972, 72). Only philosophy opens the way to what-is, because "philosophy is the knowledge of the essence of things" (Heidegger 1994, 29). This produces a leap that brings us "to a completely different place," and "what becomes clear with it can never be demonstrated, if demonstrating means: deducing propositions about a state of affairs from appropriate premisses through a chain of inferences" (Heidegger 2000, 133).

This, however, contrasts with Heidegger's above claim that the development of philosophy into the independent sciences is the legitimate completion of philosophy. Moreover, it contrasts with the fact that there is no evidence that there exists a higher source of knowledge that provides knowledge of the essence of things, let alone that philosophy is such a higher source of knowledge.

From this it seems fair to conclude that the attempts to find a new role for philosophy have failed.

1.4 Main Thesis of the Book

The failure of the attempts to find a role for philosophy raises the question: Can philosophy be useful again, and what type of philosophy can be such?

The main thesis of the book is that philosophy can be useful again if and only if it is acquisition of knowledge, that is, an inquiry aimed at acquiring knowledge of the world.

Given the great developments in the sciences since the nineteenth century, this thesis may seem peculiar. For, these developments have led to the belief that only the present sciences can be acquisition of knowledge.

But the thesis is not peculiar because, although the present sciences answer many questions about the world, they do not answer many other questions about the world, questions that are beyond the present sciences, in the sense that they are not open questions of any of the present sciences.

Only philosophy can seek and address these questions. For, scientists today are busy tackling major challenges in their own fields, so they cannot be expected to seek and address questions that are beyond their fields.

1.5 Task of the Book

The thesis that philosophy can be useful again if and only if it is acquisition of knowledge sets the task of the book: to formulate the view that philosophy is acquisition of knowledge, to show that the arguments against it are not valid, and to develop the view by considering several questions about knowledge.

This task is essential for philosophy, because philosophy has never been considered so irrelevant.

This is because the dominant philosophy today, analytic philosophy, assumes that philosophy is not and should not be acquisition of knowledge, that is, an inquiry aimed at acquiring knowledge of the world. Acquiring knowledge of the world is the aim of science, not of philosophy, philosophy aims only to understand what we already know.

This assumption has had a very negative effect on philosophy.

As Dennett says, philosophy "has become self-indulgent, clever play in a vacuum that's not dealing of problems of any intrinsic interest," it is "willfully cut off from any serious issues," it deals with questions that are "idle—just games," and therefore it is little more than a "luxury decoration on society" (Goldhill 2016).

The irrelevance of philosophy is not, however, an inevitable fate of philosophy. It is simply the result of the assumption of analytic philosophy that philosophy is not acquisition of knowledge, it aims only to understand what we already know.

Conversely, if philosophy is acquisition of knowledge, philosophy can escape from irrelevance because it can seek and address questions about the world that are beyond the present sciences. Then, if successful, philosophy can open new ways to knowledge and lead to the birth of new sciences.

1.6 Inadequacy of a Reformation

Even some eminent analytic philosophers recognize that there is something deeply wrong with analytic philosophy.

Thus, Putnam says that "the present situation in philosophy is one that calls for a revitalization, a renewal, of the subject" (Putnam 1992, ix).

Hintikka says that "the survival of analytic philosophy" is at stake, "we have to make a new start in practically all branches of philosophical studies" (Hintikka 1998, 260).

As Preston observes, "what Putnam and Hintikka are calling for" is "a reformation of analytic philosophy" (Preston 2004, 447). The "goal of reformation—as opposed to revolution, say—is to preserve and better some system or order (an institution, a tradition, and so on) by eliminating those of its characteristics that stand in the way of its being what it ought" (Preston 2004, 447).

In calling for a reformation of analytic philosophy, Putnam and Hintikka take it for granted that what is deeply wrong with analytic philosophy can be eliminated without altering the characterizing features of analytic philosophy.

But this is an illusion. For, what is deeply wrong with analytic philosophy is the assumption that philosophy is not acquisition of knowledge, it aims only to understand what we already know. Therefore this assumption must be abandoned. But this assumption is at the core of analytic philosophy, therefore, to abandon it is to abandon analytic philosophy itself.

1.7 A Special Acknowledgment

In the book I repeatedly refer to Plato. A direct reason for this is that Plato first explicitly formulated the view that philosophy is acquisition of knowledge (see Chap. 4).

An indirect reason is that I was originally a mathematical logician working on proof theory, for example, sequent natural deduction (see Cellucci 1992). But, over time, I became disillusioned with the ability of mathematical logic to account for the making of mathematics, and science in general.

Then I read an article by Cambiano, who argued that "for Plato, unlike for Aristotle, a hypothesis is not a principle and is not unprovable, a reason can be given for it" (Cambiano 1967, 147). A "hypothesis is used in the search for solutions to problems: as such the hypothesis is a condition of solvability" (ibid., 138). But the hypothesis is in its turn a problem, one must "account for the hypothesis by resorting to another hypothesis for this purpose" (ibid., 139). And so on. As a model for philosophy, Plato had "before him a method formally identical to that used by Hippocrates of Chios to solve the problem of the quadrature of the lunules: this is the hypothetical method, an analytic-reductive method, not a deductive one" (ibid., 136).

Reading Cambiano's article led me to reconsider the nature and method of mathematics, science, and philosophy.

In this process, I found several points of agreement with Plato, but also several points of disagreement, even on important issues. Therefore, by referring to Plato, I do not intend to contribute to the Platonic tradition or offer a new historically grounded interpretation of Plato. I simply want to acknowledge that certain views of Plato have been a source of inspiration for my views on the nature and method of mathematics, science, and philosophy.

1.8 Organization of the Book

The book is divided into three parts after the present Introduction, which occurs as Chap. 1.

Part I, The Ways of Philosophy, analyzes the present condition of irrelevance of philosophy, and argues that philosophy must be acquisition of knowledge.

Chapter 2 argues that analytic philosophy has special responsibility for the present condition of irrelevance of philosophy, because it assumes that philosophy is not acquisition of knowledge, it aims only to understand what we already know. Chapter 3 discusses some arguments of analytic philosophy for this view and argues that they are not valid. Chapter 4 argues that philosophy can escape from irrelevance if and only if it is acquisition of knowledge. Chapter 5 considers several arguments against the view that philosophy is acquisition of knowledge and argues that they are not valid. Chapter 6 considers the methods used in analytic philosophy and argues that they are inadequate. Chapter 7 considers the question of utility of philosophy. Chapter 8 considers the question of progress in philosophy. Chapter 9 discusses some alternative views of philosophy as aimed at acquiring knowledge of the world.

Part II, The Ways of Knowledge, considers several questions about knowledge necessary to develop the view that philosophy is acquisition of knowledge.

Chapter 10 discusses the nature of knowledge. Chapter 11 considers the relation of knowledge to reality. Chapter 12 considers the relation of knowledge to objectivity. Chapter 13 considers the relation of knowledge to mind. Chapter 14 discusses the question of whether acquisition of knowledge, in particular discovery, is a rational process. Chapter 15 argues that the method of discovery is the analytic method. Chapter 16 discusses the role of non-deductive rules and deductive rules in knowledge. Chapter 17 discusses the role of logic in knowledge. Chapter 18 argues that knowledge is problem solving by the analytic method, and therefore so is philosophy as acquisition of knowledge. Chapter 19 considers the nature of perception. Chapter 20 considers the nature of explanation. Chapter 21 considers the relation of knowledge to emotions. Chapter 22 considers the relation of knowledge to naturalization.

Part III, Conclusion, completes and concludes the book.

Chapter 23 recapitulates and makes some concluding remarks on the main thesis of the book.

1.9 General Remarks

The following are general remarks about the book.

(1) The term ‘analytic philosophy’ is not univocal. By it I mean general views held by acknowledged exponents of analytic philosophy.

(2) I make several references to Gödel’s incompleteness theorems and other limitative results. Details about them can be found in many books, and in any case in Cellucci 2022.

(3) Instead of ‘he or she’ I use the generic ‘he’, specifying that by it I mean to refer to persons of both genders.

(4) When I quote from ancient Greek philosophers, and also from some modern philosophers, translations are mine unless otherwise indicated.

References

Bacon, Francis. 2004. *The new organon*. Cambridge: Cambridge University Press.

Bona Meyer, Jürgen. 1874. *Philosophische Zeitfragen: Populäre Aufsätze*. Bonn: Adolph Marcus.

Cambiano, Giuseppe. 1967. Il metodo ipotetico e le origini della sistemazione euclidea della geometria. *Rivista di Filosofia* 58:115–149.

Cellucci, Carlo. 1992. Existential instantiation and normalization in sequent natural deduction. *Annals of Pure and Applied Logic* 58:111–148.

Cellucci, Carlo. 2022. *The theory of Gödel*. Cham: Springer.

Goldhill, Olivia. 2016. One of the most famous living philosophers says much of philosophy today is ‘self-indulgent’. *Quartz*, 28 August 2016. https://qz.com/768450/one-of-the-most-famous-living-philosophers-says-much-of-philosophy-today-is-self-indulgent.

Heidegger, Martin. 1972. *On time and being*. New York: Harper.

Heidegger, Martin. 1994. *Basic questions of philosophy: Selected ‘problems’ of ‘logic’*. Bloomington: Indiana University Press.

Heidegger, Martin. 1998. *Pathmarks*. Cambridge: Cambridge University Press.

Heidegger, Martin. 2000. *Vorträge und Aufsätze*. Frankfurt: Klostermann.

Hintikka, Jaakko. 1998. Who is about to kill analytic philosophy? In *The story of analytic philosophy: Plot and heroes*, ed. Anat Biletzki and Anat Matar, 253–269. London: Routledge.
Husserl, Edmund. 1965. Philosophy as rigorous science. In Edmund Husserl, *Phenomenology and the crisis of philosophy*, 71–147. New York: Harper.
Husserl, Edmund. 1970. *The crisis of European sciences and transcendental phenomenology*. Evanston: Northwestern University Press.
Neurath, Otto. 1983. *Philosophical papers 1913–1946*. Dordrecht: Springer.
Preston, Aaron. 2004. Prolegomena to any future history of analytic philosophy. *Metaphilosophy* 35:445–465.
Putnam, Hilary. 1992. *Renewing philosophy*. Cambridge: Harvard University Press.
Quine, Willard. 1969. *Ontological relativity and other essays*. New York: Columbia University Press.
Rorty, Richard. 1991–2007. *Philosophical papers*. Cambridge: Cambridge University Press.
Rorty, Richard. 1999. *Philosophy and social hope*. London: Penguin.

Part I
The Ways of Philosophy

Chapter 2
Philosophy at a Crossroads

Abstract There have never been so many professional philosophers as today, yet philosophy has never been considered so irrelevant. This chapter argues that analytic philosophy has special responsibility for this, because it assumes that philosophy is not acquisition of knowledge, that is, an inquiry aimed at acquiring knowledge of the world, philosophy aims only to understand what we already know. As a result, philosophy has become more and more inward-looking and self-referential, of no interest to people working in other fields, and to cultured people in general.

Keywords Irrelevance of philosophy · Multiplication of philosophers · Special responsibility of analytic philosophy · An ineffectual escape from irrelevance

2.1 The Irrelevance of Philosophy

The present condition of philosophy is a peculiar one. On the one hand, there have never been so many professional philosophers as today, on the other hand, philosophy has never been considered so irrelevant.

In fact, most of the questions addressed by professional philosophers today are of interest only to academics working in one small corner of philosophy, not to those working in other corners of philosophy, let alone to people working in other fields and to cultured people in general.

This is not healthy for philosophy because, although a discipline may exist for some time even with a limited audience, this will put its long-term survival at risk, at least in academic institutions.

Then, philosophy is at a crossroads: either continue on the present line, which relegates it to irrelevance, or analyze the reasons for irrelevance and set up an escape from it.

C. Cellucci, *A Philosophy for the Future*, Synthese Library 508,
https://doi.org/10.1007/978-3-031-98443-3_2

2.2 The Multiplication of Philosophers

That there have never been so many professional philosophers as today is clear, for example, from the fact that the American Philosophical Association, which in 1920 had 260 members, has over 10,000 members.

If we add the members of the philosophical associations of all other countries, we have a total number of philosophers of some tens of thousands, presumably greater than the number of philosophers that existed from antiquity to the nineteenth century.

That there have never been so many professional philosophers as today is also clear from the fact that philosophy conferences are held virtually every day all over the world. This explains why most philosophy books published today are edited collections rather than monographs.

2.3 Scientists and Mathematicians on the Irrelevance of Philosophy

On the other hand, that philosophy has never been considered so irrelevant is clear, for example, from the attacks of many scientists and mathematicians on contemporary philosophy.

Thus, Dirac says: "The field of philosophy has terribly declined. I feel that philosophy will never lead to important discoveries. It's just a way of talking about discoveries which have already been made" (Dirac 1963).

Weinberg says: "I know of no one who has participated actively in the advance of physics in the postwar period whose research has been significantly helped by the work of philosophers" (Weinberg 1993, 168–169). Philosophy of science "at its best seems to me a pleasing gloss on the history and discoveries of science. But we should not expect it to provide today's scientists with any useful guidance" (ibid., 167). At most, "the work of some philosophers helps us to avoid the errors of other philosophers" (ibid., 168).

Rota says: "The classical problems of philosophy have become forbidden topics in many philosophy departments," and "in this dictatorial regime we have witnessed the shrinking of philosophical activity to an impoverished 'problématique'" (Rota 1997, 98). In fact, "our latter-day philosophers are not concerned with facing up" to "any relevant features" whatsoever "of the world" (ibid., 102–103). Therefore, "like ostriches with their heads buried in the sand, they will meet the fate of those who refuse to remember the past and fail to face the challenges of our difficult present: increasing irrelevance followed by eventual extinction" (ibid., 103).

Mullis says: "Chemists always believe they're smarter than biochemists. Of course, physicists think they're smarter than chemists, mathematicians think they're smarter than physicists" (Mullis 1998, 38). And, "for a while, philosophers thought they were smarter than mathematicians, until they found out in this century that they really didn't have anything much to talk about" (ibid.).

Wolpert says: "No scientist that I know of" has "the slightest interest in the philosophy of science," because the philosophy of science "has contributed zero to the understanding of the scientific process" (Wolpert 2000). Scientists "are very ambitious. They're very competitive. If they really thought philosophy would help them, they'd learn it and use it. They don't" (ibid.).

Hersh says: "A famous mathematician said to me, 'I am willing to leave that question to the philosophers'. Which philosophers? Professional philosophers who are not mathematicians?! To obtain answers meaningful to us, I'm afraid we'll have to get to work ourselves" (Hersh 2001, 4).

Crick says: "If you ask how many cases in the past has a philosopher been successful at solving a problem, as far as we can say there are no such cases. Essentially, their main technique is the thought experiment, and here you can argue indefinitely" (Crick 2005, 74).

Gowers says: "Suppose a paper were published tomorrow that gave a new and very compelling argument for some position in the philosophy of mathematics," and the "argument caused many philosophers" to "embrace a whole new -ism" (Gowers 2006, 198). Then, "what would be the effect on mathematics? I contend that there would be almost none" (ibid.). For, "the questions considered fundamental by philosophers are the strange, external ones that seem to make no difference to the real, internal business of doing mathematics" (ibid.).

Feynman says: "Philosophy is about as useful to scientists as ornithology is to birds; philosophy of science, as ornithology, is a civilizing hobby but of no real use to its objects" (Pernu 2008, 30). Scientists discover new sites, philosophers only visit them, so "scientists are explorers, philosophers are tourists" (Sykes 1994, 260),

Hawking and Mlodinow say: "How can we understand the world in which we find ourselves? How does the universe behave? What is the nature of reality? Where did all this come from? Did the universe need a creator?" (Hawking and Mlodinow 2010, 5). Traditionally, "these are questions for philosophy, but philosophy is dead. Philosophy has not kept up with modern developments in science" (ibid.). While, "in the eighteenth century, philosophers considered the whole of human knowledge, including science, to be their field," afterwards they "reduced the scope of their inquiries" (Hawking 1988, 185). So much so that Wittgenstein "said, 'The sole remaining task for philosophy is the analysis of language'. What a comedown from the great tradition of philosophy from Aristotle to Kant!" (ibid.).

Krauss says: "Science progresses and philosophy doesn't," and "the worst part of philosophy is the philosophy of science; the only people" who "read work by philosophers of science are other philosophers of science. It has no impact on physics whatsoever," so "it's really hard to understand what justifies it" (Krauss 2012).

Dyson says: "For most of the twenty-five centuries since written history began, philosophers were important" and, "until the end of the nineteenth century, philosophers were giants playing a dominant role in the kingdom of the mind" (Dyson 2015, 243). But, "compared with the giants of the past," the present philosophers "are a sorry bunch of dwarfs," they "are historically insignificant. At some time toward the end of the nineteenth century, philosophers faded from public life" and "suddenly and silently vanished. So far as the general public was concerned, philosophers became invisible" (ibid.). This raises the question: "When and why did philosophy lose its bite? How did it become a toothless relic of past glories?" (ibid.).

These attacks of scientists and mathematicians on contemporary philosophy abundantly show the level of irrelevance to which philosophy has fallen.

2.4 Philosophers' Reactions to the Irrelevance of Philosophy

Several philosophers treat the attacks of scientists and mathematicians on contemporary philosophy with condescension.

Thus, Crane argues that, when Hawking and Mlodinow claim that philosophy is dead, they "cannot mean that the activity of philosophising is dead," because their book "itself is full of philosophy. In fact, it is probably Hawking's most philosophical book" (Crane 2017, 177). But "unfortunately, much of the book's own philosophical argument is of a very low standard, and shows a striking lack of reflection on the complexities of what is being claimed" (ibid.). Therefore, "on the evidence of Hawking and Mlodinow's book, the situation is actually the opposite of the way they describe it: it is the scientists who have not kept up with developments in philosophy" (ibid., 179).

This argument, however, is an example of ad hominem fallacy. For, instead of responding to the charge that the discipline of academic philosophy is dead, Crane seeks to discredit the philosophical abilities of Hawking and Mlodinow. But in doing so he overlooks that discrediting the messenger does not discredit the message.

While several philosophers treat the attacks of scientists and mathematicians on contemporary philosophy with condescension, some philosophers are aware that philosophy is considered irrelevant today.

Thus, Kreisel says: "A kind of paradox: high hopes for (the subject of) philosophy" and "little trust" in "the work of contemporary professional philosophers," in particular in "the logic chopping and obviously minor distinctions of which contemporary (Anglo-Saxon) philosophy is full" (Kreisel 1967, 212). Such work focuses on insignificant aspects, and "there is no evidence that careful work on insignificant aspects leads one, even indirectly, to recognize what is essential" (ibid.).

Bunge says: "There seems to be consensus that philosophy is currently at a low ebb. Some even claim that it is dead" (Bunge 2001, 207). But philosophy is not dead, "it is stagnant. In fact few if any radically new and correct philosophical ideas, let alone systems, are being proposed. Gone are the days of exciting new and grand philosophical ideas that spilled over other disciplines," today "most philosophers teach, analyze, comment on, or embellish other scholars' ideas" (ibid.). Others "play frivolous if ingenious academic games. Few philosophers think on a large scale," therefore "philosophy is in crisis" (ibid., 208).

Pinker says: "Philosophy today gets no respect. Many scientists use the term as a synonym for effete speculation" (Pinker 2003, 11). When "Ned Block told his father that he would major in the subject, his father's reply was 'Luft!'—Yiddish for 'air'. And then there's the joke in which a young man told his mother he would become a Doctor of Philosophy and she said, 'Wonderful! But what kind of disease is philosophy?'" (ibid.).

Mulligan, Simons, and Smith say: "Analytic philosophers" believe "neither that philosophy can be a science nor that it can add to the stock of positive human knowledge" (Mulligan et al. 2006, 64). They pursue philosophy "on the basis of the attitude of 'horror mundi'" (ibid.). In "science results are expected. In analytic philosophy everyone waits for the next new puzzle," analytic philosophy "is at its core a culture driven by puzzles," in fact "the quickest way to a career in modern analytic philosophy is to pick a puzzle in a trendy area" and "come up with a hitherto unsuspected twist in the dialectic" (ibid., 65). But "a philosophical tradition which suffers from the vice of 'horror mundi' in an endemic way is condemned to futility" (ibid.).

Preston says: "Analytic philosophy" presents "us with a double peculiarity," a "philosophical peculiarity" and "a sociological peculiarity" (Preston 2007, 2–3). The philosophical peculiarity is that "the analytic philosophy's original philosophical outlook initially was thought to be so promising that it generated a revolution in Anglo-American philosophy, but it came to be seen as so badly flawed that it was completely abandoned" (ibid., 3). The sociological peculiarity is that analytic philosophy "has dominated academic philosophy in at least the United States and Britain since at least the second quarter of the twentieth century," and "has maintained its dominant position even though the original reason for its dominance has long since been undermined" (ibid.).

Boghossian says: "The most interesting thing about philosophy today is how uninteresting and largely irrelevant it has become. The overwhelming majority of professional philosophers deal with issues no one outside of their sub-disciplines care about, and use language few outside of their specialties understand" (Boghossian 2016, 29). Philosophers "relentlessly pursue topics about which almost nobody cares, and professionally reward obfuscation and insularity" (ibid.). They "view pedantry not as problem to be overcome, but as a virtue to which less seasoned philosophers should aspire" (ibid.).

Haack says: "Something is rotten in the state of philosophy" (Haack 2017, 40). Over the years, "philosophy has become more and more out of touch with its own history, more and more hyper-specialized, more and more fragmented into cliques, niches, cartels, and fiefdoms, and more and more dominated by intellectual fads and fashions" (ibid.). The debates in these "niches have a shorter and shorter shelf-life, as questions once fashionable simply fade from view when people get bored with them" (Haack 2021, 41). Rarely are "such passing problems resolved; rather, the person from whose theory they derived dies, or retires, or fades from view, and people just move on to yet another passing fad" (ibid.).

Gaukroger says: "Analytic philosophy in the twentieth century increasingly became a matter of resolving self-generated problems, so that motivation collapsed into the philosophical programme itself, as the exercise became increasingly remote and fruitless, and increasingly removed from effective engagement with concerns outside its own quasi-Platonic realm" (Gaukroger 2020, 15).

Kitcher says: "Much analytic philosophy does little to advance the important goals aimed at—and sometimes achieved—in the history of western philosophy" (Kitcher 2024a, 521). Analytic philosophers "labor over questions without answering them, spend hours in heated debates about very little (if anything)," so they appear to engage in "pointless jousts" (Kitcher 2023, ix). They reshape ordinary concepts without asking what their point is. But, "when philosophers no longer ask after the point of a concept, they not only devote themselves to exploring bizarre fantasies of no practical significance; they also lose sight of possibilities of conceptual reform" (ibid., 62). As a result, "philosophy is currently an isolated discipline within a larger cluster of academic fields" (ibid., 178).

Frodeman says: "Twentieth century philosophy has been unhealthily insular" (Frodeman 2013, 1918). Philosophers "became experts like other disciplinary specialists" (Frodeman and Briggle 2016, 7). They developed "sub-specializations far from the comprehension of the person on the street" (ibid., 9). But a philosophy "where philosophers primarily work with and write for other philosophers, is in the end no philosophy at all" (Frodeman 2013, 1935). This "is part of what has led philosophy, potentially the most relevant of subjects, to become a synonym for irrelevance" (ibid., 1918).

That, as Frodeman says, philosophy, potentially the most relevant of subjects, has become a synonym for irrelevance, should seriously worry philosophers and convince them that there is something deeply wrong with the present approach to philosophy.

Aristotle hailed philosophy as "the supreme among sciences" (Aristotle, *Metaphysica*, A 2, 982 b 4). And the "most divine and most worthy" (ibid., A 2, 983 a 5). He even said: "One ought either to philosophize or say goodbye to life and depart hence," because, compared with philosophy, "all other things seem to be a lot of chatter and nonsense" (Aristotle, *Protrepticus*, fr. 110, ed. Düring).

Who would dare say anything even remotely similar about today's philosophy?

As Bobbio humorously says, the irrelevance of philosophy risks proving right "Xanthippe, Socrates' wife, who has gone down in tradition as the wife that does not understand why her husband wastes his time discussing philosophy, getting lost in discussions that are of no use. And what if she were right?" (Bobbio 2017, 45).

2.5 Alternatives to the Image of the Great Philosopher

That there have never been so many professional philosophers as today raises the question: What is the view of philosophy best suited to philosophers in this situation?

In this regard, Marconi observes that, while the traditional image of the philosopher was that of "the Great Philosopher," in the past century there was a multiplication of philosophers, so "the image of the Great Philosopher tended to tarnish" (Marconi 2014, 18). For, how could there be "dozens and then hundreds of personalities capable of so much? Wasn't there something ridiculous about this infinite multiplication of world systems, 'allgemeine Wissenschaftslehren', revelations of the essence of things?" (ibid.). So, "it was necessary to find ways to make philosophy a task within the reach of scholars perhaps educated and intelligent, but not necessarily geniuses nor astonishingly original," and to make the professional philosopher more into "a competent artisan" than into "an architect of cathedrals" (ibid., 18–19).

According to Marconi, there have been three main solutions to "the problem of transforming the Great Philosopher into a philosophy professional" (ibid., 60).

The first solution is history of philosophy, which is "a work within the reach of many people" because it does not require "Descartes' or Kant's inventiveness" (ibid., 20). Therefore, this solution "had, and still has, great fortune" (ibid.). But "it had a drawback: it deprived philosophy of its theoretical soul," because "it made philosophers into football journalists, from football players that they had been" (ibid.). Then, as Floridi says, "to maintain that historiographic research is today the best training ground to become good philosophers means to have lost sight of the fundamental difference between sports reporters and professional footballers" (Floridi 2001, 345). Therefore, "this solution has not been the most successful one among philosophers who did not intend to give up the theoretical vocation" (Marconi 2014, 20).

The second solution is hermeneutics, which in Gadamer's version requires "to reconstruct individual events, to describe specific constellations of concepts, and mostly—in practice—to narrate the use certain authors made of certain words" (ibid., 22). These tasks are surely "challenging, but within the reach of many scholars of less than Kantian or Hegelian standing" (ibid.). In this respect, "the hermeneutic solution is similar to the historico-philosophical solution; but, in addition, it has the advantage of presenting itself as a theoretical research, which fully meets that need to illuminate the present which is traditionally assigned to philosophy, and which may be thought to be frustrated by historico-philosophical research" (ibid.). This solution, however, has the drawback that it does not present philosophy as something to be done now but, as Engel says, as something that has already been done, "upon which you can only comment, propose new readings, or to which you come back" (Engel 1991, 170).

The third solution is analytic philosophy, which is "interested in philosophical problems in their current formulation," not in their "past history" (Marconi 2014, 23). This "permits one to restrict one's research area as much as desired," and "the temporal limitation to strict contemporaries drastically limits bibliographies" (ibid., 24). Analytic philosophy "is the paradigm of the artisan view of philosophy: the one that considers philosophy as an 'honest craft', not reserved for geniuses but open to many people" (ibid., 25). The analytic philosopher, being freed of "the enormous task of providing an overall view of things, can be identified—along with natural scientists—as a professional who tries to give a contribution to the solution of a problem at which many others work" (ibid., 24). So, "philosophy is a joint undertaking" (ibid., 13).

Thus, according to Marconi, with analytic philosophy the traditional image of the Great Philosopher is replaced with that of a professional who tries to give a contribution to a joint undertaking.

In fact, several eminent exponents of analytic philosophy have put forward this view.

Thus, Russell says that "most philosophies hitherto have been constructed all in one block, in such a way that, if they were not wholly correct, they were wholly incorrect, and could not be used as a basis for further investigations" (Russell 1994, 110). So, "each original philosopher has had to begin the work again from the beginning, without being able to accept anything definite from the work of his predecessors," therefore philosophy has "hitherto been unprogressive" (ibid.). Instead, with analytic philosophy "it becomes possible at last for philosophy to deal with its problems piecemeal, and to obtain, as the sciences do," partial results that "subsequent investigation can utilise" (ibid.). Philosophy becomes a cooperative enterprise that admits "of tentative, partial, and progressive advance, and of appeal to principles with which, independently of temperament, all competent students must agree" (ibid., 120).

Carnap says that the propositions of traditional philosophy "have no logical contents, but are only expressions of feeling" (Carnap 2001, 278). Indeed, "the attitude of the traditional philosopher" is "like that of a poet" (Carnap 2003, xvi). Instead, with analytic philosophy a "new type of philosophy has arisen in close contact with the work of the special sciences, especially mathematics and physics" (ibid.). The philosophers of the new type "have taken the strict and responsible orientation of the scientific investigator as their guideline for philosophical work" (ibid.). The individual philosopher "no longer undertakes to erect in one bold stroke an entire system of philosophy. Rather, each works at his special place within the one unified science" (ibid.). In fact, "if we allot to the individual in philosophical work as in the special sciences only a partial task," then "in slow careful construction insight after insight will be won. Each collaborator contributes only what he can endorse and justify before the whole body of his co-workers. Thus, stone will be carefully added to stone and a safe building will be erected at which each following generation can continue to work" (ibid., xvii).

Searle says that, while "the history of philosophy" is "largely a history of the works of a number of towering geniuses," there are "no towering geniuses alive today. This" is "not because we have less talent than our predecessors" but on the contrary because "there are more good philosophers alive now than there were in the past," and "the sheer number of hard-working, able, talented figures in the field makes it impossible for a small number of people to be recognized as standing head and shoulders above all the others" (Searle 1999, 2079). One "of the many advantages in having a field which is not dominated by a tiny number of overpowering figures, is that philosophy as a cooperative enterprise" is "more possible than it has typically been in the past. It is quite possible for people working on a common set of problems to see their enterprise as one of advancing theoretical understanding in a given domain" (ibid.),

Boghossian and Williamson say that "an adversarial methodology is not opposed to a more general conception of philosophy as a cooperative enterprise, but instead illustrates ways in which sharp disputation can advance a shared overall project. It is not a zero-sum game. Through mutual criticism, each of us has been prodded into developing, extending, and sharpening his views much further than might otherwise have happened" (Boghossian and Williamson 2020, ix).

On the basis of the assumption that analytic philosophy is a joint undertaking, Marconi argues that analytic philosophy is the "view of philosophy best suited" to philosophers in "a situation in which professional philosophers are some tens of thousands" (Marconi 2014, 25).

In a sense, Marconi is right, analytic philosophy is the solution best suited to philosophers when there are tens of thousands of professional philosophers. For, it allows philosophers to multiply their publications without having to produce really new ideas, which is very useful for the tenure struggle.

But the question is not whether analytic philosophy is the solution best suited for philosophers. It is, instead, whether analytic philosophy is the solution best suited for philosophy, or rather is itself one of the causes of the irrelevance of philosophy. Since analytic philosophy is the dominant philosophy today, there is a legitimate suspicion that analytic philosophy has special responsibility for the present irrelevance of philosophy.

2.6 The Special Responsibility of Analytic Philosophy

Analytic philosophy has brought about a fundamental change in the nature of philosophy compared to the philosophical tradition.

Indeed, Marconi says that, being freed from the enormous task of providing an overall view of things, "analytic philosophy allows one to circumscribe how much one wants the area of one's research," so, "on the one hand there is room for everyone, on the other hand, the space can be made small as desired" (Marconi 2014, 24).

But, as a result of making the space small as desired, analytic philosophy deals with marginal questions, irrelevant to our knowledge of the world.

What a fall compared to Plato, who says that philosophy "is the crowning of the other sciences, and no other science can rightly be placed above it," philosophy "is at their summit" (Plato, *Respublica*, VII 534 e 2–4).

Or compared to Aristotle, who says that philosophy deals with "the first causes and first principles of everything" (Aristotle, *Metaphysica*, A 1, 981 b 28–29).

This is because, unlike Plato and Aristotle, analytic philosophy assumes that philosophy is not acquisition of knowledge, that is, an inquiry aimed at acquiring knowledge of the world, philosophy aims only to understand what we already know. Therefore philosophy deals with questions irrelevant to our knowledge of the world.

Haack admonishes philosophers to "never forget that philosophy, like physics or biology, is about the world, not just about our language or our concepts—and especially, not just about what another member of this or that little professional circle said last year" (Haack 2016, 26–27).

But most analytic philosophers forget precisely this. They do not deal with questions relevant to our knowledge of the world, but only with what another member of this or that little professional circle said last year.

Thus, Moore says: "I do not think that the world" would "ever have suggested to me any philosophical problems. What has suggested philosophical problems to me is things which other philosophers have said about the world" (Moore 1942, 14). The problems in which "I have been (and still am) very keenly interested" are "first, the problem of trying to get really clear as to what on earth a given philosopher meant by something which he said, and, secondly, the problem of discovering what really satisfactory reasons there are for supposing that what he said was true, or, alternatively, was false" (ibid.).

Moore's claims, however, are self-defeating. For, Moore would not have been able to discuss things which other philosophers said about the world if all philosophers had been like him, namely, if they had not dealt with questions about the world. Nor Moore could discover what really satisfactory reasons there were for supposing that what other philosophers said about the world was true, or, alternatively, was false, without dealing with questions about the world.

That most analytic philosophers do not deal with questions relevant to our knowledge of the world but only with what another member of this or that little professional circle said last year, has given rise to a type of philosophy that is concerned only with logical rigour in dealing with questions, rather than with the significance of the questions themselves.

Thus, Williamson says: "To give an example of the kind of problem that I'm interested in, we can say that J. F. Kennedy and Marilyn Monroe could have had a child, and the issue arises of whether one can say that there is such a thing as a possible child of JFK and Marilyn Monroe. If so, what sort of thing is it? This is an area which is susceptible of quite rigorous logical analysis" (Williamson 2002, 162).

Williamson is a distinguished representative of contemporary analytic philosophy. Kitcher even says that "nobody is a more distinguished exponent of contemporary analytic philosophy than Timothy Williamson" (Kitcher 2024b, 571). So, his views are significant expressions of contemporary analytic philosophy. And yet, the question is not whether a problem like that of the status of a possible child of JFK and Marilyn Monroe is susceptible of quite rigorous logical analysis. It is, rather, whether this type of problem is significant and relevant to our knowledge of the world.

In this regard, Dennett suggests the following significance test.

If you are tempted to deal with a problem and wonder whether the problem is significant and worth addressing, "don't count on the validation of your fellow graduate students or your favorite professors" or any other member of the reference philosophical community, for "they all have a vested interest in keeping the enterprise going. It's what they know how to do; it's what they are good at" (Dennett 2006, 40). Try "seeing if folks outside philosophy," or uninitiated "undergraduates, can be made to care" (ibid., 41). If they cannot be made to care, you should consider the possibility that "you're following a self-supporting community of experts into an artifactual trap" (ibid., 40).

Would the type of problem Williamson is interested in pass Dennett's significance test? It seems unlikely, even in light of Williamson's answer to his problem: "Since Marilyn Monroe and John F. Kennedy could have had a child (although they actually didn't), there is something that could have been a child of Marilyn Monroe and John F. Kennedy," and this something "is neither a child nor a collection of atoms, but rather something that merely could have been a child, made out of atoms, but actually has no location in space and time" (Williamson 2014, 232).

In general, many problems considered by analytic philosophers would not pass Dennett's significance test because, as Dennett says, "many projects" in analytic "philosophy are artifactual puzzles of no abiding significance" (Dennett 2006, 39).

As Midgley says, analytic philosophers "build up a set of games out of simple oppositions and elaborate them until, in the end, nobody else can see what they are talking about. All this can go on until somebody from outside the circle finally explodes it by moving the conversation onto a quite different topic, after which the games are forgotten" (Midgley 2013).

An extreme example of this is provided by Capra, Garrett and Joaquin who, apparently with utmost seriousness, dispute over the question: "What exactly is a fart?" (Capra 2022, 39). They regret that "farts have not received the metaphysical attention they deserve" (Garrett and Joaquin 2022, 19).

Therefore, as Boghossian and Lindsay say, "philosophy isn't accorded much respect outside the academy—it twiddles away and seriously entertains the hyper-esoteric and inconsequential. One need look no further than to a philosophy conference" to "marvel at the obscurity and irrelevance of what's become of the discipline" (Boghossian and Lindsay 2017, 64–65). Indeed, "it's almost as if philosophers have forgotten how to speak to people not just outside their discipline, but also outside their niche" (ibid., 65).

A parody of analytic philosophers is made by Feynman, who speaks of the "paralysis of thought that comes to" analytic "philosophers, who sit opposite each other, one saying to the other, 'You don't know what you are talking about!' The second one says, 'What do you mean by know? What do you mean by talking? What do you mean by you?' and so on" (Feynman et al. 2010, 8-1).

All this confirms that analytic philosophy has special responsibility for the present irrelevance of philosophy.

2.7 The Claim That Analytic Philosophy Is a Turning Point

However, the vast majority of analytic philosophers would disagree that analytic philosophy has special responsibility for the present irrelevance of philosophy, and would indeed deny the present irrelevance of philosophy itself. For, many analytic philosophers claim that analytic philosophy is a turning point in philosophy.

Already Schlick claimed: "An experience of more than two thousand years seems to teach us that all attempts to put an end to the chaos of systems, and to alter the destiny of philosophy, can no longer be taken seriously" (Schlick 1979, 155). But this is not the case. For, with the rise of analytic philosophy, "we are at present in the midst of an altogether final change in philosophy, and are justly entitled to consider the fruitless conflict of systems at an end" (ibid.). This situation "has no parallel with any earlier one," it "is really unique, and the change now in progress" is "a really final one" (ibid.). Of course, "many will still continue for centuries to pursue the accustomed paths; philosophical authors will go on discussing the old pseudo-problems for a long time yet, but in the end they will no longer be listened to and will resemble actors who go on performing for a time before noticing that the audience has gradually stolen away" (ibid., 159–160).

That analytic philosophy is a turning point in philosophy has been reaffirmed even later.

For example, Casati claims that "the work done in philosophy in the last century, if not in the last twenty years, is incomparably more interesting and clear than the work done by the masters of ages past" (Casati 2001, 75). It "is thus better, if one wants to get a grip on what preoccupies a professional epistemologist, to read, rather than the *Theaetetus*, the works" of "BonJour" (ibid.).

But this claim is not valid. For, BonJour says: "The following quite venerable answer" to the question 'What is knowledge?' "is, I believe, at least approximately correct. For a person A to know that P, where P is some proposition, three conditions must be satisfied: (1) A must believe confidently that P, (2) P must be true, and (3) A's belief that P must be adequately justified" (BonJour 1985, 3–4).

Thus, Bonjour says he believes that the view that knowledge is justified true belief is at least approximately correct.

But this contrasts with the fact that, as will be seen in Chap. 10, precisely in the *Theaetetus* Plato had shown that the view that knowledge is justified true belief is untenable. It is thus better, if one wants to get a grip on what knowledge is, to read the *Theaetetus* rather than the works of BonJour.

As another example, Marconi claims that, in contrast to previous philosophy, with analytic philosophy "philosophical problems are solvable and several of them have in fact been solved" (Marconi 2014, 91). Thus, analytic philosophy has solved the problem of the relation between truth and provability, demonstrating that "it is impossible to reduce truth to provability" (ibid., 110).

But this claim is not valid. For, it is unfounded to say that analytic philosophy has solved the problem of the relation between truth and provability, demonstrating that it is impossible to reduce truth to provability, Indeed, this problem has been solved by Gödel's first incompleteness theorem, which cannot be ascribed to analytic philosophy but only to mathematical logic, which is a conventional, even though somewhat marginal, branch of mathematics.

Gödel himself forcefully emphasizes that mathematical logic and his first incompleteness theorem owe nothing to analytic philosophy.

In fact, Gödel says that "Wittgenstein's negative attitude toward symbolic language is a step backwards. Those who, like Carnap, misuse symbolic language want to discredit mathematical logic; they want to prevent the appearance of philosophy" (Wang 1996, 174). In general, "the whole movement of the positivists wants to destroy philosophy; for this purpose they need to destroy mathematical logic as a tool" (ibid.). As to Gödel's first incompleteness theorem, "Wittgenstein did not understand it," in fact "he interpreted it as a kind of logical paradox, while in fact it is just the opposite, namely a mathematical theorem within an absolutely uncontroversial part of mathematics (finitary number theory or combinatorics)" (ibid., 179).

From this, it is clear that the claim that analytic philosophy is a turning point in philosophy is unfounded.

As Kitcher says, "for two millennia, philosophers tried to provide their readers with tools for making sense of important facets of their lives and the world in which they lived" (Kitcher 2024b, 571–572). But today, "although a much larger population of people are now paid for writing philosophy than at any previous stage in history, we are (largely) failing to give the world what our great predecessors provided" (ibid., 572).

2.8 The Ideal of Clarity

Why do many analytic philosophers claim that analytic philosophy is a turning point in philosophy? There are several reasons for this.

One reason is that they believe that, unlike other types of philosophy, analytic philosophy aims at clarity.

Thus, Wittgenstein says that "without philosophy thoughts are, as it were, cloudy and indistinct," therefore "philosophy aims at the logical clarification of thoughts," its "task is to make them clear and to give them sharp boundaries," so "philosophy does not result in 'philosophical propositions', but rather in the clarification of propositions," a "philosophical work consists essentially of elucidations" (Wittgenstein 2001, 4.112)

Ayer says that "it's perfectly true that people who write very clearly may be superficial," but "it isn't at all true that someone who writes clearly has to be superficial. On the contrary," a "good philosophical writer is someone who can put difficult theories" in "a clear fashion. One of the great dangers in philosophy is woolliness, and woolliness, particularly among Germans, is always marked by very unclear writing" (Ayer 1991, 224–225)

Williams says that "the aim of analytical philosophy" is "to be clear" (Williams 1985, viii). Indeed, "what distinguishes analytical philosophy from other contemporary philosophy" is "a certain way of going on, which involves argument, distinctions," and the fact that "it distinguishes sharply between obscurity and technicality. It always rejects the first, but the second it sometimes finds a necessity. This feature peculiarly enrages some of its enemies. Wanting philosophy to be at once profound and accessible, they resent technicality but are comforted by obscurity" (ibid.).

But the belief that, unlike other types of philosophy, analytic philosophy aims at clarity, is unfounded.

As Glock argues, clarity "cannot serve as a distinctive mark of analytic philosophy, not even vis-á-vis the much derided continental thinkers. It is neither necessary nor sufficient for qualifying as an analytic philosopher," it "is not even a characteristic of some of its most prominent manifestations. Wittgenstein springs to mind. There is a striking irony about his work," because "Wittgenstein devoted his philosophy to the pursuit of clarity, yet in a fashion which is at times extremely obscure," so there is a "striking discrepancy between design and result" (Glock 2024, 346).

Moreover, clarity is not an end in itself, because it can be empty and trivial. Clarity is worth pursuing only when it has a relevant content.

2.9 The Ideal of Logical Rigour

Another reason why many analytic philosophers claim that analytic philosophy is a turning point in philosophy is that they believe that, unlike other types of philosophy, analytic philosophy aims at logical rigour.

Thus, Williamson says that "analytic philosophy at its best uses logical rigor" to "achieve a sharpness of philosophical vision unobtainable by other means" (Williamson 2022, 48).

Analytic philosophy draws its ideal of logical rigour in philosophy from Frege's ideal of logical rigour in mathematics.

Frege aims to achieve the "maximum logical precision" of concepts and "the gapless rigour of demonstration" (Frege 1990, 223). Mathematics needs definitions of concepts that "determine unambiguously for every object whether it falls under the concept or not" (Frege 2013, Volume II, 69). And it needs proofs "drawn up in a logically rigorous way, with no gaps in the chains of inference" (Frege 1979, 217). So, mathematics needs logical rigour.

In particular, Frege aims to achieve logical rigor in arithmetic. He presents his work as a natural continuation of the work of rigorization of calculus of Cauchy, Weierstrass, Dedekind, and Cantor. Indeed, he says that, "proceeding along these lines, we are hound eventually to come to the concept of number and to the simplest propositions holding of positive whole numbers, which form the foundation of the whole of arithmetic" (Frege 1960, 2). The "fundamental propositions of arithmetic should be proved, if in any way possible, with the utmost rigour; for only if every gap in the chain of deductions is eliminated with the greatest care can we say with certainty upon what primitive truths the proof depends" (ibid., 4).

But Frege's work is not really a natural continuation of the work of rigorization of the analysis of Cauchy, Weierstrass, Dedekind and Cantor. The latter responded to the need to remove difficulties relating to issues such as Fourier series, convergence or the existence of derivatives and integrals, which hindered the development of mathematics because they caused uncertainty about the domain of validity of theorems or the conditions of existence of limits. On the contrary, Frege's aim to achieve logical rigor in arithmetic did not respond to any real need of mathematical activity, it was an end in itself, it was only inspired by an abstract ideal of logical hygiene.

In particular, this is clear from the importance that Frege placed on giving an answer to the question of "what the number one is" (Frege 1960, xiii). He says: "Is it not a scandal that our science should be so unclear about the first and foremost among its objects, and one which is apparently so simple?" (ibid., xiv). If "a concept fundamental to a mighty science gives rise to difficulties, then it is surely an imperative task to investigate it more closely until those difficulties are overcome" (ibid.). We "shall hardly succeed in finally clearing up negative numbers, or fractional or complex numbers, so long as our insight into the foundation of the whole structure of arithmetic is still defective" (ibid.).

But this contrasted with the fact that none of the results of number theory depended on an answer to the question of what the number one is. Giving an answer to this question did not respond to any real need of mathematical research.

Logical rigour has a positive function when it is fruitful for research. If it is an end in itself, it is only a tight dress that prevents one from moving freely, and therefore is pedantry. What matters in mathematics is not logical rigour but fruitfulness. Therefore, logical rigour must be replaced with fruitfulness.

The same can be said of logical rigour in philosophy. What matters in philosophy is not logical rigour but fruitfulness.

In fact, the limitations of Frege's ideal of logical rigour for mathematics are significant also for analytic philosophy because, as Dummett says, Frege is "the true father of analytical philosophy" (Dummett 2007, 27).

In particular, analytic philosophy considers the logic created by Frege, mathematical logic, to be part of philosophy.

Indeed, Wittgenstein says: "Philosophy consists of logic and metaphysics" (Wittgenstein 1979, 106).

By 'logic' Wittgenstein means mathematical logic. Therefore, Wittgenstein says that mathematical logic is part of philosophy.

But the claim that mathematical logic is part of philosophy is not valid. Mathematical logic is a conventional, although somewhat marginal, branch of mathematics, with no special relevance to philosophy. This has been underlined by several mathematicians and philosophers.

Thus, Rota says: "The fake philosophical terminology of mathematical logic has misled philosophers into believing that mathematical logic deals with the truth in the philosophical sense" (Rota 1997, 93). But this is not the case. The "snobbish symbol-dropping found nowadays in philosophical papers raises eyebrows among mathematicians, like someone paying his grocery bill with Monopoly money" (ibid.).

Kripke says: "Logical investigations can obviously be a useful tool for philosophy," but it "should not be supposed that the formalism can grind out philosophical results in a manner beyond the capacity of ordinary philosophical reasoning. There is no mathematical substitute for philosophy" (Kripke 1976, 416).

2.10 The Ideal of Professionalism

Another reason why many analytic philosophers claim that analytic philosophy is a turning point in philosophy is that they conceive of themselves as professionals.

Thus, Marconi says that analytic philosophers "conceive of themselves as professionals" and, "as professionals," they "are concerned with conforming to certain standards" that "require clarity and correctness of argument" (Marconi 2014, 85). They also require that, "when dealing with topic X, one takes into account what has been published on X, at least in the last twenty years and in the most important journals, or by the most prestigious scientific publishing houses" (ibid., 86). A philosopher "who intends to respect these requirements can hardly deal with many disparate subjects: he will usually deal with no more than two or three philosophical questions, and will in any case move within the scope of a subdiscipline: he will be a philosopher of the mind or a moral philosopher or a metaphysician etc. Therefore, a professional philosopher" cannot be "anything other than a specialist" (ibid., 86–87). Analytic philosophers are "conscious of these characteristics of their work," and "are also proud of them," because "they think that it is precisely by virtue of these characteristics that philosophy qualifies as a respectable intellectual activity" (ibid., 87). In contrast, they think that non-analytic philosophers, in particular "what they call 'continental philosophers', do not conform to these standards, in fact, they are incapable of conforming to them" (ibid.).

But the ideal of professionalism, instead of producing highly relevant philosophy, has produced irrelevant philosophy.

As Rescher says, the ideal of professionalism has led to "the turning of philosophy from global general, large-scale issues to more narrowly focused investigations of matters of microscopically fine-grained detail" (Rescher 1993, 731). So, "philosophy has become increasingly technical in character," philosophical investigations "make increasingly extensive use of the formal machinery of semantics, modal logic, compilation theory, learning theory, and so forth. Ever heavier theoretical armaments are brought to bear on ever smaller problem-targets in ways that" make one "wonder whether the important principle that technicalities should never be multiplied beyond necessity has been lost sight of" (ibid.). The "increasing specialization has impelled" analytic "philosophy toward the ivory tower" (ibid., 733). Consequently, "outside the academy," even main analytic "philosophers exert no influence" (ibid., 732).

The characteristics of analytic philosophy that Rescher points out—technicality and focus on matters of microscopically fine-grained detail—are particularly evident today, when analytic philosophy is in a phase of exhaustion. The elephantiasis of technique and the sophistry of content are characteristics of a philosophical tradition in the process of exhaustion.

Moreover, as Gaukroger says, "the way in which analytic philosophy" has "trivialized problems so that they have become merely technical, as if one were solving scientific or economic problems," instead of "clearing away the rubbish in the way of knowledge," has "in fact put rubbish in its way" (Gaukroger 2020, 284). And "the response to trivializing" problems "has perhaps been even worse: the promotion of an 'analytical metaphysics' in which philosophical questions are pasted into a scientific framework" (ibid.).

An example of 'analytical metaphysics' is Williamson's problem of the status of a possible child of JFK and Marilyn Monroe.

In fact, professionalism in philosophy is impossible. A philosopher cannot be a professional in the same sense as a mathematician, or a physicist, or a biologist, because philosophy has no special field of its own.

Moreover although, as will be argued in Chap. 18, the method of philosophy is the same as that of mathematics and natural sciences, that is, the analytic method, the latter is only a general framework and its application requires much experience in the relevant field. Mathematicians and scientists have much experience in their fields, not so philosophers who move on an unexplored ground, of which they have very limited experience.

2.11 Lack of Interest in Questions of Analytic Philosophy

In contrast to the claim that analytic philosophy is a turning point in philosophy, most questions of analytic philosophy do not elicit any interest outside analytic philosophy.

As Fodor says, today "nobody reads philosophy," in particular "it's mainly the laity that seems to have lost interest. And it's mostly Anglophone analytic philosophy that it has lost interest in" (Fodor 2004, 17).

Why this lack of interest in questions of analytic philosophy?

Marconi says that it is due to the enormous "growth in the number of those who can be considered professional philosophers" (Marconi 2014, 10). This has led "to a redefinition of the figure of the philosopher" as "a normal researcher who controls a limited area of study and aims to contribute to it" (ibid., 60). A philosopher is a specialist, to be "a specialist is to limit one's research to a small sector of a discipline" (ibid., 15). This has "the great advantage of allowing even ordinary scholars to do honest and sensible research (and therefore to increase the amount of honest and sensible research)" (ibid.). On the other hand, however, specialism "has produced a drastic fall in the ability to communicate and therefore in the cultural influence of philosophy, both with respect to other disciplines and with respect to the cultured public at large" (ibid., 61).

According to Marconi, this is "sufficient to explain the poor communication capacity of today's philosophy. For, specialism implies that someone who writes about philosophy must necessarily take a lot for granted" (ibid., 25). This is inevitable because, "by the very nature of specialism, the objects to which philosophical research is applied are often impenetrable to non-experts" (ibid., 61). But "we have long been accustomed to the specialism of the natural sciences," and "why should it be different in the case of philosophy? Why is philosophy made a request for communicativeness that no one dreams of making of molecular biology?" (ibid., 28). And "why, if this request is not satisfied, should philosophy be attributed a lack of legitimacy which, once again, no one dreams of attributing to the 'hard' sciences, with all their incomprehensibility? Why, as just said, does it seem that philosophy cannot afford specialism?" (ibid.).

Marconi's explanation of the lack of interest in questions of analytic philosophy, however, is unconvincing. The specialism of natural sciences does not lead to a lack of interest in scientific questions.

Thus, according to Funk, Gottfried, and Mitchell, "about seven-in-ten Americans express at least some interest in science news," one-in-three adults get "science news regularly," and "one-in-six adults" both "get science news regularly and seek it out" (Funk et al. 2017, 13).

Nothing of the kind can even remotely be said of philosophy. What is at issue here is not specialism but the lack of interest in questions of analytic philosophy.

Marconi claims that the problem of the poor communication capacity of analytic philosophy can be solved through the "popularization of philosophy" (Marconi 2014, 61). Much "popularization of natural sciences consists in the presentation (in simplified form) of theories and the experiments and observations that confirm them" in "this or that area of science" (ibid.). Then "why couldn't something similar be done in the case of philosophy?" (ibid.). Indeed, "if we believe that professional philosophical research has produced, in recent decades, results of some interest, we should be able to do so; not only that, we should want to do so" (ibid., 62).

This view is endorsed by several analytic philosophers.

For example, Williamson says: "A civilized society has popular philosophy just as it has popular physics, popular psychology, popular history," so "one might expect the relation between popular and academic philosophy to resemble the corresponding relations for other disciplines. Thus, popular philosophy would communicate recent research in academic philosophy to a wider audience" (Williamson 2020). In fact, "recent philosophical research has produced lots of fascinating new ideas, which deserve to be better known. Now there's a task for popular philosophy" (ibid.).

But this solution to the problem of the poor communication capacity of analytic philosophy would not work. The popularization of natural sciences is successful because there is interest in scientific questions. The popularization of analytic philosophy would not be successful because there is lack of interest in questions of analytic philosophy.

Williamson claims: "Throughout my life in academic philosophy, a recurrent event has been reading" the "latest contemptuous dismissal of analytic philosophy" (Williamson 2024, 527). But these dismissals are completely unjustified because "contemporary analytic philosophy has plenty to offer other disciplines," and "it will have even more to offer other disciplines if it gets better at answering its own questions" (ibid., 534).

But this claim is in sharp conflict with the lack of interest in questions of analytic philosophy. Only analytic philosophers find them interesting, and their interest in them has no objective justification.

In fact, when the problem arises of whether a question is a philosophical question and it is interesting, Williamson's answer is: "I am a philosopher, I find" the "question interesting" (Williamson 2022, 26).

Contrary to Williamson's claim, analytic philosophy has nothing to offer to other disciplines, as it is clear from the attacks by many scientists and mathematician on contemporary philosophy, considered earlier in this chapter. Therefore, if analytic philosophy gets better at answering its own questions, this will make no difference to other disciplines.

The lack of interest in questions of analytic philosophy is a consequence of the fact that analytic philosophy assumes that philosophy is not acquisition of knowledge, that is, an inquiry aimed at acquiring knowledge of the world, it aims only to understand what we already know. As a result, analytic philosophy deals with marginal questions, irrelevant to our knowledge of the world.

2.12 An Inadequate Escape from Irrelevance

While the vast majority of analytic philosophers would disagree that analytic philosophy has special responsibility for the present irrelevance of philosophy, and would indeed deny the present irrelevance of philosophy itself, some rare analytic philosophers admit it and seek an escape from irrelevance.

Thus, Unger says that analytic philosophy "offers little or nothing substantial about concrete reality" (Unger 2014, 21). Indeed, "the ideas offered by analytic philosophers are, almost without exception, concretely empty ideas, making or meaning no difference, at all, for how anything ever is with concrete reality" (ibid., 23). Many analytic philosophers take it "as a point in favor of their quite limited activity that it" avoids "making any grand claims as to concrete reality's general nature, structure, or character" (ibid., 2). According to them, "by contrast with the natural sciences, intellectually responsible philosophy should offer no substantial thoughts about the general nature of concrete reality" (ibid., 1). This contrasts with the fact that, "throughout most of its history, philosophy has aspired to offer substantial thoughts about concrete reality," and in fact "it's proper for philosophy to offer quite substantial ideas about concrete reality" (ibid., 21). Therefore, analytic philosophy is a "terribly deficient philosophy" as "compared with much philosophy saliently offered in earlier eras (or, at the very least, as compared with what has been so widely and plausibly presumed for a good deal of earlier philosophy)" (ibid.). As a result, analytic philosophy is "heavily pervaded with concretely empty ideas" (ibid., 25).

But what escape from irrelevance does Unger propose for philosophy?

He argues that, "for there to be any significant hope as regards philosophy's prospects, at least two things must come to pass" (Unger 2014, 239). First, "some of the most intelligent and philosophically talented young people must become serious scientists, contributing a great deal to the science in which they are involved" (ibid.). Second, "we must not require so very much of a proposed idea, by way of depth," for "the idea to count as a properly philosophical thought" a "little leeway" must be "tolerated" (ibid., 240). Indeed, "as regards matters of depth," it must be tolerated that there will be "a significant difference between what will be allowed to count as a philosophical idea and, on the other side, very nearly all other concretely substantial ideas, including very nearly all scientific substantial propositions" (ibid.).

Unger's proposed escape from irrelevance, however, would not work. Unger himself admits that "it will be extremely difficult and, perhaps, it may be very unlikely," that the two things in question will come to pass; but he insists that, "short of that, there is no realistic chance, at all, for there being offered, during the next century or so, any new concretely substantial philosophical ideas that amount to anything much more than idle speculation" (ibid.).

But even if the two things in question came to pass, this would not be very rewarding for philosophy, because it would reduce philosophy to an entertainment for scientists.

For, the image of the philosopher of the future that Unger proposes is that of a scientist who, to take a break from his demanding work as a scientist, occasionally relaxes by philosophizing about it, but without too many pretensions. For, one cannot be too demanding, one cannot expect his philosophy to be characterised by depth.

Since Unger's proposed escape from irrelevance would not work, an alternative escape is necessary. This will be presented in Chap. 4.

References

Ayer, Alfred. 1991. An interview with A. J. Ayer, by Ted Honderich. In *A. J. Ayer: Memorial essays*, Phillips Griffiths, ed. 209–226. Cambridge University Press.

Bobbio, Norberto. 2017. *La filosofia e il bisogno di senso*. Brescia: Morcelliana.

Boghossian, Peter. 2016. Philosophy that matters. *The Philosophers' Magazine* 72 (1): 29–30.

Boghossian, Peter, and James Lindsay. 2017. Abiding significance, and rabbit holes. In *Philosophy's future: The problem of philosophical progress*, ed. Russell Blackford and Damien Broderick, 63–74. Hoboken: Wiley.

Boghossian, Paul, and Timothy Williamson. 2020. *Debating the a priori*. Oxford: Oxford University Press.

BonJour, Laurence. 1985. *The structure of empirical knowledge*. Cambridge: Harvard University Press.

Bunge, Mario. 2001. *Philosophy in crisis: The need for reconstruction*. Amherst: Prometheus Books.

Capra, Bill. 2022. The metaphysics of farts. *Think* 21:39–43.

Carnap, Rudolf. 2001. *Logical syntax of language*. London: Routledge.

Carnap, Rudolf. 2003. *The logical structure of the world—Pseudoproblems in philosophy*. Chicago: Open Court.

Casati, Roberto. 2001. Review of Paolo Rossi (ed.), *La filosofia*. *Dialectica* 55:74–84.

Crane, Tim. 2017. Philosophy, science and the value of understanding. In *Sapere aude: The future of the humanities in British universities*, ed. Kurt Almqvist and Isabella Thomas, 175–183. Stockholm: Axel and Margaret Ax:son Johnson Foundation.

Crick, Francis. 2005. You're just a pack of neurons. In *Conversations on consciousness*, ed. Susan Blackmore, 68–78. Oxford: Oxford University Press.

Dennett, Daniel. 2006. Higher-order truths about Chmess. *Topoi* 25:39–41.

Dirac, Paul. 1963. Interview by Thomas Kuhn, 6 May 1963. College Park: American Institute of Physics, Niels Bohr Library and Archive.

Dummett, Michael. 2007. The place of philosophy in European culture. *European Journal of Analytic Philosophy* 3:21–30.
Dyson, Freeman. 2015. *Dreams of earth and sky*. New York: The New York Review of Books.
Engel, Pascal. 1991. French and American philosophical dispositions. *Stanford French Review* 15:165–181.
Feynman, Richard, Robert Leighton, and Matthew Sands. 2010. *Lectures of Physics*. Vol. 1. New York: Basic Books.
Floridi, Luciano. 2001. I filosofi: calciatori o giornalisti? In *La filosofia italiana in discussione*, ed. Francesco Firrao, 337–361. Milano: Bruno Mondadori.
Fodor, Jerry. 2004. Water's water everywhere. *London Review of Books* 26 (20): 17–19.
Frege, Gottlob. 1960. *The foundations of arithmetic: A logico-mathematical enquiry into the concept of number*. New York: Harper.
Frege, Gottlob. 1979. *Posthumous writings*. Oxford: Blackwell.
Frege, Gottlob. 1990. *Kleine Schriften*. Hildesheim: Olms.
Frege, Gottlob. 2013. *Basic laws of arithmetic*. Oxford: Oxford University Press.
Frodeman, Robert. 2013. Philosophy dedisciplined. *Synthese* 190:1917–1936.
Frodeman, Robert, and Adam Briggle. 2016. *Socrates tenured*. London: Rowman and Littlefield.
Funk, Cary, Jeffrey Gottfried, and Amy Mitchell. 2017. *Science news and information today*. Pew Research Center, September 2017.
Garrett, Brian, and Jeremiah Joaquin. 2022. Letting Rip: Rebutting Capra on the metaphysics of farts. *Think* 21:19–22.
Gaukroger, Stephen. 2020. *The failures of philosophy: A historical essay*. Princeton: Princeton University Press.
Glock, Hans-Johann. 2024. Clarity-rigour-progress: Three ideals of analytic philosophy. *History of Philosophy & Logical Analysis* 27:337–369.
Gowers, Timothy. 2006. Does mathematics need a philosophy? In *18 unconventional essays on the nature of mathematics*, ed. Reuben Hersh, 182–200. New York: Springer.
Haack, Susan. 2016. The fragmentation of philosophy, the road to reintegration. In *Susan Haack: Reintegrating philosophy*, ed. Julia Göhner and Eva-Maria Jung, 3–32. Cham: Springer.
Haack, Susan. 2017. The real question: Can philosophy be saved? *Free Inquiry* 37:40–43.
Haack, Susan. 2021. Philosophy as a profession, and as a calling. *Syzetesis* 8:33–51.
Hawking, Stephen. 1988. *A brief history of time: From the Big Bang to black holes*. London: Bantam Books.
Hawking, Stephen, and Leonard Mlodinow. 2010. *The grand design*. New York: Bantham Books.
Hersh, Reuben. 2001. Reply to Martin Gardner. *The Mathematical Intelligencer* 23 (2): 3–5.
Kitcher, Philip. 2023. *What's the use of philosophy?* Oxford: Oxford University Press.
Kitcher, Philip. 2024a. Précis of What's the use of philosophy? *Philosophia* 52:521–525.
Kitcher, Philip. 2024b. Reply to commentators. *Philosophia* 52:571–583.
Krauss, Lawrence. 2012. Interview by Ross Andersen. *The Atlantic*, 23 April 2012.
Kreisel, Georg. 1967. Mathematical logic: What has it done for the philosophy of mathematics? In *Bertrand Russell, philosopher of the century*, ed. Ralph Schoenman, 201–272. London: Allen & Unwin.
Kripke, Saul. 1976. Is there a problem about substitutional quantification? In *Truth and meaning: Essays in semantics*, ed. Gareth Evans and John Henry McDowell, 325–419. Oxford: Oxford University Press.
Marconi, Diego. 2014. *Il mestiere di pensare: La filosofia nell'epoca del professionismo*. Turin: Einaudi.
Midgley, Mary. 2013. The golden age of female philosophy. *The Guardian*, 28 November 2013. https://www.theguardian.com/world/2013/nov/28/golden-age-female-philosophy-mary-midgley
Moore, George. 1942. An autobiography. In *The philosophy of G. E. Moore*, ed. Paul Schilpp, 3–39. La Salle: Open Court.
Mulligan, Kevin, Peter Simons, and Barry Smith. 2006. What's wrong with contemporary philosophy? *Topoi* 25:63–67.

Mullis, Kary. 1998. *Dancing naked in the mind field.* New York: Vintage Books.
Pernu, Tuomas. 2008. Philosophy and the front line of science. *The Quarterly Review of Biology* 83:29–36.
Pinker, Steven. 2003. *The blank slate: The modern denial of human nature*. New York: Penguin Books.
Preston, Aaron. 2007. *Analytic philosophy: The history of an illusion*. New York: Continuum.
Rescher, Nicholas. 1993. American philosophy today. *The Review of Metaphysics* 46:717–745.
Rota, Gian-Carlo. 1997. *Indiscrete thoughts*. Boston: Birkhäuser.
Russell, Bertrand. 1994. *Mysticism and logic*. London: Routledge.
Schlick, Moritz. 1979. The turning point in philosophy. In Moritz Schlick, *Philosophical papers*, vol. 2, 154–160. Dordrecht: Reidel.
Searle, John. 1999. The future of philosophy. *Philosophical Transactions of the Royal Society: Biological Sciences* 354 (1392): 2069–2080.
Sykes, Christopher, ed. 1994. *No ordinary genius: The illustrated Richard Feynman*. New York: Norton.
Unger, Peter. 2014. *Empty ideas: A critique of analytic philosophy*. Oxford: Oxford University Press.
Wang, Hao. 1996. *A logical journey: From Gödel to philosophy*. Cambridge: The MIT Press.
Weinberg, Steven. 1993. *Dreams of a final theory*. New York: Vintage.
Williams, Bernard. 1985. *Ethics and the limits of philosophy*. London: Routledge.
Williamson, Timothy. 2002. On vagueness. In *New British philosophy*, ed. Julian Baggini and Jeremy Stangroom, 147–162. London: Routledge.
Williamson, Timothy. 2014. Classical investigations. In *Philosophy at 3:AM: Questions and answers with 25 top philosophers*, ed. Richard Marshall, 222–234. Oxford: Oxford University Press.
Williamson, Timothy. 2020. Popular philosophy and populist philosophy. *Daily Nous*, 8 June. http://dailynous.com/2020/06/08/popular-philosophy-populist-philosophy-guest-post-timothy-williamson/
Williamson, Timothy. 2022. *The philosophy of philosophy*. Hoboken: Wiley-Blackwell.
Williamson, Timothy. 2024. Philip Kitcher's purge on philosophy. *Philosophia* 52:527–537.
Wittgenstein, Ludwig. 1979. *Notebooks 1914–1916*. Oxford: Blackwell.
Wittgenstein, Ludwig. 2001. *Tractatus logico-philosophicus*. London: Routledge.
Wolpert, Lewis. 2000. Round table debate: Science versus philosophy? *Philosophy Now*, Issue 27, June/July 2000.

Chapter 3
Philosophy and Understanding

Abstract In Chap. 2 it has been said that analytic philosophy assumes that philosophy is not acquisition of knowledge, that is, an inquiry aimed at acquiring knowledge of the world, philosophy aims only to understand what we already know. This chapter considers the arguments of analytic philosophy for this view and argues that they are not valid.

Keywords Poverty of philosophy as cognitive discipline · Midwife explanation · Philosophy as only aiming at understanding · Philosophy as conceptual analysis

3.1 The Arguments of Analytic Philosophy

In Chap. 2 it has been said that analytic philosophy assumes that philosophy is not acquisition of knowledge, that is, an inquiry aimed at acquiring knowledge of the world, philosophy aims only to understand what we already know.

This chapter considers the arguments of analytic philosophy for this view and argues that they are not valid.

A clear formulation of these arguments is provided by Hacker. Therefore, the chapter discusses the arguments of analytic philosophy as formulated by Hacker.

3.2 Cognitive Poverty of Philosophy and Midwife Explanation

Hacker claims that philosophy is not acquisition of knowledge, therefore philosophy is not an inquiry aimed at acquiring knowledge of the world.

Against the view that philosophy is acquisition of knowledge, Hacker argues that philosophy is poor as a cognitive discipline. For, philosophy has so little to show for more than two thousand years of endeavour.

C. Cellucci, *A Philosophy for the Future*, Synthese Library 508,
https://doi.org/10.1007/978-3-031-98443-3_3

Indeed, Hacker says that, "throughout its history, philosophy has" been "held to be a participant in the quest for knowledge—a cognitive discipline" (Hacker 2013a, 3). But this conflicts with "the poverty of philosophy qua cognitive discipline" (ibid., 7). That is, it conflicts with the fact that, "conceived as a part of our quest for knowledge, philosophy, unlike the sciences, has so little to show for more than two thousand years of endeavour" (ibid., x). In fact, philosophy is not "engaged, like" the "sciences, in the pursuit of knowledge of the world" (ibid., xx). It does not advance knowledge of the world, "indeed, one might say, with only a little exaggeration, that in philosophy, 'if it's news, it's wrong'" (ibid., 19). This raises the challenging question: "How is the poverty of philosophy, construed as a cognitive discipline, to be explained?" (ibid., 5).

Supporters of the view that philosophy is acquisition of knowledge respond to this question with the midwife explanation.

According to it, the poverty of philosophy as a cognitive discipline is only apparent because, "once questions are sufficiently sharply formulated to be answerable, they are handed over to an independent science, which then contributes to the extension of human knowledge" (ibid., 16). Thus, "many questions that were opened by philosophers were subsequently handed over to scientists," so physics "became independent of philosophy in the seventeenth" century, and "psychology broke free of philosophy at the end of the nineteenth century" (ibid.).

3.3 Against the Midwife Explanation

According to Hacker, however, the midwife explanation of the poverty of philosophy as a cognitive discipline is not valid. He advances the following arguments against it.

(1) "Although independence was achieved" by physics or psychology, "new areas of philosophical investigation were thereby generated, e.g. philosophy of physics or philosophy of the psychological sciences" (Hacker 2013a, 6). These new areas of philosophical investigation may contribute to enlightening the corresponding sciences, and "it would be misguided to suppose that questions in the philosophies of the special sciences remain philosophical only because they are insufficiently clearly understood to be handled by a new metascience" (ibid.).

(2) "Although" physics or psychology "achieved independence, it would be mistaken to suppose that they achieved freedom from conceptual confusion" (ibid., 6). In fact, "the sciences (and to a lesser degree mathematics), in our times, are the primary source of misguided metaphysics—which it is the task of philosophy to curb" (ibid., 10). Philosophy "is a Tribunal of Reason, before which scientists and mathematicians may be arraigned for their transgressions" (ibid.). So philosophy can "restrain their tendency to transgress the bounds of sense" (ibid.).

(3) "The birth of an independent science does not free philosophy from a host of questions which have always been on the philosophical agenda associated with the subject matter of that special science" (ibid., 6). Such questions remain "on the

agenda of philosophy" (ibid., 7). They cannot be answered by that science, because between philosophy and science there is a categorial gulf that is "as deep as any categorial gulf" (ibid., 16). Therefore, on the one hand, "no philosophical question can be answered by scientific enquiry," and, on the other hand, "no scientific discovery can be made by philosophical investigation" (ibid.).

(4) "The suggested explanation is implausible when we turn to practical philosophy (in Kant's sense of the term)—to ethics, political and legal philosophy" (ibid., 7). For, "moral, legal, and political philosophy do not give birth to new sciences, but contribute to the emergence of new moral, legal, and political distinctions, principles and constitutional arrangements" (ibid.).

On the basis of these arguments, Hacker concludes that "the poverty of philosophy qua cognitive discipline cannot be explained as a consequence of the fact that once knowledge is achievable the subject becomes a science" (ibid.). It can be explained only by recognizing that "philosophy is not a contribution to human knowledge" (ibid., 9). Therefore, philosophy is not acquisition of knowledge.

3.4 Non-Validity of Arguments Against the Midwife Explanation

Hacker's arguments against the midwife explanation, however, are not valid.

(1) It is not valid to say that, although independence was achieved by physics or psychology, new areas of philosophical investigation were thereby generated, such as the philosophy of physics or the philosophy of the psychological sciences, which may contribute to enlightening the corresponding sciences without becoming the midwives of new sciences.

This argument assumes that the philosophy of physics or the philosophy of the psychological sciences are new philosophical disciplines that have originated in the modern period from the rise of physics or psychology as independent sciences. But this assumption is unfounded, because the philosophy of physics and the philosophy of the psychological sciences have been part of philosophy from the very beginning, as it is clear from the fact that Aristotle dedicated two of his most important works to them, *Physica* and *De anima*.

For Aristotle, however, the philosophy of physics and the philosophy of the psychological sciences were not separate disciplines, they were an integral part of general philosophy.

On the contrary, the philosophy of physics and the philosophy of the psychological sciences, as Hacker and analytic philosophy understand them, are not part of general philosophy but separate disciplines.

The philosophy of physics and the philosophy of the psychological sciences, understood as separate disciplines, have the typical defects of analytic philosophy. They make pedantic distinctions about small inconsequential matters and are unable to make any significant contribution to our knowledge of the world. Evidence for this is provided by the fact that, as we have seen in Chap. 2, many mathematicians and scientists consider them insignificant.

That the philosophy of physics and the philosophy of the psychological sciences, understood as separate disciplines, are unable to make any significant contribution to our knowledge of the world, depends on the fact that it is impossible to treat philosophical problems sectorially, independently of a general approach to knowledge. Philosophical problems concerning physics or the psychological sciences are part of the general problem of knowledge, therefore they are incomprehensible if treated independently of a general approach to knowledge. Philosophy is not limited to sectorial questions, it offers a global view. So, there cannot be a philosophy of physics alone, or a philosophy of the psychological sciences alone. For, the question of the nature of physics, or of the psychological sciences, cannot be adequately approached locally, in isolation from all other knowledge, but only globally, as an integral part of a general approach to knowledge.

(2) It is not valid to say that, although physics or psychology achieved independence, it would be mistaken to suppose that they achieved freedom from conceptual confusion, in fact the sciences (and to a lesser degree mathematics), in our times, are the primary source of misguided metaphysics which is the task of philosophy to curb.

This argument is based on what Locke calls the view of the philosopher as an under-labourer of the sciences.

Indeed, Locke says: "In an age that produces such masters as the great Huygenius and the incomparable Mr Newton," the philosopher must "be employed as an under-labourer in clearing the ground a little, and removing some of the rubbish that lies in the way to knowledge" which is "cumbered with the learned but frivolous use of uncouth, affected, or unintelligible terms, introduced into the sciences" (Locke 1975, 10). In fact, "vague and insignificant forms of speech, and abuse of language, have so long passed for mysteries of science" that to free science from them will be "some service to human understanding" (ibid.).

The view of the philosopher as an under-labourer of the sciences has been inherited by analytic philosophy, but it is illusory to think that philosophy can curb conceptual confusions of mathematics and sciences. Mathematicians and scientists do not have to wait for philosophers to clear up conceptual confusions, this is an integral part of their work, and they are much more competent to the task than philosophers.

Even an analytic philosopher like Ryle says: "I have long since learned to doubt the native sagacity of philosophers when discussing technicalities which they have not learned to handle on the job" (Ryle 1964, 12).

Moreover, a philosopher conceived as an under-labourer of the sciences could not open new ways to knowledge and lead to the birth of new sciences, because he would not formulate new ideas about questions that are beyond the existing sciences.

(3) It is not valid to say that the birth of an independent science does not free philosophy from a host of philosophical questions associated with the subject matter of that special science because, on the one hand, no philosophical question can

be answered by scientific enquiry, and, on the other hand, no scientific discovery can be made by philosophical investigation.

For, on the one hand, the claim that no philosophical question can be answered by scientific enquiry conflicts with the fact that, since antiquity, many philosophical questions about the world have been answered by sciences. On the other hand, the claim that no scientific discovery can be made by philosophical investigation conflicts with the fact that philosophy has led to the birth of several new sciences and, through them, to several scientific discoveries.

(4) It is not valid to say that the midwife explanation is implausible when we turn to practical philosophy, namely ethics, political and legal philosophy, because the latter do not give birth to new sciences.

This argument assumes that ethics, legal, and political philosophy are about values while knowledge is about facts, and values and facts are completely independent of each other. But this assumption is unjustified because, as will be argued in Chap. 5, ethics, legal, and political philosophy are also concerned with knowledge, although not theoretical knowledge but practical knowledge, that is, knowledge as a means to action. Therefore, the midwife explanation is plausible even when we turn to practical philosophy.

Since Hacker's arguments against the midwife explanation are not valid, Hacker's claim that philosophy is not acquisition of knowledge is not valid.

3.5 Philosophy as Only Aiming at Understanding

While, negatively, Hacker argues that philosophy is not acquisition of knowledge, positively, he argues that philosophy aims only to understand what we already know.

He says that this is "a conception of philosophy that is wholly inspired by Wittgenstein," although it "is not quite the same as his" (Hacker 2013a, x).

Hacker puts forward the following arguments for the view that that philosophy aims only to understand what we already know.

(1) "Philosophy is not a contribution to human knowledge but to human understanding" (Hacker 2013a, 9). It achieves understanding through "conceptual analysis" (Hacker 2013b, 446). Conceptual analysis does not require new knowledge, to perform it "one need only assemble and marshal what one already knows" (ibid., 454). Through conceptual analysis, philosophy is able "to solve or dissolve conceptual unclarity or misunderstanding, and to answer conceptual questions" (ibid., 462). The concepts that philosophy can clarify are, on the one hand, the "ordinary ones, familiar to any mature speaker of the language" (Hacker 2013a, 17). On the other hand, they are the concepts of the sciences and mathematics. Of course, philosophy cannot "introduce novel concepts for the purposes of other disciplines (e.g. for mathematics, physics or biology)—for that is not its business," but it "may show other disciplines what is awry with their concepts or with their use of their concepts" (Hacker 2013b, 463). For example, philosophy clarifies the nature of non-Euclidean geometries by showing that, contrary to a widespread opinion, they "are

not alternative theories of space but alternative grammars for the description of spatial relationships" (Hacker 2013a, 11).

(2) The conceptual analysis through which philosophy achieves understanding is an investigation into "the uses of words, phrases, and sentences" (ibid., 17). For, "to possess a concept is to have mastered the use of a word or phrase" (Hacker 2007, 16). And "there is no way to characterize a concept other than by describing the relevant features of the uses of expressions that express that concept or belong to the category of concepts it subsumes" (Hacker 2013a, 13). The aim of an investigation into the uses of words, phrases, and sentences is to determine "their correct use" (Hacker 2013b, 460). But "we do not need to engage in sociolinguistic surveys to establish how the expressions that concern us are used" (Hacker 2007, 14). What is needed for this purpose is only "one's competence as a mature language-user (aided by the reminders of the *Oxford English Dictionary* and by the etymology and history of words it provides)" (Hacker 2013b, 451). For, an investigation into the uses of words, phrases, and sentences is an investigation into the rules for such uses, and the latter are "identical with rules for the uses of words in a dictionary (e.g. 'A vixen is a female fox')," although "they are not crafted for lexicographical purposes" (Hacker 2013a, xiv). Such an investigation is independent of the "differences between languages," because "a philosophical investigation into the use of a word is an investigation into the concept expressed" (ibid., 12). And a concept is independent of the differences between languages. Thus, a conceptual analysis of the concept of knowledge is an "investigation into the use of 'know'" and such an investigation "will yield much the same results as a philosophical investigation into the uses of 'wissen' and 'kennen'" (ibid.). For, "the investigation, whether conducted in English or in German, is an investigation into those features of usage that determine the common concept of knowledge" (ibid.).

(3) Through conceptual analysis, philosophy can study the nature of things. Indeed, what philosophy is "ultimately interested in is the nature of things! So it is this that we should be investigating" (Hacker 2013b, 442). In fact, while "the study of the nature of things, in one sense, belongs to the empirical sciences," this study, "in another sense, belongs to philosophy" (Hacker 2007, 7). Philosophy studies the nature of things through conceptual analysis, because "to give such a conceptual analysis is to describe the nature of what is signified" (Hacker 2013b, 452). The nature of a thing "is surely independent of the forms of natural languages, the logical adequacy of which is to be judged precisely by the extent to which they mirror the nature of the things they describe" (ibid., 442). The independence of the nature of a thing of the forms of natural languages derives from the fact that "the nature of something (that has a nature) consists of those attributes without which the thing would not be what it is" (ibid., 452). Nevertheless, although the philosophical investigation aims at the study of the nature of a thing thus understood, it concerns "rules for the use of the words signifying things" (ibid., 446). The rules determine the meaning of these words. For example, the study of the nature of red concerns the "rules for the use of the word 'red' that determine its meaning" (ibid., 143). The rules state "that red is a colour, that if any object is red it is extended, that red is darker than pink, that nothing can simultaneously be red all over and green all over,

that red is seen and not heard, and so forth" (ibid.). These rules determine the meaning of the word 'red' and hence the nature of red.

3.6 Against Philosophy as Only Aiming at Understanding

Hacker's arguments for the view that philosophy aims only to understand what we already know, however, are not valid.

(1) It is not valid to say that philosophy is not a contribution to human knowledge but to human understanding, and achieves understanding through conceptual analysis, which can clarify both ordinary concepts and the concepts of the sciences and mathematics.

Indeed, on the one hand, conceptual analysis cannot clarify ordinary concepts. For, many unclarities concerning ordinary concepts depend on prejudices or lack of knowledge. Dealing with them generally requires knowledge. This contrasts with the fact that, as Hacker himself admits, conceptual analysis permits only to assemble and marshal what one already knows.

On the other hand, conceptual analysis cannot clarify the concepts of mathematics and the sciences. Therefore, it is not valid to say that philosophy clarifies the nature of non-Euclidean geometries by showing that, contrary to a widespread opinion, they are not alternative theories of space but alternative grammars for the description of spatial relationships. If non-Euclidean geometries were merely alternative grammars for the description of spatial relationships, there would have been no urge to formulate them. On the contrary their creators, from Lobatchevsky to Riemann, formulated them because they felt that Euclidean geometry was inadequate to deal with the physical space, and that new geometries were needed for that purpose.

For example, Riemann formulated his elliptic geometry because he felt that Euclidean geometry was inadequate to deal with the infinitely small, and that a new geometry was needed for that purpose. He was moved by the conviction that, if bodies do not "exist independently of position," then "we cannot draw conclusions from metric relations of the great, to those of the infinitely small" (Riemann 2016, 39–40). We "are therefore quite at liberty to suppose that the metric relations of space in the infinitely small do not conform to the hypotheses of" Euclidean "geometry; and we ought in fact to suppose it, if we can thereby obtain a simpler explanation of phenomena" (ibid., 40).

(2) It is not valid to say that the conceptual analysis through which philosophy achieves understanding is an investigation into the uses of words, phrases, and sentences. For, philosophical problems are not questions about the use of language but questions about the world, and such questions cannot be dealt with through an investigation into the uses of words, phrases, and sentences.

This has been emphasized by many philosophers, from Plato to Hume and Kant.

Thus, Plato says that a philosopher who only investigates into the uses of words, phrases, and sentences is only "a maker of words, incapable of doing any" worthwhile "work" (Plato, *Epistulae*, VII 328 c 5–6). Indeed, "things are to be learned and sought for not from names, but from things themselves much more than from names" (Plato, *Cratylus*, 439 b 6–8).

Hume says that there is always the risk "for philosophers to encroach upon the province of grammarians; and to engage in disputes of words, while they imagine, that they are handling controversies of the deepest importance and concern" (Hume 1983, 98).

Kant says that, "in matters over which one has quarreled over a long period of time, especially in philosophy, there has never been at the basis a quarrel of words but always a true quarrel over things" (Kant 2007, 179). And "it is to be feared that in artificially fathoming 'logomachia'," namely verbal disputes, "everywhere" one "falls prey to 'logodaedalia'," namely quibbling over words, "beyond which nothing more disadvantageous can happen to philosophy" (ibid., 180). Thus, Kant strongly emphasizes that "the work to be done in philosophy is not exhausted at the linguistic level" (Capozzi 2011, 334–335).

It is unfounded to think that philosophy can deal with questions about the world through an investigation into the uses of words, phrases, and sentences. For, the sciences change our image of the world time and time again, and the uses of words, phrases, and sentences can only chase these changes, but will always lag behind. An investigation into the uses of words, phrases, and sentences, either ends up reducing problems about the world to verbal trivialities, or bases the investigation into questions about the world upon prejudices or mistaken opinions.

An example of this is Hacker's claim that, to understand what knowledge is, it suffices one's competence as a mature language-user (aided by the reminders of the *Oxford English Dictionary* and by the etymology and history of words it provides).

Contrary to this claim, already Plato warned that, when someone asks you to tell him "what you think knowledge is," he does not want to know what the word 'knowledge' means but "what knowledge in itself is" (Plato, *Theaetetus*, 146 c 3–146 e 10).

Anyway, suppose that, disregarding Plato's warning, you interpret the request to tell 'what you think knowledge is' as a request to explain what the word 'knowledge' means, and, in order to answer this request, you look up the *Oxford English Dictionary*. There you find: '*Philos*. true, justified belief; certain understanding, as opp. to opinion'. This will not let you know what knowledge is. For, on the one hand, saying that knowledge is 'true, justified belief', conflicts with the fact that, as already Plato argued, this is not valid (see Chap. 10). On the other hand, saying that knowledge is 'certain understanding, as opp. to opinion', conflicts with the fact that, by Gödel's second incompleteness theorem, even mathematical knowledge is not certain understanding.

In fact, looking up the *Oxford English Dictionary* does not tell you what knowledge is, it only gives you information about the opinions, or the prejudices, of the author of the dictionary entry 'knowledge'. An investigation into the use of the word 'knowledge' is totally irrelevant to the philosophical question of determining what knowledge is. As will be argued in Chap. 10, what is required to answer this question is an investigation into the role knowledge plays in human life. But this is not an investigation into the ordinary use of the word 'knowledge'.

Hacker, and the whole Oxford analytic philosophy, worship the use of words, phrases, and sentences. Even some analytic philosophers outside of Oxford analytic philosophy worship this use.

Thus Cappelen and McKeever say: "What role does ordinary language play in philosophical theorizing today? One might think: little. After all, analytic philosophy has moved past its 'ordinary language' phase" (Cappelen and McKeever 2022, 221). But this is not the case. For, "contemporary analytic philosophy pays more attention to ordinary language use than the implicit historiography mentioned above suggests" (ibid.). In fact, "ordinary language is not just a tool or instrument for seeing the world better. It also determines what it is we look at, and—sometimes—it tells us things about what we look at" (ibid., 236).

But this use strongly depends, if not on prejudices or mistaken opinions, at least on current knowledge. And it can become soon obsolete because of scientific, technological, or social developments that are too fast for the use of words, phrases, and statements to keep pace with them.

By assuming that philosophy is not acquisition of knowledge but only a contribution to understand what we already know, analytic philosophy has led philosophy to irrelevance.

Even Hacker admits that analytic philosophy "has split into numerous rivulets meandering aimlessly through the marshlands. It has" become "over-specialized. It is imbued with the spirit of scholasticism, conflating pedantry with precision, and confusing technicality with clarity" (Hacker 2013a, xxi–xxii). Hacker says that "whether all this represents the dying embers of a once great movement of thought, or whether this phase is merely a pause before the further development of something that can be deemed a continuation of the analytic tradition, only time can tell" (ibid., xxii). But time has already told, all this represents the dying embers of the analytic tradition.

(3) It is not valid to say that, through conceptual analysis, philosophy can study the nature of things. For, Hacker asserts that the nature of something consists of those attributes without which the thing would not be what it is. But the attributes without which a thing would not be what it is define the essence of the thing. Therefore, by saying that, through conceptual analysis, philosophy can study the nature of things, Hacker claims that, through conceptual analysis, philosophy can study the essence of things.

This means that Hacker claims that philosophy can perform a task that, as we will see in Chap. 4, Galileo says to be an impossible and vain undertaking for science. Hacker also claims that philosophy can perform that task by the poor means of the investigation into the ordinary uses of words, phrases, and sentences afforded by our competence as mature language-users (aided by the reminders of the *Oxford English Dictionary* and by the etymology and history of words it provides). Given the disproportion between penetrating the essence of things and the poverty of the means envisaged by Hacker to that aim, his claim that by conceptual analysis philosophy can study the nature of things is utterly implausible.

In particular, as we have seen, Hacker asserts that the essence of red is given by the rules of use of the word ‘red’, namely that red is a colour, that if any object is red it is extended, that red is darker than pink, that nothing can simultaneously be red all over and green all over, that red is seen and not heard, and so forth. But these rules do not give the essence of red, they only state some phenomenal properties of red. The same can be said of any other colour. If we want to know what red is, or what colour itself is, we must give up the demand to know their essence, and content ourselves with knowing some of their phenomenal properties, such as those investigated by physics and the physiology of human visual perception.

Since Hacker’s arguments for the view that philosophy aims only to understand what we already know are not valid, we may conclude that this view is untenable.

References

Capozzi, Mirella. 2011. Philosophy and writing: The philosophical book according to Kant. *Quaestio* 11:307–350.

Cappelen, Herman, and Matthew McKeever. 2022. In defense of ordinary language philosophy. *Metaphilosophy* 53:221–237.

Hacker, Peter. 2007. *Human nature: The categorical framework*. Malden: Blackwell.

Hacker, Peter. 2013a. *Wittgenstein: Comparisons and context*. Oxford: Oxford University Press.

Hacker, Peter. 2013b. *The intellectual powers: A study of human nature*. Chichester: Wiley-Blackwell.

Hume, David. 1983. *An enquiry concerning the principles of morals*. Indianapolis: Hackett, Indianapolis.

Kant, Immanuel. 2007. *Anthropology, history, and education*. Cambridge: Cambridge University Press.

Locke, John. 1975. *An essay concerning human understanding*. Oxford: Oxford University Press.

Riemann, Bernhard. 2016. *On the hypotheses which lie at the bases of geometry*. Cham: Springer.

Ryle, Gilbert. 1964. *Dilemmas*. Cambridge: Cambridge University Press.

Chapter 4
Philosophy as Acquisition of Knowledge

Abstract That philosophy has never been considered so irrelevant raises the question of how to escape from irrelevance. This chapter argues that philosophy can escape from irrelevance if and only if it is acquisition of knowledge, that is, an inquiry aimed at acquiring knowledge of the world.

Keywords Philosophy as acquisition of knowledge · Philosophy and the creation of new sciences · The heuristic view of philosophy · Differences between philosophy and the present sciences

4.1 What Is Philosophy?

That philosophy has never been considered so irrelevant raises the question of how to escape from irrelevance. In Chap. 2 it has been argued that Unger's proposed escape from irrelevance would not work, so an alternative escape is necessary. This chapter argues that philosophy can escape from irrelevance if and only if it is acquisition of knowledge, that is, an inquiry aimed at acquiring knowledge of the world.

To deal with the question of how philosophy can escape from irrelevance, we must begin by asking: What is philosophy?

Against this, Popper argues that "a philosopher should philosophise: he should try to solve philosophical problems, rather than talk about philosophy" (Popper 2002, 90).

But this argument is self-refuting. For, Popper is the author of an article entitled 'How I see philosophy' in which he says that "all philosophy must start from the dubious and often pernicious views of uncritical common sense. Its aim is to reach enlightened, critical common sense: to reach a view nearer to the truth; and with a less pernicious influence on human life" (Popper 1994, 179–180).

Therefore, Popper talks about philosophy.

What is more important, Popper's argument conflicts with the fact that which problems one wants to solve and how one solves them depend on the answer to the question: What is philosophy?

C. Cellucci, *A Philosophy for the Future*, Synthese Library 508,
https://doi.org/10.1007/978-3-031-98443-3_4

From its very beginning as a discipline, philosophy has called into question all human knowledge including itself, so it has been a reflection on the nature of philosophy. Therefore, philosophers have always talked about philosophy.

As Illetterati and Miolli say, "'metaphilosophy' is a recent word for an ancient theoretical practice. At least from Plato's speculation onwards, philosophy has questioned its own 'nature', making it somewhat redundant today to emphasize this aspect" (Illetterati and Miolli 2022, 2), On the contrary, "the need to isolate a specific name for philosophy's investigation of itself and to clearly circumscribe the scope of this self-reflective activity is new and symptomatic of the extraordinary fragmentation of twentieth-century philosophical discourse" (ibid., 3).

Sellars even says that "a philosopher could scarcely be said to have his eye on the whole in the relevant sense, unless he has reflected on the nature of philosophical thinking. It is this reflection on the place of philosophy itself, in the scheme of things which is the distinctive trait of the philosopher," indeed, "in the absence of this critical reflection on the philosophical enterprise, one is at best but a potential philosopher" (Sellars 1991, 3).

A philosopher must reflect on the nature of philosophy especially today because, in order to find an escape from irrelevance, it is essential to give an adequate answer to the question: What is philosophy? Without such an answer, it is impossible to know in which direction philosophy should go.

And to know in which direction philosophy should go is essential because, as Seneca says: "Ignoranti, quem portum petat, nullus suus ventus est [For a man who does not know what port to make for, no wind is the right wind]" (Seneca, *Ad Lucilium Epistulae morales*, LXXI, 3).

4.2 Origin of Philosophy as Acquisition of Knowledge

An important answer to the question 'What is philosophy?' is that philosophy is acquisition of knowledge, that is, an inquiry aimed at acquiring knowledge of the world.

The view that philosophy is acquisition of knowledge is implicit in the assertion of several Presocratics that philosophers must be inquirers of the world.

Thus, Pythagoras says that "every human being has been made up by the god to know and inquire" (Aristotle, *Protrepticus*, fr. 20, ed. Düring). When Pythagoras was asked what "is the thing for the sake of which nature and the god engendered us," he "replied: to inquire the heavens," and "he himself was an inquirer of nature and had come to life for this purpose" (ibid., fr. 18).

Xenophanes says that "the gods have not revealed all things to men from the beginning but, by inquiring, men find out, in time, what is better" (Xenophanes 21 B 18). They find out what is better by making hypotheses, so "all is but a woven web of hypotheses" (ibid., 21 B 34).

Heraclitus says that "nature loves to hide" (Heraclitus 22 B 123, ed. Diels-Kranz). Therefore, "philosophers must be good inquirers into many things" (ibid.,

22 B 35). In their inquiries, philosophers are always liable to fail, they are like those "seekers of gold" who "dig up much earth, but find little gold" (ibid., 22 B 22). However, they can also succeed, because "what encompasses us," namely nature, "is rational and endowed with intelligence" (ibid., 22 A 16). And the human mind "recognizes the intelligence which rules all things through all things" (ibid., 22 B 41).

Parmenides says that the philosopher "must inquire into everything" (Parmenides 28 B 1 Diels–Kranz). So the philosopher "will know the nature of the aether, and in the aether all the stars, and the resplendent works of the glowing sun's clear torch and whence they arose" (ibid., 28 B 10). He will know "the wandering deeds of the round-faced moon and her origin," and "the heaven that surrounds everything, whence it arose, and how Necessity bound it to keep the limits of the stars" (ibid.).

4.3 The Rise of Philosophy as Discipline

While implicit in the assertion of several Presocratics that philosophers must be inquirers of the world, the view that philosophy is acquisition of knowledge is first explicitly formulated by Plato.

Indeed, Plato says: "Philosophy is acquisition of knowledge [Philosophia ktesis epistemes]" (Plato, *Euthydemus*, 288 d 8).

The view that philosophy is acquisition of knowledge is at the origin of philosophy as discipline.

As Nightingale says, with his assertion that philosophy is acquisition of knowledge, "Plato appropriated the term 'philosophy' for a new and specialized discipline—a discipline that was constructed in opposition to the many varieties of 'sophia' or 'wisdom' recognized by Plato's predecessors and contemporaries" (Nightingale 1995, 14).

Plato explains the view that philosophy is acquisition of knowledge by saying that, although the name 'philosophia' means 'love of wisdom', wisdom is in no way "different from knowledge," because people are "wise in just those things of which they have knowledge," so "knowledge and wisdom are the same thing" (ibid., 145 e 1–6). Therefore, "to be a philosopher is the same thing as to be a lover of knowledge" (Plato, *Respublica*, II 376 b 9–10). Only "the one who is wholeheartedly ready to taste every sort of knowledge, who is eager to know and is insatiable for it, can be rightly called a philosopher" (ibid., V 475 c 6–8).

Philosophy as discipline, based on the view that philosophy is acquisition of knowledge, was first institutionalized in Plato's Academy.

As Moore says, Plato "propounded" the view "of 'philosophia' as the pursuit of a comprehensive and deep understanding of the world" in "the Academy, and, indeed, its institutional structure, encouraging research in mathematics and methodology, assumed it" (Moore 2020, 221). Plato founded the Academy to have "an institution of research and teaching that would not only outlive him but also, even while he lived, produce materials independent of his hand" (ibid.).

Thus, not only Plato's assertion that philosophy is acquisition of knowledge is at the origin of philosophy as discipline, but philosophy as discipline was first institutionalized in Plato's Academy, based on the view that philosophy is acquisition of knowledge.

4.4 The Tradition of Philosophy as Acquisition of Knowledge

The view that philosophy is acquisition of knowledge has been repeatedly reaffirmed since Plato.

Thus, Aristotle says: "By nature, all humans desire to know" (Aristotle, *Metaphysica*, A 1, 980 a 21). This especially holds of philosophers, who aim to know "the first causes and the principles of things" (ibid., A 1, 981 b 28–29). This type of knowledge is "the ultimate thing for the sake of which we have come to be" (Aristotle, *Protrepticus*, fr. 17, ed. Düring).

Cicero says: "Philosophy, to interpret the word, is nothing else but the study of wisdom. And wisdom, as old philosophers defined it, is knowledge of divine and human things and of the causes by which these things are connected together" (Cicero, *De officiis*, II.ii.5).

Bacon says: "I have taken all knowledge to be my province" (Bacon 1857–1874, VIII, 109). Philosophy is a "universal science, to be as the mother of the rest, and to be regarded in the progress of knowledge as portion of the main and common way, before the ways separate and divide" (ibid., IV, 337).

Descartes says: "This word 'philosophy' means the study of wisdom," where "by wisdom is meant not only prudence in everyday affairs, but also a perfect knowledge of all things that mankind is capable of knowing, both for the conduct of his life, and for the conservation of his health, and for the discovery of all the arts" (Descartes 1996, IX–2, 2). Indeed, "to try to acquire" this type of knowledge "is properly termed philosophizing" (ibid.).

Hobbes says: "Philosophy" is "the study of wisdom" (Hobbes 1839–1845, I, xiv). But "wisdom, properly called, is nothing else" but "the perfect knowledge of the truth in all matters whatsoever" (ibid., II, iii).

Berkeley says: "Philosophy" is "nothing else but the study of wisdom and truth" (Berkeley 2009, 68).

Kant says: "We would do well to leave" the word 'philosophy' "in its ancient sense" as "the doctrine of wisdom," because this "would suit the Greek expression (which signifies love of wisdom) while yet sufficing to embrace under the name of philosophy love of science" (Kant 2015, 88). For, science is the "organon of wisdom" and as such is "indispensable for it, so that one may well maintain that wisdom without science is" only "a silhouette of a perfection to which we shall never attain" (Kant 1992, 539). In fact, "without cognitions one will never become a philosopher" (ibid., 538).

Mach says: "Scientific thought presents itself in two seemingly different forms: as philosophy and as specialist research," namely research of "the specialist scientist" (Mach 1976, 2). But "the ultimate end of all research is just the same. This shows itself also in the fact that the greatest philosophers" have "opened up new ways of specialist enquiry" for the specialist scientist, and "philosophy has furnished science with some positive notions of value too" (ibid., 3).

The view that philosophy is acquisition of knowledge also underlies a multi-volume history of philosophy by Chiaradonna and Pecere, who argue that "philosophy represents, in a privileged way, the moment of the search for knowledge, the creation of new forms of knowledge" (Chiaradonna and Pecere 2018, vol. 1A, v). Over the centuries, "scientific knowledge has often had a philosophical genesis" (ibid.). On the other hand, "studying philosophy reminds us that there is still much to be written on the path of knowledge" (ibid.).

4.5 Analytic Philosophy and Acquisition of Knowledge

While the view that philosophy is acquisition of knowledge has been repeatedly reaffirmed since Plato, in the last century the dominant philosophy, analytic philosophy, has abandoned it. According to analytic philosophy, philosophy is not and should not be acquisition of knowledge, that is, an inquiry aimed at acquiring knowledge of the world, because acquiring knowledge of the world is the purpose of science, not philosophy. Philosophy aims only to understand what we already know.

This view has been strongly affirmed by eminent exponents of analytic philosophy.

Thus, Wittgenstein says that "philosophers constantly see the method of science before their eyes, and are irresistibly tempted to ask and answer questions in the way science does," but this "leads the philosopher into complete darkness" (Wittgenstein 1969, 18). In fact, it is "essential to our investigation that we do not seek to learn anything new by it" but only "to understand something that is already in plain view" (ibid.). In philosophy, "it is not that a new building has to be erected, or that a new bridge has to be built," but only "that the geography, as it now is, has to be described" (Wittgenstein 1978, 302). Philosophy has no impact on the growth of knowledge, it "leaves everything as it is" (Wittgenstein 2009, 124). It "just puts everything before us, and neither explains nor deduces anything" (ibid., 126). In philosophy "there are no great essential problems in the sense of science" (Wittgenstein 2005, 301). Philosophy "is a tool which is useful only against the philosophers" (Wittgenstein 1932–1933, 11).

Ryle says that "philosophy is not a sister science or a parent science," its "business is not to add to the number of scientific statements" (Ryle 2009b, I, 261–262). Philosophy is "intended not to increase what we know," but only to determine "the logical geography of the knowledge which we already possess" (Ryle 2009a, lix). The aim of philosophy is only "the clarification of ideas" (Ryle 2009b, II, 212). Indeed, "philosophers do not make known matters of fact which were unknown before. The sense in which they throw light is that they make clear what was unclear before, or make obvious things which were previously in a muddle" (ibid., II, 166). Thus, "the philosopher throws new light, but he does not give new information" (ibid., II, 173).

Dummett says that "philosophy is indeed concerned with reality, but not to discover new facts about it: it seeks to improve our understanding of what we already know. It does not seek to observe more, but to clarify our vision of what we see" (Dummett 2010, 10). Philosophy "concerns our view of reality" only "by seeking to clarify the concepts in terms of which we conceive of it, and hence the linguistic expressions by means of which we formulate our conception" (ibid.).

Sellars says that "the aim of the philosopher is not to discover new truths, but to 'analyse' what we already know" (Sellars 1991, 3). Indeed, "the aim of philosophy" is "to understand how things in the broadest possible sense of the term hang together in the broadest possible sense of the term" (ibid., 1). Therefore, philosophy "makes no substantive contribution to what we know," it is only "concerned in some way to improve the manner in which we know it" (ibid., 3).

Brandom says that "philosophers aim" not "at a kind of knowledge" but "at a kind of understanding" (Brandom 2009, 113). Specifically, the kind of understanding that is the characteristic goal of philosophy is "the explication of concepts—or, put slightly more carefully, the development and application of expressive tools with which to make explicit what is implicit in the use of concepts" (ibid., 114). The "point of explicating concepts" is "opening them up to rational criticism," because "defective concepts distort our thought and constrain us" (ibid.).

The view of analytic philosophy that philosophy is not acquisition of knowledge but aims only to understand what we already know, marks a sharp discontinuity between analytic philosophy and the philosophical tradition starting with Plato.

As Glymour says, "Plato and Aristotle thought the goal of philosophy" was "to provide methods for coming to have knowledge. This same conception utterly dominated philosophy in the 17th century," but "after 1925 or thereabouts," a "tradition that joined together much of the classical philosophical literature simply vanished" (Glymour 1991, 75).

This has produced disorientation and bewilderment among analytic philosophers themselves.

Thus, Quine says that by now 'philosophy' is only "one of a number of blanket terms used by deans and librarians in their necessary task of grouping the myriad topics and problems" under "a manageable number of headings" (Quine 2008, 293–294). By this "I am not alluding to the fragmentation of specialties; I speak of the insignificance of a certain verbal grouping" (ibid., 294).

But above all, the view of analytic philosophy that philosophy is not acquisition of knowledge has led to the irrelevance of philosophy. For, if philosophy is not acquisition of knowledge, then philosophy is like "a wheel that can be turned though nothing else moves with it" (Wittgenstein 2009, 271).

4.6 Analytic Philosophy, Experience and Intuition

The view of analytic philosophy that philosophy is not acquisition of knowledge, implies that experience has no part in solving philosophical problems.

Thus, Wittgenstein says that "philosophical problems are not solved by experience" (Wittgenstein 2001, 3).

Since, according to analytic philosophy, experience has no part in solving philosophical problems, analytic philosophy is bound to appeal to intuition rather than to experience as the source for their solution.

Kripke even says: "I think" that "something's having intuitive content" is "very heavy evidence in favor of anything, myself. I really don't know, in a way, what more conclusive evidence one can have about anything, ultimately speaking" (Kripke 1980, 42).

But, being subjective and arbitrary, intuition cannot be really evidence in favor of anything. Asserting that a statement is true on the basis of intuition amounts to saying: I feel it is so, therefore it must be so. But two philosophers can say 'I feel it is so' of two opposing views, and this leaves no criterion on how to choose between them.

In fact, as Lewis says, "'intuitions' are simply opinions," where "some are commonsensical, some are sophisticated; some are particular, some are general, some are firmly held, some less. But they are all opinions" (Lewis 1983, x).

4.7 An Incongruous Claim

Some eminent exponents of analytic philosophy claim that, from time to time, philosophy throws off some portion of itself to take station as a science.

Thus, Russell says that, "as soon as definite knowledge concerning any subject becomes possible, this subject ceases to be called philosophy, and becomes a separate science" (Russell 1998, 90). In fact, "the whole study of the heavens, which now belongs to astronomy, was once included in philosophy; Newton's great work was called 'the mathematical principles of natural philosophy'. Similarly, the study of the human mind, which was a part of philosophy," has now "become the science of psychology" (ibid.). So, "those questions which are already capable of definite answers are placed in the sciences" (ibid.).

Austin says that "from time to time" philosophy "throws off some portion of itself to take station as a science" (Austin 1970, 232). This "happened long ago at the birth of mathematics, and again at the birth of physics," in "the last century we have witnessed the same process once again" in the birth "of mathematical logic," and it is possible that "the next century may see the birth of" a "science of language" (ibid.). So philosophy shall have rid itself of one more portion of itself by "kicking it upstairs" (ibid.).

Dummett says that "philosophy has budded off" several sciences, which then "have declared their independence from it" (Dummett 2010, 3). Thus "psychology, pursued by rational methods alone, was already solidly established as a branch of philosophy;" and "logic was formerly regarded as a branch of philosophy," but with mathematical logic it "sought a home within mathematics;" and "general linguistics, likewise, parted company with philosophy, which had nurtured it" (ibid.). Also, "cognitive science has raided philosophical territory and set itself up as a science in its own right" (ibid., 4).

The claim that, from time to time, philosophy throws off some portion of itself to take station as a science, is incongruous with the basic assumption of analytic philosophy that philosophy is not acquisition of logic. For, the portion of itself that philosophy throws off is knowledge that philosophy has acquired.

4.8 Another Incongruous Claim

Some eminent exponents of analytic philosophy also claim that the denial that philosophy is acquisition of knowledge has had a very negative effect on philosophy.

Thus, Russell says that, while "philosophers from Thales onwards have tried to understand the world" and, "even when they have failed, they have supplied material to their successors and an incentive to new efforts," analytic philosophy is not "carrying on this tradition" (Russell 1995a, 170). It seems "to have abandoned, without necessity, that grave and important task which philosophy throughout the ages has hitherto pursued," and "to concern itself, not with the world and our relation to it, but only with the different ways in which silly people can say silly things" (ibid.). It "cares only about language, and not about the world" (Russell 1960, 15). Thus, Wittgenstein "seems to have grown tired of serious thinking and to have invented a doctrine which would make such an activity unnecessary" (Russell 1995b, 161). His "positive doctrines seem to me trivial," and his "negative doctrines, unfounded" (ibid., 160–161). Wittgenstein makes philosophy, "at best, a slight help to lexicographers, and at worst, an idle tea-table amusement" (ibid., 161). But, "if this is all that philosophy has to offer, I cannot think that it is a worthy subject of study" (ibid., 170).

Dummett says that, if philosophy is taught today in universities, it is only because of the "historical precedent" that, "when the first Western universities came into being," philosophy "was not sharply differentiated from what we call "natural science"," and "the quest for truth was a single activity" (Dummett 2010, 2). Then it was easy to find a justification for philosophy. But in the twentieth century "the distinction between the natural sciences and the humanities came to be universally recognized" (ibid., 3). So, it has become difficult to find a justification for philosophy. Thus, "if universities had been an invention of the second half of the twentieth century, would anyone have thought to include philosophy among the subjects that they taught and studied? It seems very doubtful," therefore "it would be easy to conclude that" philosophy "is an anachronism" (ibid., 2). The "layman or non-professional expects philosophers to answer deep questions of great import for an understanding of the world," and "the layman is quite right: if philosophy does not aim at answering such questions, it is worth nothing" (Dummett 1991, 1).

Ayer says that many analytic philosophers "do not set out to describe, or even to explain, the world, still less to change it. Their concern is only with the way in which we speak about the world," so for them philosophy is merely "talk about talk" (Ayer 1963, 3). This has driven "analytic philosophy" into "the scholasticism which has been threatening to overtake it" (ibid., 35).

The claim that the denial that philosophy is acquisition of knowledge has had a very negative effect on philosophy, is incongruous with the basic assumption of analytic philosophy that philosophy is not acquisition of knowledge. For, it admits that the denial is self-refuting.

4.9 Analytic Philosophy and Scholasticism

As we have seen above, Ayer says that the fact that, for many analytic philosophers, philosophy is merely talk about talk, has driven analytic philosophy into scholasticism.

The scholasticism to which Ayer refers is a result of the conservatism and conformism of analytic philosophy.

As Vintiadis says, analytic "philosophers are under a lot of pressure to publish in specific peer-reviewed journals and in a very narrow range of topics in order to advance their careers" (Vintiadis 2021, 88). As a result, "current publication standards have bred tunnel vision and conservatism and have dangerously narrowed the scope of what" philosophers do, "most philosophers do not address many issues" because "they are not the kind of thing that would help them advance their careers" (ibid., 88–89).

Peer review is instrumental to conservatism and conformism.

As Hanna says, "peer review" is both "an effective straitjacket that massively stifles creativity and guarantees in-crowd orthodoxy," and "an effective bottleneck that massively slows down and chokes the dissemination and sharing of creative unorthodox ideas" (Hanna 2023).

The tenure system is also instrumental to conservatism and conformism.

As Frodeman says, "ironically, tenure seems to have made" analytic philosophers "less rather than more willing to challenge the status quo. It was thought that by safeguarding jobs, tenure would make" analytic philosophers "bolder, giving them the liberty to develop and enact new ideas. Instead, tenure has diminished the incentive for" analytic philosophers "to change the system that provides them with job security" (Frodeman 2019, 105).

Even the analytic philosophers' training system is instrumental to conservatism and conformism.

As Feyerabend says, analytic philosophers "define their tasks as the analysis and preservation of what is" (Feyerabend 1999, 384). For, they think "that it is good to conform, and that it is bad not to conform," to the "standards of philosophy" which are determined by the "consensus of those in the discipline" (ibid., 383). Therefore, analytic philosophers claim that students "must be trained to repeat the tricks after" their trainers "so that sometime in the future they may perhaps be able to become trainers themselves, modifying the tricks a little here and there (this is called 'original research'), and being equally stern in the propagation of their knowledge (this is called 'professional conscience')" (ibid., 386).

This contrasts with the fact that "philosophy has very often turned against the status quo," in fact "this is one of the most eminent functions of philosophy: to critically examine existing standards, to expose their vicious character, to overthrow them by the suggestion of new standards" (Feyerabend 1999, 383). Against this, analytic philosophers might argue: If you do not conform to the standards that are determined by the consensus of those in the discipline, "the 'standards of philosophy'" will "suffer from this" (ibid., 387). But to this argument we can answer: "Well, Plato says that all things change for the worse; the worst thing alone will change for the better. The standards of philosophy, then, can only improve" (ibid.).

Philosophy lives with the great problems that confront humanity and require new and bold thoughts, it becomes irrelevant with the scholasticism of analytic philosophy.

4.10 Philosophical Revolutions and Counter-Revolutions

Plato's formulation of the view that philosophy is acquisition of knowledge, that is, an inquiry aimed at acquiring knowledge of the world, has been a revolution in philosophy, because it has given rise to philosophy as discipline.

On the other hand, as Preston says, analytic philosophy is "widely believed to have brought about a revolution in philosophy. That is to say, the birth of analytic philosophy is ordinarily thought of as being not merely the beginning of yet another philosophical school but the beginning of a new era in philosophy at large" (Preston 2004, 449).

But this is confusing counter-revolutions with revolutions. For, Plato's formulation of the view that philosophy is acquisition of knowledge has been a revolution in philosophy, because it has given rise to philosophy as discipline. On the contrary, analytic philosophy has been a counter-revolution because, by denying that philosophy is acquisition of knowledge, it has led to the irrelevance of philosophy.

4.11 Reaffirming Philosophy as Acquisition of Knowledge

Against analytic philosophy, it must be reaffirmed that philosophy is acquisition of knowledge.

Human beings are faced all the time with new problems in the world and solving them requires acquiring knowledge about the world. This is the task of both philosophy and science, therefore philosophy is acquisition of knowledge.

Nozick says: "Once upon a time, philosophy promised more than simply contents of thought" (Nozick 1989, 19).

What philosophy promised more than simply contents of thought is acquisition of knowledge. Many human problems and difficulties depend on inadequate knowledge, therefore, acquisition of knowledge is a most important promise of philosophy.

4.12 The Aim of Philosophy as Acquisition of Knowledge

According to the view that philosophy is acquisition of knowledge, the aim of philosophy is to acquire knowledge of the world.

For, philosophy is continuous with the present sciences, in the sense that it aims at a type of knowledge that is not essentially different from that of the present sciences. Therefore, the objectives of philosophy are not different from those of the present sciences, and philosophy is an activity that in many respects is not different from the present sciences.

However, philosophy differs from the present sciences in one important respect. While the present sciences are what we already know about the world, philosophy is about what we do not yet know about it, so philosophy is concerned with questions that are beyond the present sciences. They are beyond the present sciences in the sense that they are not open questions of any of the present sciences.

Only philosophy can seek and address these questions. For, scientists today are busy tackling major challenges in their own fields.

For example, physicists are busy tackling the challenge that, "even though the Standard Model" of particle physics "is exceptionally successful, it leaves a number of important questions unanswered. Among them, some are of cosmological nature. These include the origin of dark matter, inflation as well as matter-antimatter asymmetry. Addressing these fundamental questions requires physics beyond the Standard Model in some form" (Lebedev 2021).

Facing serious challenges in their own fields, scientists cannot be expected to seek and address questions that are beyond their field. It is up to philosophy to seek and address them.

In seeking and addressing these questions, philosophy tries unexplored paths and is always exposed to the risk of failure. But, if successful, it can open new ways to knowledge and lead to the birth of new sciences.

Philosophy has done so even in the past century, when it has led to the birth, for example, of computer science, cognitive science, artificial intelligence, and linguistics.

There are plenty of things we still don't know, including ninety-five percent of the universe which is still a mystery to us. There is no reason to suppose that new sciences will not arise in the future and that philosophy will not lead to their birth.

A stimulus in this direction comes from Seneca's warning: "Veniet tempus quo posteri nostri tam aperta nos nescisse mirentur [A time will come when our posterity will marvel that we did not know such obvious things]" (Seneca, *Naturales Quaestiones* 7.25).

4.13 Philosophy Is Not the Residue of the Sciences

That philosophy can open new ways to knowledge and lead to the birth of new sciences does not mean, as Austin says, that when philosophy does so, it throws off some portion of itself to take station as a science, so philosophy rids itself of one portion of itself by kicking it upstairs.

As Overgaard, Gilbert, and Burwood argue, what Austin says implies that philosophy is the "'residue' of the sciences," so "philosophy 'shrinks' as problems hitherto considered philosophical are, in Austin's phrase, 'kicked upstairs'" (Overgaard et al. 2013, 30). Then philosophy will eventually be finished off. In fact, some analytic "philosophers in Austin's day" did talk "about how long it would take to finish off the subject" (ibid.)

On the contrary, the view that philosophy is acquisition of knowledge implies that philosophy is not the residue of the sciences. According to it, by opening new ways to knowledge and leading to the birth of new sciences, philosophy does not shrink because it does not lose a part of itself, it simply opens new ways to knowledge of the world. So, philosophy will not be finished off to the extent that knowledge will not be finished off.

4.14 The Role of Philosophy in Galileo's Revolution

Not only philosophy can open new ways to knowledge and lead to the birth of new sciences, but it has played a crucial role in the birth of modern science itself, which has been the result of a philosophical revolution, Galileo's revolution.

Galileo says: "Either, by speculating, we seek to penetrate the true and intrinsic essence of natural substances, or we content ourselves with coming to know some of their" phenomenal "properties" (Galilei 1968, V, 187). But trying to penetrate the essence of natural substances is "a not less impossible and vain undertaking with regard to the closest elemental substances than with the remotest celestial things" (ibid.). Therefore, we must content ourselves with dealing with "some" phenomenal "properties of them, such as their location, motion, shape, size, opacity, mutability, generation, and dissolution" (ibid., V, 188). Such phenomenal properties of natural substances are all mathematical in kind, including mutability, generation, and dissolution, because they are changes in size and shape. While we cannot know the essence of natural substances, "we need not despair of our ability" to come to know such phenomenal properties "even with respect to the remotest bodies, just as those close at hand" (ibid.).

As we will argue below, this is a philosophical revolution with respect to the philosophical tradition going back to Aristotle.

4.15 The Two Components of Galileo's Revolution

Galileo's revolution has two components. The first component is the change of the object of science with respect to Aristotle.

According to Aristotle, "to know a thing is to know its essence" (Aristotle, *Metaphysica*, Z 6, 1031 b 20–21). For, "any single thing and its pure essence coincide" (ibid., Z 6, 1031 b 19–20). The "essence of a thing is what the thing is said to be in itself" (ibid., Z 4, 1029 b 14–15).

Since to know a thing is to know its essence, according to Aristotle the object of science is to know the essence of things.

On the contrary, according to Galileo's revolution, the object of science cannot be to know the essence of things, it can only be to know some phenomenal properties of things, mathematical in kind.

The second component of Galileo's revolution is the assignment of a central role to mathematics in science. This is another change with respect to Aristotle.

According to Aristotle, natural science and mathematics are different sciences, because "natural science treats of certain movable things," while mathematics treats of "things which are immovable" (Aristotle, *Metaphysica*, E 1, 1026 a 12, 14–15). Now, different sciences have different genera, and the principles of a science "must be in the same kind as the things demonstrated" (Aristotle, *Analytica Posteriora*, A 28, 87 b 2–3). Therefore, "what pertains to one science cannot be demonstrated by another science" (ibid., A 7, 75 b 14). For, "the underlying genus is different" (ibid., A 9, 76 a 12).

According to Aristotle, since natural science and mathematics are different sciences and what pertains to one science cannot be demonstrated by another science, it follows that what pertains to natural science cannot be dealt with by mathematics.

On the contrary, according to Galileo's revolution, since the object of science is to know some phenomenal properties of things mathematical in kind, what pertains to natural science can and must be dealt with by mathematics. For, dealing with phenomenal properties of things mathematical in kind necessarily requires mathematics.

In fact, Galileo says that "wanting to deal with natural questions without geometry is attempting to do what is impossible to be done" (Galilei 1968, VII, 229).

As Kant argues, with Galileo's revolution, "in any special doctrine of nature there can be only as much proper science as there is mathematics therein" (Kant 2002, 184). That is, there can be only "as much proper science as there is mathematics capable of application there" (ibid., 185).

Galileo's revolution has played an essential role in the birth of modern science because the latter is based on Galileo's philosophical assumptions.

Thus, Newton says that "the moderns, rejecting substantial forms and occult qualities, have endeavoured to subject the phenomena of nature to the laws of mathematics" (Newton 1962, I, xvii).

So, Newton reaffirms the two components of Galileo's revolution: the rejection of substantial forms and occult qualities, thus the essence of things, and the attempt to subject the phenomena of nature to the laws of mathematics.

4.16 The Argument of the Great Book of the World

An argument could be raised against the claim that, according to Galileo's revolution, the object of science cannot be to know the essence of things, it can only be to know some phenomenal properties of things, mathematical in kind.

Galileo asserts that natural "philosophy is written in this very great book that is continually open to us before our eyes (I say the universe)" (Galilei 1968, VI, 232). It "is written in mathematical language, and the characters are triangles, circles, and other geometrical figures, without the use of which it is impossible to humanly understand any word of it" (ibid.).

Husserl claims that Galileo' assertion means that, for him, "nature is, in its 'true being-in-itself', mathematical" (Husserl 1970, 54).

This implies that, by knowing properties of things mathematical in kind, science knows nature in its true being-in-itself, thus in its essence. This contradicts the claim that, according to Galileo's revolution, the object of science cannot be to know the essence of things.

But this argument is not valid. Contrary to Husserl's claim, Galileo's assertion does not mean that, for him, nature is, in its true being-in-itself, mathematical. Otherwise, by making this assertion, Galileo would contradict the basic tenet of his philosophical revolution, that trying to penetrate the essence of natural substances is a not less impossible and vain undertaking with regard to the closest elemental substances than with the remotest celestial things.

Galileo's assertion is only meant to be polemical with contemporary Aristotelians, according to whom natural philosophy is written in Aristotle's books, and "philosophizing is and can be nothing but to work Aristotle's texts over and over" (Galilei 1968, V, 190). They never want to lift their eyes from Aristotle's texts, "as if this great book of the world had been written by nature to be read by nobody but Aristotle, and his eyes had been destined to see for all posterity" (ibid.).

On the contrary, according to Galileo, the great book of the world has been written by nature to be read by anybody, but what can be read in it is not nature in its true being-in-itself, but only some phenomenal properties of things, mathematical in kind. This confirms that the object of science cannot be to know the essence of things, it can only be to know some phenomenal properties of things, mathematical in kind.

That this is the object of science has been repeatedly reasserted after Galileo.

For example, Heisenberg says that "what we observe is not nature in itself but nature exposed to our method of questioning. Our scientific work in physics consists in asking questions about nature in the language we possess and trying to get an answer from experiment by the means that are at our disposal" (Heisenberg 1958, 58).

4.17 The Recognition of the Role of Philosophy in Science

As a result of Galileo's revolution, both Galileo and Newton recognize the role of philosophy in science.

Thus, Galileo says that there is an intimate connection between science and "the true and good philosophy, especially in that part which concerns the constitution of the universe" (Galilei 1968, V, 102). So, there is a substantial difference between the mere astronomers, who formulate their hypotheses only "to facilitate their calculations" and "the philosophical astronomers," who "seek to investigate, as the greatest and most admirable problem, the true constitution of the universe" (ibid.). Copernicus, "wearing the philosopher's dress," engaged himself "in the investigation of this constitution, recognizing" that "in philosophy one would have gained a knowledge as eminent as that which consists in knowing the true disposition of the parts of the world" (ibid., V, 297–298). Galileo himself "studied for a greater number of years in philosophy than months in pure mathematics" (ibid., X, 353).

Newton says: "I have in this treatise cultivated mathematics as far as it relates to philosophy" (Newton 1962, I, xvii). I "offer this work as the mathematical principles of philosophy, for the whole burden of philosophy seems to consist in this—from the phenomena of motions to investigate the forces of nature, and then from these forces to demonstrate the other phenomena" (ibid., xvii–xviii). And "I hope the principles here laid down will afford some light either to this or some truer method of philosophy" (ibid., I, xviii).

Because of the role of philosophy in science, in the modern period those who studied the natural world were called 'natural philosophers' until 1834, when Whewell coined the term 'scientist'.

Indeed, Whewell says: "There was no general term by which" those who studied the natural world "could describe themselves with reference to their pursuits. 'Philosophers' was felt to be too wide and too lofty a term," while "'savans' was rather assuming, besides being French instead of English;" then "some ingenious gentleman," that is, Whewell himself, "proposed that, by analogy with 'artist', they might form 'scientist'" (Whewell 1834, 59).

4.18 Philosophy as Acquisition of Knowledge and Method

If philosophy is acquisition of knowledge, then philosophy needs a method. This has been asserted by eminent philosophers from the antiquity to the modern period.

Thus, Plato says that, proceeding to acquire knowledge without a method "would be like walking with the blind. But someone who goes about his subject skillfully must not be like the blind" (Plato, *Phaedrus*, 270 d 9–e 2).

Aristotle says that, to acquire knowledge, "one must have been educated in the method by which each thing should be produced" (Aristotle, *Metaphysica*, A 3, 995 a 12–13). For, "by having a method we shall be able more easily to attack any proposed question" (Aristotle, *Topica*, A 2, 101 a 29–30).

Bacon says that, to acquire knowledge, "we need a thread to guide our steps; and the whole road, right from the first perceptions of sense, has to be made with a sure method" (Bacon 2004, 10).

Descartes says that, to acquire knowledge, a method is necessary because "it is far better to never think of investigating the truth of anything than to do so without a method" (Descartes 1996, X, 371). To have no method is to be like a person who "is burning with such a stupid desire to find a treasure, that he constantly roams about the streets to see if by chance he might find one lost by a passer-by" (ibid.). In fact, a method exists and "is twofold, one by analysis, the other by synthesis" (ibid., VII, 155).

Regarding the fact that eminent philosophers from the antiquity to the modern period have asserted that philosophy requires a method, Grice even says: "The greatest philosophers have been the greatest, and the most self-conscious, methodologists; indeed, I am tempted to regard this fact as primarily accounting for their greatness as philosophers" (Grice 1986, 66).

4.19 The Heuristic View of Philosophy

In particular, if philosophy is acquisition of knowledge, then philosophy needs a method of discovery.

As will be argued in Chap. 18, there is, in fact, a method of discovery not only for mathematics and science but also for philosophy, namely the analytic method. So, if philosophy is acquisition of knowledge, then its method is the analytic method.

Since the analytic method is a method of discovery, the view that philosophy is acquisition of knowledge can also be called the heuristic view of philosophy.

Heuristics is often understood in a reductive way, which diminishes its status.

Thus, Williamson says that "a heuristic is a rule of thumb for solving problems of some type" (Williamson 2024, 2).

On the contrary, the name 'heuristic view of philosophy' is intended to emphasize that philosophy does not have a mere rule of thumb but a proper method of discovery.

4.20 An Open Set of Inference Rules

A method of discovery cannot be based on a closed set of inference rules, that is, a set of inference rules given once and for all, it must be based on an open set of inference rules, that is, a set of inference rules that can always be extended.

For, nothing guarantees that new knowledge can be acquired using the existing inference rules.

Every extension of the existing inference rules is a development of the method.

Thus, as Bacon says, "the art of discovery may improve with discoveries" (Bacon 2004, 101).

In particular, as we will see in Chap. 15, in the analytic method the existing inference rules can be extended in the course of solving a problem.

4.21 Differences of Philosophy from the Present Sciences

It has been said above that, if philosophy is acquisition of knowledge, then philosophy is continuous with the present sciences. However, philosophy differs from the present sciences in at least two respects.

A difference has been already pointed out above. While the present sciences are what we already know about the world, philosophy is concerned with what we do not yet know about the world, therefore philosophy is concerned with questions that are beyond the present sciences.

Comparing what we already know about the world with what we do not yet know about it provides guidance on what to investigate.

Therefore, Plato says that in philosophy it is important to "know what one knows and what one does not know" (Plato, *Charmides*, 167 a 7–8).

Another difference is that philosophy is concerned with everything that exists. For, as Russell says, "what concerns philosophy is the universe" (Russell 1995b, 189).

Each of the present sciences has a specific object, different from the object of the other sciences. On the contrary, philosophy has no specific object of its own, because it is concerned with what we do not yet know about the world, and what we do not yet know about it can be everything that exists.

4.22 Influence of Philosophy on the Present Sciences

As said above, philosophy is concerned with questions that are beyond the present sciences.

This does not mean, however, that philosophy has no influence on the present sciences. In fact, after the birth of modern science, philosophy has influenced subsequent developments of physical sciences.

As Rovelli says, the influence of philosophy on physics "can be recognised in every major step. Take the twentieth century. Both major advances made by twentieth century physics," relativity and quantum mechanics, "were strongly influenced by philosophy. They would have been inconceivable without the philosophy of the time" (Rovelli 2018, 483). Einstein "explicitly recognised his debt to the philosophical writings of Mach and Poincaré. Without these inputs, his special relativity would have been inconceivable" (ibid., 483–484). Even with regard to his general relativity, Einstein "was explicit in recognising his debt to philosophy, this time to the critical thinking of Leibniz, Berkeley, and Mach" (ibid., 484).

In fact, Einstein says that "science without epistemology is—insofar as it is thinkable at all—primitive and muddled" (Einstein 1970, 684).

On the other hand, as Rovelli also says, "quantum mechanics springs from an intuition due to Heisenberg, grounded in the strongly positivist philosophical atmosphere in which he found himself: one gets knowledge by restricting oneself to what is observable" (Rovelli 2018, 483).

In fact, Heisenberg says that his aim was "to set a basis for a theory of quantum mechanics based exclusively on relations between quantities which in principle are observable" (Heisenberg 1925, 879).

Philosophy has also influenced developments of non-physical sciences. For example, Berkeley's attack on Newton's and Leibniz's infinitesimal calculus influenced the development of the calculus.

As Grabiner says, Berkeley's attack "pointed to the questions which had to be answered if a successful foundation were to be given. D'Alembert and Lazare Carnot both used some of Berkeley's arguments in their own discussions of foundations, and Lagrange took Berkeley's criticisms with the utmost seriousness" (Grabiner 2005, 27).

4.23 Characters of Philosophy as Acquisition of Knowledge

From what has been said above it is clear that philosophy as acquisition of knowledge has the following characters.

(1) *Philosophy aims at acquiring knowledge*. For, human beings are faced all the time with new problems in the world, and solving them requires knowledge about the world.

(2) *Philosophy is continuous with the present sciences*. For, philosophy aims at a type of knowledge that is not essentially different from that of the present sciences.

(3) *Philosophy can make use of the results of the present sciences*. For, philosophy starts from existing knowledge, and this is provided by the present sciences.

(4) *Philosophy is concerned with questions that are beyond the present sciences.* They are beyond the present sciences, in the sense that they are not open questions of any of the present sciences.

(5) *Philosophy is concerned with everything that exists.* For, philosophy is concerned with what we do not yet know about the world, and what we do not yet know about the world can be everything that exists.

(6) *Philosophy can open new ways to knowledge and lead to the birth of new sciences.* For, philosophy is concerned with questions that are beyond the present sciences.

(7) *Philosophy needs a method of discovery.* For, philosophy aims at acquiring knowledge of the world, and this requires a method of discovery,

(8) *The method of discovery must be based on a set of inference rules that can always be extended.* For, nothing guarantees that new knowledge can be acquired using the existing inference rules.

(9) *The method of discovery is the analytic method.* For, knowledge is acquired by solving problems, and the analytic method is the best available method for solving problems.

(10) *Philosophy can also influence the development of the present sciences.* For, philosophy can help analyze their basic concepts and improve their method.

References

Austin, John. 1970. *Philosophical papers*. Oxford: Oxford University Press.

Ayer, Alfred. 1963. *The concept of a person and other essays*. New York: St Martin's Press.

Bacon, Francis. 1857–1874. *Works*. London: Longmans.

Bacon, Francis. 2004. *The new organon*. Cambridge: Cambridge University Press.

Berkeley, George. 2009. *Philosophical writings*. Cambridge: Cambridge University Press.

Brandom, Robert. 2009. *Reason in philosophy: Animating ideas*. Cambridge: Harvard University Press.

Chiaradonna, Riccardo, and Paolo Pecere. 2018. *Filosofia: La ricerca della conoscenza*. Milan: Mondadori.

Descartes, René. 1996. *Oeuvres*. Paris: Vrin.

Dummett, Michael. 1991. *The logical basis of metaphysics*. Cambridge: Harvard University Press.

Dummett, Michael. 2010. *The nature and future of philosophy*. New York: Columbia University Press.

Einstein, Albert. 1970. Autobiographical notes—Remarks concerning the essays brought together in this co-operative volume. In *Albert Einstein: Philosopher-scientist, 3–95*, ed. Paul Schilpp, 665–688. New York: MJF Books.

Feyerabend, Paul. 1999. Letters to the Director of the Department of Philosophy. In Imre Lakatos and Paul Feyerabend, *For and against method*, 382–393. Chicago: The University of Chicago Press.

Frodeman, Robert. 2019. A robot took my boyfriend and my job: Positioning philosophy for a resurgence. In *Why philosophy?* ed. Paolo Bubbio and Jeff Malpas, 105–121. Berlin: de Gruyter.

Galilei, Galileo. 1968. *Opere*. Florence: Barbera.

Glymour, Clark. 1991. The hierarchies of knowledge and the mathematics of discovery. *Minds and Machines* 1:75–95.

Grabiner, Judith. 2005. *The origins of Cauchy's rigorous calculus*. Mineola: Dover.
Grice, Paul. 1986. Reply to Richards. In *Philosophical grounds of rationality: Intentions, categories, ends*, ed. Richard Grand and Richard Warner, 45–106. Oxford: Oxford University Press.
Hanna, Robert. 2023. The end of peer review and the matrix of ideas. Unpublished manuscript. https://www.academia.edu/109130045/The_End_of_Peer_Review_and_The_Matrix_of_Ideas_November_2023_version_
Heisenberg, Werner. 1925. Über quantentheoretische Umdeutung kinematischer und mechanischer Beziehungen. *Zeitschrift fur Physik* 33:879–893.
Heisenberg, Werner. 1958. *Physics and philosophy: The revolution in modern science*. New York: Harper and Brothers.
Hobbes, Thomas. 1839–1845. *English works*. London: Longman.
Husserl, Edmund. 1970. *The crisis of European sciences and transcendental phenomenology*. Evanston: Northwestern University Press.
Illetterati, Luca, and Giovanna Miolli. 2022. Why Hegel's metaphilosophy matters: An introduction. In *The relevance of Hegel's concept of philosophy: From classical German philosophy to contemporary metaphilosophy*, ed. Luca Illetterati and Giovanna Miolli, 1–14. London: Bloomsbury.
Kant, Immanuel. 1992. *Lectures on logic*. Cambridge: Cambridge University Press.
Kant, Immanuel. 2002. *Theoretical philosophy after 1781*. Cambridge: Cambridge University Press.
Kant, Immanuel. 2015. *Critique of practical reason*. Cambridge: Cambridge University Press.
Kripke, Saul. 1980. *Naming and necessity*. Cambridge: Harvard University Press.
Lebedev, Oleg. 2021. The Higgs portal to cosmology. *Progress in Particle and Nuclear Physics* 120:103881.
Lewis, David. 1983. *Philosophical papers: Volume I*. New York: Oxford University Press.
Mach, Ernst. 1976. *Knowledge and error: Sketches on the psychology of enquiry*. Dordrecht: Springer.
Moore, Christopher. 2020. *Calling philosophers names: On the origin of a discipline*. Princeton: Princeton University Press.
Newton, Isaac. 1962. *Mathematical principles of natural philosophy—The system of the world*. Berkeley: University of California.
Nightingale, Andrea. 1995. *Genres in dialogue: Plato and the construct of philosophy*. Cambridge: Cambridge University Press.
Nozick, Robert. 1989. *The examined life: Philosophical meditations*. New York: Simon and Schuster.
Overgaard, Søren, Paul Gilbert, and Stephen Burwood. 2013. *Introduction to metaphilosophy*. Cambridge: Cambridge University Press.
Popper, Karl. 1994. How I see philosophy. In Karl Popper, *In search of a better world: Lectures and essays from thirty years*, 173–187. London: Routledge.
Popper, Karl. 2002. *Conjectures and refutations: The growth of scientific knowledge*. London: Routledge.
Preston, Aaron. 2004. Prolegomena to any future history of analytic philosophy. *Metaphilosophy* 35:445–465.
Quine, Willard. 2008. *Quine in dialogue*. Cambridge: Harvard University Press.
Rovelli, Carlo. 2018. Physics needs philosophy. Philosophy needs physics. *Foundations of Physics* 48:481–491.
Russell, Bertrand. 1960. Introduction. In *Words and things*, ed. Ernest Gellner, 13–15. Boston: Beacon Press.
Russell, Bertrand. 1995a. *My philosophical development*. London: Routledge.
Russell, Bertrand. 1995b. *An outline of philosophy*. London: Routledge.
Russell, Bertrand. 1998. *The problems of philosophy*. Oxford: Oxford University Press.
Ryle, Gilbert. 2009a. *The concept of mind*. London: Routledge.
Ryle, Gilbert. 2009b. *Collected papers*. London: Routledge.
Sellars, Wilfrid. 1991. *Science, perception and reality*. Atascadero: Ridgeview.

Vintiadis, Elly. 2021. What philosophy is and what it could be. In *Philosophy by women: 22 philosophers reflect on philosophy and its value*, ed. Elly Vintiadis, 83–90. London: Routledge.
Whewell, William. 1834. Review of Mary Sommerville, *On the Connexion of the Physical Sciences*. *The Quarterly Review* 51:54–68.
Williamson, Timothy. 2024. *Overfitting and heuristics in philosophy*. Oxford: Oxford University Press.
Wittgenstein, Ludwig. 1932–1933. *Muss sich denn nicht....* In *Wittgenstein's Nachlass*, TS 219.
Wittgenstein, Ludwig. 1969. *The blue and brown books: Preliminary studies for the 'Philosophical investigations'*. Oxford: Blackwell.
Wittgenstein, Ludwig. 1978. *Remarks on the foundations of mathematics*. Oxford: Blackwell.
Wittgenstein, Ludwig. 2001. *Lectures, Cambridge, 1932–1935*, from the notes of Alice Ambrose and Margaret Macdonald. Amherst: Prometheus Books.
Wittgenstein, Ludwig. 2005. *The big typescript TS 213*. Oxford: Blackwell.
Wittgenstein, Ludwig. 2009. *Philosophische Untersuchungen—Philosophical investigations*. Oxford: Blackwell.

Chapter 5
Against Philosophy as Acquisition of Knowledge

Abstract Chapter 3 considered one argument that can be raised against the view that philosophy is acquisition of knowledge, the cognitive poverty of philosophy, and argued that it is not valid. This chapter considers several other arguments that can be raised against the view and argues that they are not valid. On this basis, the chapter confirms that philosophy is acquisition of knowledge, and that this is the only view capable of restoring vitality to philosophy.

Keywords Philosophy and science · Philosophy and understanding · Philosophy and scientism · Philosophy and truth · Philosophy and consensus · Philosophy and scepticism · Philosophy and meaning · Philosophy and mathematics · Philosophy and language · Philosophy and values · Philosophy and history of philosophy

5.1 Questioning Philosophy as Acquisition of Knowledge

Chapter 3 considered one argument that can be raised against the view that philosophy is acquisition of knowledge, the cognitive poverty of philosophy, and argued that it is not valid.

Several other arguments can be raised against this view. This chapter considers them and argues that they are not valid.

On this basis, the chapter confirms that philosophy is acquisition of knowledge, and that this is the only view capable of restoring vitality to philosophy.

5.2 Only Science Is Acquisition of Knowledge

Another argument that can be raised against the view that philosophy is acquisition of knowledge is that only science is acquisition of knowledge because science provides all significant truths about reality and, when 'complete', what science tells us will not be surprisingly different from what it tells us today.

C. Cellucci, *A Philosophy for the Future*, Synthese Library 508,
https://doi.org/10.1007/978-3-031-98443-3_5

Thus, Rosenberg says that "science provides all the significant truths about reality, and knowing such truths is what real understanding is all about" (Rosenberg 2011, 7). Also, "when 'complete', what science tells us will not be surprisingly different from what it tells us today" (ibid., 6–7).

This argument, however, is not valid. For, it depends on the assumption that 'science' means 'the present sciences', therefore the latter are the only possible source of knowledge. This assumption is unwarranted because there are areas of experience that the present sciences cannot deal with. There is no evidence that in the future there will not be new sciences capable of dealing with them, nor that philosophy will not lead to their birth. Moreover, it is highly implausible that, when 'complete', what science tells us will not be surprisingly different from what it tells us today, considering that ninety-five percent of the universe is still a mystery to us.

5.3 Philosophy Only Asks Questions

Another argument that can be raised against the view that philosophy is acquisition of knowledge is that philosophy only asks questions, not answers them, only science answers questions. Beyond the territories conquered by the scientific enterprise, there are only questions without answers.

Thus, Bobbio says that, while "science gives partial answers," and yet answers, "philosophy only asks questions without giving answers" (Bobbio 2017, 46). So, "one must not expect of philosophy what one expects of science, namely answers, not even partial ones" (ibid., 31). For, "beyond the territories conquered by the scientific enterprise, there are only questions without answer" (ibid., 67). This is "a territory into which one ventures, knowing that there is no way out" (ibid.). Indeed, "at that point the sphere of knowledge ends and that of non-knowledge begins" (ibid.).

This argument is an expression of what Marconi calls "the jogger's view" of philosophy because, for the jogger, "the important thing is to run, not to get somewhere" (Marconi 2014, 92). Similarly, according to the jogger's view of philosophy, for the philosopher the important thing is to ask questions, not to get somewhere.

Sometimes the jogger's view of philosophy is ascribed to Socrates. Thus, Black says that "a major innovation" in "Socrates was asking questions without giving answers" (Black 2008, 30).

This, however, is misleading. For, the questions Socrates asked were 'leading' questions, that is, questions phrased in a manner that suggested the desired answer to the discussant.

What is more important, the jogger's view of philosophy is not valid. Already Plato said that the philosopher is not the man who only knows how to ask questions, but "the man who knows how to ask and answer questions" (Plato, *Cratylus*, 390 c 10). Indeed, if the philosopher only asks questions but does not answer them, how can he be useful? For a layman, not receiving an answer to a question from the philosopher is just as frustrating as it is, for a child, not receiving an answer to a question from his parents.

Moreover, it is unwarranted to say that, beyond the territories conquered by the scientific enterprise, there are only questions without answers. Beyond those territories there is the vast land of what we do not yet know, and there is no reason to suppose that it is unknowable and that philosophy will not lead to know it.

Indeed, Rota says that "the awesome complexities of the phenomena that are being studied" in "experimental psychology, neurophysiology and computer science" have convinced scientists that "progress in science will depend on philosophical research in the most classical vein" (Rota 1997, 99).

5.4 Philosophy Aims Only at Understanding

Another argument that can be raised against the view that philosophy is acquisition of knowledge is that philosophy aims only at understanding philosophical questions, not at solving them.

Thus, Hannon and Nguyen say that "philosophy does not aim at discovering true answers to philosophical questions" (Hannon and Nguyen 2022). For, philosophical questions "are perennials for which it is unreasonable to expect a solution" (ibid.). Philosophy aims only at "increasing our understanding of philosophical issues," which will "deepen our thinking about philosophical issues" (ibid.).

This argument, however, is not valid. For, there is no point in increasing our understanding of philosophical questions, deepening our thinking about philosophical issues, if we cannot solve them anyway. Increasing our understanding would only increase our sense of impotence.

5.5 Philosophy Is Not Scientism

Another argument that can be raised against the view that philosophy is acquisition of knowledge is that it is scientism. For, it involves that philosophy must be continuous with the natural sciences, and scientism is the disposition to regard the natural sciences as the only true channel of knowledge. This reduces the task of philosophy to that of adding ornamentation to the theories of the scientists.

Thus, Dummett says that the view that philosophy is acquisition of knowledge implies that "the idea that philosophy has a subject matter or a method of its own must be discarded: if it is to contribute to knowledge at all, it must be continuous with the natural sciences" (Dummett 2010, 35). But this is scientism, because "scientism is the disposition to regard the natural sciences as the only true channel of knowledge" (ibid.). In fact, this "is not philosophy; but it is not science either. It is the result of the bedazzlement of those who have undertaken one manner of intellectual inquiry via the successes of another" (ibid., 36). It reduces the task of philosophy "to that of adding ornamentation to the theories of the scientists" (ibid.).

This argument, however, is not valid. The view that philosophy is acquisition of knowledge is not scientism, because scientism is not the disposition to regard the natural sciences as the only true channel of knowledge. Instead, scientism is the disposition to regard the present sciences as the only true channel of knowledge. Scientism arises from overrating the scope of the present sciences. It assumes that the present sciences cover all areas of experience, so it ignores that there are areas of experience that the present sciences are unable to address. Philosophy is concerned with them. It is in this sense that the present sciences are what we already know about the world, philosophy is concerned with what we do not yet know about it. This does not reduce the task of philosophy to that of adding ornamentation to the theories of the scientists.

Philosophy does not reduce to the present sciences. On the other hand, it is not opposed to them. Philosophy and the present sciences are part of a common enterprise, which aims at acquiring knowledge of the world. Therefore, philosophy and the present sciences are distinct, but not distant. They are a continuum not in the sense of scientism, but in the sense that the type of knowledge at which philosophy aims does not differ essentially from scientific knowledge and is not limited to any area.

5.6 Philosophy Aims at Necessary Truths

Another argument that can be raised against the view that philosophy is acquisition of knowledge is that it implies that the answer to a philosophical question is a contingent not necessary truth. But philosophy aims at necessary truths.

Thus, Bealer says that, if philosophy is acquisition of knowledge, then "most philosophical knowledge must rely on empirical evidence," so "philosophy would have no significant independence from empirical science" (Bealer 1987, 293). Then, the answer to a philosophical question is a contingent not necessary truth. But "philosophers have nearly all agreed that, at least in most cases, the answer to a philosophical question should be a necessary not contingent truth" (ibid., 289). For, "philosophers do not want to know" what "such things as the nature of mind, intelligence, the virtues, and life," and so forth, "just happen to be, but rather what those things must be" (ibid.). Philosophy aims at necessary truths. On the contrary, "no such necessity ever holds for science" (Bealer 1996, 19).

This argument, however, is not valid. There is no evidence that, unlike the answer to a scientific question, the answer to a philosophical question can be a necessary not contingent truth.

Indeed, as will be argued in Chap. 18, even a priori knowledge is contingent not necessary knowledge.

5.7 Philosophy Stands in Complete Contrast with Sciences

Another argument that can be raised against the view that philosophy is acquisition of knowledge is that philosophy stands in complete contrast with sciences. For, the objectives, methods, and results of philosophy wholly diverge from those of the sciences.

Thus, Dummett says that "philosophy stands in complete contrast with sciences" (Dummett 2010, 7). For, "its methods wholly diverge from those of science, and its objective differs to an equal extent," and "the results of philosophy differ fundamentally in character from those of the sciences" (ibid.).

This argument, however, is not valid. For, as argued in Chap. 4, the methods of philosophy do not wholly diverge from those of science, its objective does not differ from that of science, and the results of philosophy do not differ in character from those of the sciences.

That philosophy stands in complete contrast with sciences is also claimed, for opposite reasons, by several scientists, who believe that philosophy is, at best, a parasitic commentary on the achievements of the sciences, and the sciences are philosophy-free. But, as Dennett observes, "there is no such thing as philosophy-free science; there is only science whose philosophical baggage is taken on board without examination" (Dennett 1996, 21).

5.8 Sciences Are Alien Elements in Philosophy

Another argument that can be raised against the view that philosophy is acquisition of knowledge is that, by budding off a number of sciences, philosophy has liberated itself from alien elements. Because of the presence of these alien elements, philosophy has remained for so long in its early stages, but with the rise of analytic philosophy it has reached maturity.

Thus, Dummett says that "no practicing philosopher would explain the value of the subject merely as a matrix out of which new disciplines could develop" (Dummett 2010, 4). If philosophy has budded off a number of sciences, it is because philosophy contained alien elements, indeed, philosophy is "what is left when the" sciences "to which it gave birth have left the parental home" (ibid., 3–4). That is, philosophy is what is left when it is liberated from alien elements. Genuine philosophical problems are those that are left after this liberation. Because of the presence of alien elements, philosophy has remained "for so long 'in its early stages'," that is, "those in which its practitioners" had "not yet attained a clear view of its subject-matter and its goals." (Dummett 1978, 457). Philosophy got rid of the alien elements only in the nineteenth century, so "it was not until the nineteenth century that it made sense to ask for an example of a philosophical problem, as opposed to a problem of some other kind" (Dummett 2010, 8). At last, with the rise of analytic philosophy, philosophy has "struggled out of its early stage into maturity" (Dummett 1978, 457).

This argument, however, is not valid. It is unjustified to say that, by budding off a number of sciences, philosophy has liberated itself from alien elements. For, the objective of philosophy does not differ from that of the present sciences, and the results of philosophy do not differ fundamentally in character from those of the present sciences.

Moreover, it is unjustified to say that, because of the presence of these alien elements, philosophy has remained for so long in its early stages, but with the rise of analytic philosophy it has reached maturity.

This is based on the assumption that, while in philosophy for two and a half millennia everything has been questioned and every stance has been just a matter of individual convictions, analytic philosophy has marked a turning point. It has started an entirely new approach to philosophy in which each philosopher, adding stone to stone, contributes to erecting a shared building in cooperation with other philosophers.

But this assumption is not valid because, as van Inwagen observes, "there is almost no thesis in philosophy about which" analytic "philosophers agree" (van Inwagen 2017, 129).

In fact, surveys show that analytic philosophers have contrasting views even on key questions (see Bourget and Chalmers 2014, 2023).

5.9 In Philosophy There Is No Consensus

Another argument that can be raised against the view that philosophy is acquisition of knowledge is that in philosophy there is no consensus over solutions to problems. Even if philosophy has given birth to new fields, the results of the new sciences are answers to non-philosophical questions.

Thus, Brennan says that the "radical dissensus" between philosophers over solutions to problems shows that "philosophy is an unreliable instrument for finding truth" (Brennan 2010, 3). It is not valid to argue that "philosophy gives birth to new fields" and hence "is indirectly truth-tracking, as it produces other fields that find the truth" (ibid., 12). For, the results of the new fields are answers to non-philosophical questions, and there are "questions that cannot be made non-philosophical," so there is no "good reason to think that all questions" of philosophy "will one day be turned over to other fields" (ibid.).

This argument, however, is not valid because several results of the new fields are answers to what were previously considered genuine philosophical questions, and there is consensus on them.

5.10 Philosophy Is Opposed to Philosophical Scepticism

Another argument that can be raised against the view that philosophy is acquisition of knowledge is that it leads to philosophical scepticism. For, the methods of philosophy are not truth-conducive, therefore we have no reason to think that the philosophical theories that those methods deliver are true. At the origin of philosophical scepticism there is the assumption that the aim of philosophy is knowledge, so this assumption must be abandoned.

Thus, Beebee says that the view that that philosophy is acquisition of knowledge leads to "philosophical scepticism," because it implies that "philosophers do not and cannot know many of the substantive philosophical claims that they make or implicitly assume" (Beebee 2018, 1). For, the methods of philosophy "can stake no claim to truth-conduciveness; thus we have no particular reason to think that" the philosophical "theories that those methods deliver are true" (ibid., 2). At the origin of philosophical scepticism there is the assumption that "the aim of philosophy" is "knowledge" (ibid., 2). Therefore, "the right response to philosophical scepticism is to abandon the view that philosophy aims at knowledge" (ibid., 1).

This argument, however, is not valid. For, it depends on the assumption that the aim of philosophy is truth. But the assumption is unwarranted. As will be argued in Chap. 11, the aim of philosophy cannot be truth because a criterion of truth is impossible, so it is generally impossible to distinguish true propositions from false ones. Instead, the aim of philosophy is plausibility.

5.11 Philosophy Is the Midwife of Fewer and Fewer Sciences

Another argument that can be raised against the view that philosophy is acquisition of knowledge is that philosophy is the midwife of fewer and fewer new sciences, so the view that philosophy is acquisition of knowledge implies that pretty soon we would stop practicing philosophy.

Thus, Brennan says that "the birth rate" of the new sciences to whose birth philosophy contributes "appears to be dropping. Philosophy is not founding new fields as often as it used to," therefore "the maieutic defence suggests that pretty soon we should stop practicing philosophy, because the expected utility (in terms of founding new fields) is too low" (Brennan 2010, 12).

This argument, however, is not valid. There is no evidence that philosophy is the midwife of fewer and fewer new sciences. As already pointed out in Chap. 4, even in the past century philosophy has led to the birth, for example, of computer science, cognitive science, artificial intelligence, and linguistics. There are plenty of things we still don't know, including ninety-five percent of the universe which is still a mystery to us.

5.12 Questions About Knowledge Are No Longer Central

Another argument that can be raised against the view that philosophy is acquisition of knowledge is that the purpose of a theory of knowledge is not to acquire knowledge, but to inquire into the possibility of certain, objective, universal knowledge. But, while the existence of this type of knowledge seemed problematic in the seventeenth century, it is no longer so, therefore questions about knowledge are no longer central in philosophy.

Thus, Searle says that "in the seventeenth century" Descartes, Bacon and others "took epistemology as the central element of the entire philosophical enterprise, because" then "the possibility of certain, objective, universal knowledge seemed problematic," and "the very existence of knowledge was in question" (Searle 2008, 4). It "was not at all clear how their various beliefs could be established with certainty, and it was not even clear how they could be made consistent" (ibid., 4–5). So "we had three and a half centuries in which epistemology was at the centre of philosophy" (ibid., 5). But now, "because of the sheer growth of certain, objective, and universal knowledge, the possibility of knowledge is no longer a central question in philosophy," so questions about knowledge no longer "lie at the heart of the philosophical enterprise" (ibid.).

This argument, however, is not valid. Being acquisition of knowledge, philosophy aims primarily at knowledge, so a theory of knowledge, necessarily has a central place in philosophy. The purpose of a theory of knowledge is not to inquire into the possibility of certain, objective, universal knowledge, because there is no such knowledge. By Gödel's second incompleteness theorem, even mathematical knowledge is not certain, objective, universal knowledge. The purpose of a theory of knowledge is, rather, to inquire into how to get more knowledge, fallible knowledge and yet knowledge.

As Kitcher says: "What might we want from a theory of knowledge? There's an obvious answer. A theory of knowledge should enable us to get more of it" (Kitcher 2011, 508).

This transfers the theory of knowledge from the context of justification to the context of discovery, which includes that of justification because, in the process of discovery, hypotheses are accepted only when they are plausible, that is, such that the arguments for them are stronger than the arguments against them, on the basis of experience (see Chap. 11).

5.13 Questions About Meaning Are Now Central

Another argument that can be raised against the view that philosophy is acquisition of knowledge is that, rather than questions about knowledge, questions about meaning are now central in philosophy. In fact, Frege made the theory of meaning the foundation of all philosophy.

Thus, Dummett says that Descartes made the theory of knowledge "the foundation of philosophy because he had conceived the task of philosophy as being that of introducing rigour into science," intended as "the whole body of things we conceive ourselves to know" (Dummett 1973, 676). But Descartes was wrong. For, the theory of knowledge concerns questions of justification of knowledge, and "until we have first achieved a satisfactory analysis of the meanings of the relevant expression, we cannot so much as raise questions of justification, since we remain unclear about what we are attempting to justify" (ibid., 667). Therefore, Frege made "the theory of meaning" the "foundation of all philosophy, and not theory of knowledge, as Descartes misled us into believing" (ibid.). Thus Frege "effected a revolution in philosophy," so we can "date a whole epoch in philosophy as beginning with the work of Frege" (ibid., 669). As a result of Frege's philosophical revolution, "the theory of meaning" is "the fundamental part of philosophy which underlies all others" (ibid., 669).

This argument, however, is not valid. Frege did not make the theory of meaning the foundation of all philosophy. For, his main purpose was to provide a secure foundation for mathematics so as to "place the truth of a proposition beyond all doubt" (Frege 1960, 2). This was a question of justification within Descartes' tradition.

Moreover, if Descartes made the theory of knowledge the foundation of philosophy because he had conceived the task of philosophy as being that of introducing rigour into science, then Frege did the same with regard to mathematics. Indeed, Frege assumed that the theory of knowledge is the foundation of philosophy because he had conceived the task of philosophy as being that of introducing rigour into mathematics by proving "the fundamental propositions of arithmetic" with "the utmost rigour" (ibid., 4).

In fact, the theory of meaning cannot be the foundation of philosophy because it concerns questions about language, while the main philosophical problems are questions about the world.

5.14 Philosophy Does Not Start from Existing Knowledge

Another argument that can be raised against the view that philosophy is acquisition of knowledge is that, while mathematics and science build on previous results and therefore start from existing knowledge, philosophy does not start from existing knowledge, in particular from the discoveries of mathematics and science.

Thus, Wittgenstein says that philosophy does not start from existing knowledge because "the name 'philosophy' might also be given to what is possible before all new discoveries" (Wittgenstein 2009, 126). In particular, "no mathematical discovery can advance it" (ibid., 124). In fact, "no new words have to be used in philosophy—the old, ordinary words of language suffice" (Wittgenstein 2005, 309).

This argument, however, is not valid. For, if philosophy does not start from existing knowledge, in particular from the discoveries of mathematics and science, then philosophy ends up repeating old idioms, based on obsolete worldviews or simply prejudices.

Instead, philosophy must start from existing knowledge, introducing new idioms, suitable to the problems addressed.

5.15 Philosophy Cannot Have a Method of Discovery

Another argument that can be raised against the view that philosophy is acquisition of knowledge is that philosophy cannot have a method of discovery because there exists no such method, discoveries are made by intuition. An example of this is provided by Kekulé's account of his discovery of the structure of benzene.

Thus, Hempel argues that hypotheses for solving problems cannot be obtained by "any process of systematic inference" (Hempel 1966, 15). They "are not derived from observed facts, but invented in order to account for them" (ibid.). They "are invented by an exercise of creative imagination" (Hempel 2001, 32). Their discovery "requires inventive ingenuity; it calls for imaginative, insightful guessing" (Hempel 1966, 17). That is, it calls for intuition.

Thus, Kekulé "had long been trying unsuccessfully to devise a structural formula for the benzene molecule," but one evening, "while he was dozing in front of his fireplace," he "found a solution to his problem" (ibid., 16). For, "gazing into the flames, he seemed to see atoms dancing in snakelike arrays. Suddenly, one of the snakes formed a ring by seizing hold of its own tail" (ibid.). Then Kekulé "awoke in a flash: he had hit upon" the "idea of representing the molecular structure of benzene by a hexagonal ring" (ibid.).

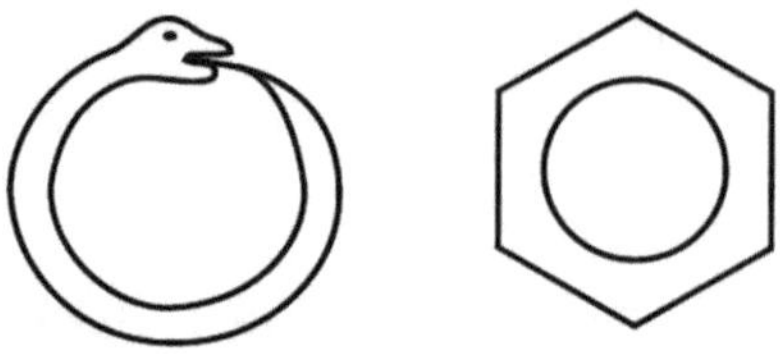

This argument, however, is not valid. According to Hempel, Kekulé's account indicates that he made his discovery by intuition. But this is unjustified, instead, Kekulé's account indicates that he made his discovery by analogy. Indeed, from the fact that the snake formed a ring, a stable structure, by analogy Kekulé arrived at the hypothesis that the structure of benzene could be a hexagonal ring, a stable structure. Then Kekulé checked that the hypothesis was compatible with the existing data. Therefore, Kekulé formed the hypothesis through a rational process.

5.16 Philosophy Is an Armchair Discipline Like Mathematics

Another argument that can be raised against the view that philosophy is acquisition of knowledge is that philosophy cannot give knowledge of the world because it is an armchair discipline like mathematics, so it needs no input from experience, it is the product of thought alone.

Thus, Dummett says that philosophy cannot give knowledge of the world because it "is a discipline that makes no observations, conducts no experiments, and needs no input from experience: an armchair subject, requiring only thought" (Dummett 2010, 4). It is like "another armchair discipline: mathematics. Mathematics likewise needs no input from experience: it is the product of thought alone" (ibid.).

This argument, however, is not valid. It is unjustified to say that mathematics is an armchair discipline, which needs no input from experience and is the product of thought alone.

As Rota observes, this amounts to saying that mathematicians "spew out solutions of one problem after another by dint of pure brain power, simply by staring long enough at a blank piece of paper" (Rota 1997, 100–101). But, while this "may appropriately describe the way to solve the linguistic puzzles of today's impoverished philosophy," namely analytic philosophy, it "is wide of the mark in describing the work of mathematicians, or any kind of serious work" (ibid., 101).

Mathematics essentially involves interactions with the world beyond the armchair. For, most mathematical problems arise from the world and hence from experience, and are solved using hypotheses that are justified on the basis of experience (see Cellucci 2025).

In the same way, it is unjustified to say that philosophy is an armchair discipline, which needs no input from experience and is the product of thought alone. For, most philosophical problems arise from the world and hence from experience, and are solved using hypotheses that are justified on the basis of experience.

If philosophy is to escape from irrelevance, it must answer questions about the world and, to this purpose, philosophy needs inputs from experience, not only thought. By thought alone, philosophy could only reformulate what is already known in other terms, not answer questions about the world.

5.17 Philosophy Is Based on Analysis of Language

Another argument that can be raised against the view that philosophy is acquisition of knowledge is that philosophy cannot give knowledge of the world because it is based on analysis of language, performed by a dictionary.

Thus, Austin says that, to conduct a philosophical inquiry, "first we may use the dictionary—quite a concise one will do, but the use must be thorough" (Austin 1970, 186). We can "read the book through, listing all the words that seem relevant," or we can "start with a widish selection of obviously relevant terms," and "consult the dictionary under each: it will be found that, in the explanations of the various meanings of each, a surprising number of other terms occur" (ibid., 186–187). We then "look up each of these, bringing in more for our bag from the 'definitions' given in each case; and when we have continued for a little, it will generally be found that the family circle begins to close, until ultimately it is complete and we come only upon repetitions" (ibid., 187). So we will "arrive at the meanings of large numbers of expressions" and we "shall comprehend clearly much that, before, we only made use of ad hoc. Definition, I would add, explanatory definition, should stand high among our aims" (ibid., 189).

This argument, however, is not valid. For, through analysis of language, performed by a dictionary, we will only get at the opinions, or prejudices, of the authors of the dictionary entries.

Moreover, dictionary entries soon become obsolete due to scientific, technological, or social developments, which are too fast for a dictionary to keep pace with them.

5.18 Philosophy Is About Values

Another argument that can be raised against the view that philosophy is acquisition of knowledge is that it excludes ethics and politics. For, knowledge is about facts while ethics and politics are about values. Facts and values are absolutely independent of each other and, while facts can be an object of inquiry, values cannot be an object of inquiry. Therefore, ethics and politics are not acquisition of knowledge.

Thus, Carnap says that ethics, meant as "the philosophy of moral values or moral norms," is "not an investigation of facts, but a pretended investigation of what is good and what is evil, what it is right to do and what it is wrong to do," so its purpose is "to state norms for human action or judgments about moral values" (Carnap 1935, 23). To state norms for human action or judgments about moral values "is merely a difference of formulation" because a norm "has an imperative form, for instance: 'Do not kill!' The corresponding value judgment would be: 'Killing is evil'" (ibid., 23–24). Since the value judgment 'Killing is evil' has the grammatical form of an assertive proposition, "most philosophers have been deceived by this form into thinking that a value statement is really an assertive proposition, and must be either true or false" (ibid., 24). But "a value statement is nothing else than a command in a misleading grammatical form," so "it is neither true nor false. It does not assert anything and can neither be proved nor disproved" (ibid.). Therefore, norms or value statements "are not scientific propositions (taking the word scientific to mean any assertive proposition)" (ibid., 25). They have "no theoretical sense" (ibid., 26).

This argument, however, is not valid for two reasons.

(1) The argument assumes that values and facts are absolutely independent of each other and, while facts can be an object of inquiry, values cannot be an object of inquiry. But this assumption is not valid.

For, on the one hand, values depend on what we know about the world, including human beings, and may change as our knowledge changes, so values depend on facts. On the other hand, values guide scientists in selecting the scientific problems to work on and the hypotheses to solve them, and in evaluating the acceptability of the hypotheses, so the selection of facts depends on values. Therefore, as Rudner says, "the scientist qua scientist makes value judgments" (Rudner 1953, 4).

This is clear if, as will be argued in Chap. 11, the aim of science is to make plausible hypotheses about the world, that is, hypotheses such that the arguments for them are stronger than the arguments against them, on the basis of experience. For, determining whether the arguments for a hypothesis are stronger than the arguments against it requires value criteria. In particular, every scientific revolution leads to a change in the values that guide scientists in selecting the scientific problems to work on and the hypotheses to solve them, and in evaluating the plausibility of hypotheses.

Then, it is not valid to say that facts and values are absolutely independent of each other and, while facts can be an object of inquiry, values cannot be an object of inquiry.

As Dewey says, "inquiry, discovery take the same place in morals that they have come to occupy in sciences and nature" (Dewey 2004, 100). If the claim is made that, in ethics and politics, "'values' are involved and that inquiry as 'scientific' has nothing to do with values, the inevitable consequence is that inquiry in the human area is confined to what is superficial and comparatively trivial" (ibid., xvi).

(2) The argument assumes that ethics and politics are not acquisition of knowledge. But this assumption is not valid. Ethics and politics are acquisition of knowledge, although not of theoretical knowledge but of practical knowledge, that is, knowledge as a means to action.

Already Plato makes it clear that the view that philosophy is acquisition of knowledge by no means excludes ethics and politics. For, good decisions require knowledge, therefore ethics and politics must be acquisition of knowledge. There are two forms of knowledge, theoretical knowledge and practical knowledge, and philosophy is acquisition of both theoretical and practical knowledge.

Indeed, Plato argues that "it is not by ignorance but by knowledge that men will make good decisions" (Plato, *Respublica*, IV 428 b 7–8). Making good decisions requires that men have "knowledge of all good things, and of their production in the present, the future, and the past, and of all bad things likewise" (Plato, *Laches*, 199 d 5–7). Therefore, ethics and politics must be acquisition of knowledge. We must "divide all knowledge" into two forms, "theoretical knowledge," and "practical knowledge" (Plato, *Politicus*, 258 e 4–5). Theoretical knowledge is proper to arts which are "stripped of action and furnish only cognition" (ibid., 258 d 5–6). Practical knowledge is "knowledge which naturally inheres in actions" (ibid., 258 d 9–e 1).

The division of knowledge into two forms is not an opposition, theoretical and practical knowledge are simply "the two forms of knowledge, which is one in its entirety" (ibid., 258 e 6–7). And "the philosopher desires the whole of knowledge, not just one part of it and not the other" (Plato, *Respublica*, V 475 b 8–9). So, philosophy is acquisition of both theoretical and practical knowledge.

That philosophy is acquisition of both theoretical and practical knowledge implies that philosophy is acquisition not only of propositional knowledge but also of non-propositional knowledge. For, much practical knowledge is non-propositional.

Again, already Plato makes this clear. For, he says that, "when you talk about cobbling," you mean "the knowledge of making shoes" (Plato, *Theaetetus*, 146 d 6–8). When you talk "about carpentering," you mean "the knowledge of making wooden furniture" (ibid., 146 e 1–2). When you talk about farming, you mean things such as "the knowledge of distributing seed over land" (Plato, *Minos*, 317 d 3–4). Now, the knowledge of making shoes, of making wooden furniture, of distributing seed over land, are all instances of practical and non-propositional knowledge.

5.19 Philosophy Does Not Need History of Philosophy

Another argument that can be raised against the view that philosophy is acquisition of knowledge is that it needs history of philosophy. But philosophy does not need history of philosophy.

Thus, Wittgenstein says: "What has history to do with me? Mine is the first and only world! I want to report how I found the world. What others in the world have told me about the world is a very small and incidental part of my experience of the world" (Wittgenstein 1979, 82, 2.9.16). The "reason why I give no sources is that it is a matter of indifference to me whether the thoughts that I have had have been anticipated by someone else" (Wittgenstein 2001, 4). A philosopher is like "a king" who has been "brought up in the belief that the world began with him" (Wittgenstein 1969, 14e, § 92).

In fact, as Beaney says, "analytic philosophy has had an uneasy relationship with the discipline of history of philosophy throughout its life. Analytic philosophers often either scorn or simply ignore history of philosophy" (Beaney 2013, 231). Admittedly, in recent years some analytic philosophers "have begun to look at the history of analytic philosophy itself more seriously and history of philosophy has been brought closer to home," nevertheless, "as a subdiscipline of philosophy, history of philosophy has often been derided by analytic philosophers, and at a crude level" (Beaney 2013, 231).

Thus, on the door of his office at Princeton university, Harman put the sign: "History of philosophy: Just say No!" (Sorell 2005, 44). So Harman wanted to say that "students of philosophy need not be required to study the history of philosophy and that a study of the history of philosophy tends not to be useful to students of philosophy" (ibid.).

The argument that philosophy does not need history of philosophy, however, is not valid.

Philosophy as acquisition of knowledge needs the history of philosophy because, positively, the history of philosophy presents and analyzes questions that philosophy has succeeded in investigating and that, in some cases, have even given rise to new sciences. The analysis helps to understand why and how philosophy has succeeded. Negatively, the history of philosophy presents and analyzes questions that philosophy has failed to investigate and which therefore require new investigations. The analysis helps to understand why and how philosophy has failed.

Philosophy as acquisition of knowledge also needs the history of philosophy because, as Antognazza says, the history of philosophy "keeps us alert to the fact that latest is not always best" (Antognazza 2015, 165). So, it "helps to avoid taking current orthodoxy for granted" (ibid., 180). Moreover, the history of philosophy keeps us alert to the fact that "a genuinely new perspective often means embracing and developing an old insight" (ibid., 165).

5.20 Confirming Philosophy as Acquisition of Knowledge

Since the arguments that can be raised against the view that philosophy is acquisition of knowledge are not valid, we can confirm that philosophy is acquisition of knowledge.

In fact, the view that philosophy is acquisition of knowledge is the only one capable of restoring vitality to philosophy. For, philosophy can be vital only if it responds to problems with which we are confronted in the world, and a solution to them essentially requires that philosophy be acquisition of knowledge of the world.

Therefore, philosophy must aim at knowledge. There are areas of experience that the present sciences are unable to address, philosophy aims at them and, if successful, it can open new ways to knowledge and lead to the birth of new sciences. This is what philosophy has repeatedly done until recently, and there is no denying that a philosophy capable of doing so is vital.

References

Antognazza, Maria Rosa. 2015. The benefit to philosophy of the study of its history. *British Journal for the History of Philosophy* 23:161–184.

Austin, John. 1970. *Philosophical papers*. Oxford: Oxford University Press.

Bealer, George. 1987. The limits of scientific essentialism. *Philosophical Perspectives* 1:289–365.

Bealer, George. 1996. On the possibility of philosophical knowledge. *Philosophical Perspectives* 10:1–34.

Beaney, Michael. 2013. Analytic philosophy and history of philosophy: The development of the idea of rational reconstruction. In *The historical turn in analytic philosophy*, ed. Erich Reck, 231–260. New York: Palgrave Macmillan.

Beebee, Helen. 2018. Philosophical scepticism and the aims of philosophy. *Proceedings of the Aristotelian Society* 118 (1): 1–24.

Black, Antony. 2008. The 'axial period': What was it and what does it signify? *The Review of Politics* 70:23–39.

Bobbio, Norberto. 2017. *La filosofia e il bisogno di senso*. Brescia: Morcelliana.

Bourget, David, and David Chalmers. 2014. What do philosophers believe? *Philosophical Studies* 170:465–500.

Bourget, David, and David Chalmers. 2023. Philosophers on philosophy: The 2020 philpapers survey. *Philosophers' Imprint* 23 (1): 11.

Brennan, Jason. 2010. Scepticism about philosophy. *Ratio* 23:1–16.

Carnap, Rudolf. 1935. *Philosophy and logical syntax*. London: Kegan Paul.

Cellucci, Carlo. 2025. Mathematics and experience. *Foundations of Science* 30:473–487.

Dennett, Daniel. 1996. *Darwin's dangerous idea: Evolution and the meaning of life*. London: Penguin.

Dewey, John. 2004. *Reconstruction in philosophy*. Mineola: Dover.

Dummett, Michael. 1973. *Frege: Philosophy of language*. New York: Harper and Row.

Dummett, Michael. 1978. *Truth and other enigmas*. Cambridge: Harvard University Press.

Dummett, Michael. 2010. *The nature and future of philosophy*. New York: Columbia University Press.

Frege, Gottlob. 1960. *The foundations of arithmetic: A logico-mathematical enquiry into the concept of number*. New York: Harper.

Hannon, Michael, and James Nguyen. 2022. Understanding philosophy. *Inquiry*. https://doi.org/10.1080/0020174X.2022.2146186.

Hempel, Carl. 1966. *Philosophy of natural science*. Englewood Cliffs: Prentice-Hall.

Hempel, Carl. 2001. *The philosophy of Carl G. Hempel: Studies in science, explanation, and rationality*. Oxford: Oxford University Press.

Kitcher, Philip. 2011. Epistemology without history is blind. *Erkenntnis* 75:505–524.

Marconi, Diego. 2014. *Il mestiere di pensare: La filosofia nell'epoca del professionismo*. Turin: Einaudi.

Rosenberg, Alex. 2011. *The atheist's guide to reality: Enjoying life without illusions*. New York: Norton.

Rota, Gian-Carlo. 1997. *Indiscrete thoughts*. Boston: Birkhäuser.

Rudner, Richard. 1953. The scientist qua scientist makes value judgments. *Philosophy of Science* 20:1–6.

Searle, John. 2008. *Philosophy in a new century*. Cambridge: Cambridge University Press.

Sorell, Tom. 2005. On saying no to history of philosophy. In *Analytic philosophy and history of philosophy*, ed. Tom Sorell and Graham Rogers, 43–59. Oxford: Oxford University Press.

van Inwagen, Peter. 2017. *Thinking about free will*. Cambridge: Cambridge University Press.

Wittgenstein, Ludwig. 1969. *On certainty*. Oxford: Blackwell.

Wittgenstein, Ludwig. 1979. *Notebooks 1914–1916*. Oxford: Blackwell.

Wittgenstein, Ludwig. 2001. *Tractatus logico-philosophicus*. London: Routledge.

Wittgenstein, Ludwig. 2005. *The big typescript TS 213*. Oxford: Blackwell.

Wittgenstein, Ludwig. 2009. *Philosophische Untersuchungen—Philosophical investigations*. Oxford: Blackwell.

Chapter 6
Methods of Philosophy

Abstract Several methods are used in contemporary philosophy. This chapter, however, considers them and argues that they are inadequate. They are the methods of analytic philosophy, which assumes that philosophy is not acquisition of knowledge. Philosophy as acquisition of knowledge needs a different method.

Keywords Grammatical analysis · Conceptual analysis · Common sense · Explication · Verificationism · Axiomatic method · Thought experiments · Inference to the best explanation · Intuition · Reflecting equilibrium · Modelling · Transcendental method · Experimental philosophy

6.1 Methods Used in Today's Philosophy

Several methods are used in contemporary philosophy: grammatical analysis, conceptual analysis, common sense, explication, verificationism, axiomatic method, thought experiments, inference to the best explanation, intuition, reflecting equilibrium, modelling, transcendental method, and experimental philosophy.

This chapter, however, considers them and argues that they are inadequate. This does not mean that philosophy does not have a method, instead, it means that philosophy needs a different method.

Indeed, the methods used in contemporary philosophy are the methods of analytic philosophy, which assumes that philosophy is not acquisition of knowledge. Philosophy as acquisition of knowledge needs a different method.

6.2 Grammatical Analysis

A method of philosophy is grammatical analysis.

Thus, Wittgenstein says that the method of philosophy is grammatical analysis because "philosophy gives no pictures of reality" (Wittgenstein 1979, 106). It

C. Cellucci, *A Philosophy for the Future*, Synthese Library 508,
https://doi.org/10.1007/978-3-031-98443-3_6

"arises neither from an interest in the facts of nature, nor from a need to grasp causal connections" (Wittgenstein 2009, 89). Philosophy is only concerned with our thoughts, indeed "the object of philosophy is the logical clarification of thoughts" Wittgenstein (2001, 4.112). This means that the object of philosophy is the logical analysis of propositions, because "in the proposition the thought is expressed perceptibly through the senses" (ibid., 3.1). But "a proposition is completely logically analysed if its grammar is made completely clear" Wittgenstein (1975, 51). Therefore, ultimately a philosophical "investigation is a grammatical one" (Wittgenstein 2009, 90). A grammatical investigation gives us the essence of things, because "grammar tells what kind of object anything is" (ibid., 373). Thus "essence is expressed in grammar" (ibid., 371). And "philosophy as custodian of grammar can in fact grasp the essence of the world" Wittgenstein (1975, 85). Therefore, "the harmony between thought and reality is to be found in the grammar of the language" (Wittgenstein 1974, Part I, 112).

This method, however, is inadequate. It is blatantly contradictory to assert, on the one hand, that philosophy gives no pictures of reality and does not arise from an interest in the facts of nature, and on the other hand, that a philosophical investigation is a grammatical one, a grammatical investigation gives us the essence of things, and through it philosophy can grasp the essence of the world. Philosophy cannot give us the essence of things if it gives no pictures of reality.

In fact, philosophy cannot give us the essence of things anyway. For, as made clear by Galileo's revolution, science cannot give us the essence of things. And there is no evidence that there exists a superior source of knowledge that provides knowledge of the essence of things, let alone that philosophy is such a superior source of knowledge.

6.3 Conceptual Analysis

Another method of philosophy is conceptual analysis.

Thus, Shaffer says that, "according to a significant number of philosophers," conceptual analysis "is the only method of philosophy" (Shaffer 2015, 555). Conceptual analyses "take the form of proposed definitions (that is, sets of necessary and sufficient conditions)," the "adequacy of any" definition "can be tested against concrete and/or imagined cases" (ibid.).

This method, however, is inadequate. For, in general it is impossible to give a definition of a philosophical concept that really covers all its cases. For example, many analytic philosophers consider the attempts to give an analysis of the concept of knowledge as an important example of conceptual analysis. According to them, "the justified true belief account of knowledge is a leading example of the thinking behind conceptual analysis" (Schukraft 2017, 23). But this is unwarranted. For, as will be argued in Chap. 10, all analyses of the concept of knowledge are subject to counterexamples.

6.4 Common Sense

Another method of philosophy is common sense.

Thus, Lewis says that you should "never put forward a philosophical theory that you yourself cannot believe in" your "most commonsensical moments" (Lewis 1986a, 135). For, "philosophy cannot credibly challenge either the positive convictions of common sense or the established theses of the natural sciences and mathematics" (Lewis 2020, 617). Therefore, "when common sense delivers a firm and uncontroversial answer about a not-too-far-fetched case, theory had better agree," only "when common sense falls into indecision or controversy, or when it is reasonable to suspect that farfetched cases are being judged by false analogy to commonplace ones, then theory may safely say what it likes" (Lewis 1986b, 194). Of course, "it's not that common sense speaks with the voice of some infallible faculty of 'intuition'. It's just that" the "only sensible policy" is the "reluctance to accept theories that fly in the face of common sense" (Lewis 1986a, 134).

This method, however, is inadequate. For, it is unwarranted to say that the only sensible policy is the reluctance to accept theories that fly in the face of common sense. Several mathematical and scientific theories fly in the face of common sense, therefore common sense leads to several errors that mathematics and the sciences must correct.

6.5 Explication

Another method of philosophy is explication.

Thus, Carnap says that the method of explication "consists in transforming a given more or less inexact concept," the explicandum, "into an exact one," the explicatum, "or, rather, in replacing the first by the second" (Carnap 1962, 3). The "explicandum may belong to everyday language or to a previous stage in the development of scientific language. The explicatum must be given by explicit rules for its use, for example, by a definition which incorporates it into a well-constructed system of scientific either logico-mathematical or empirical concepts" (ibid.). Between the explicatum and the explicandum "close similarity is not required, and considerable differences are permitted" (ibid., 7).

This method, however, is inadequate. According to it, the explicatum is not obtained through an analysis of the explicandum that makes all aspects of the explicandum explicit and clarifies it. The explicatum merely replaces the explicandum with something that might just have, but need not have, a vague similarity to the explicandum. Therefore, the explicatum does not provide a conceptual clarification of the explicandum.

6.6 Verificationism

Another method of philosophy is verificationism.

Thus, Ayer says that verificationism is the method based on "the criterion of verifiability" (Ayer 1952, 35). According to it, "we say that a sentence is" meaningful "to any given person, if, and only if, he knows how to verify the proposition which it purports to express—that is, if he knows what observations would lead him, under certain conditions, to accept the proposition as being true, or reject it as being false" (ibid.).

This method, however, is inadequate. For, the same observations may be compatible with several hypotheses that are mutually incompatible, so the observations cannot lead any given person to accept a hypothesis as being true, or reject it as being false. Therefore, the criterion of verifiability implies that several hypotheses are meaningless.

Moreover, the criterion of verifiability implies that philosophical propositions are meaningless.

For, Ayer says that "philosophical theories are not tested by observation. They are neutral with respect to particular matters of fact" (Ayer 1956, 1).

The criterion of verifiability itself is meaningless by its own standards, because no observations could lead any given person to accept the criterion of verifiability as being true or to reject it as being false.

6.7 Axiomatic Method

Another method of philosophy is the axiomatic method.

Thus, Lycan says that the axiomatic method "consists in the construction" of "deductively valid arguments whose premises are obvious, self-evident, indisputably true, or at least uncontroversial. This is the way" in which "philosophers arrive at conclusions that are guaranteed to be as indisputably true as the original premises, once the ingenious deductive arguments have been hit upon" (Lycan 1988, 116). The axiomatic method "has pervaded the rationalist tradition and survives" among "analytic philosophers" (ibid.). The "quality of self-evidence that" the "premises are supposed to have and to transmit to" the "conclusion has been variously described," for example, it has been described as "analyticity, a prioricity, clarity and distinctness, mere obviousness" (ibid.).

That the axiomatic method survives among analytic philosophers is clear, for example, from Ramsey.

Indeed, Ramsey says that "in philosophy we take the propositions we make in science and everyday life, and try to exhibit them in a logical system with primitive terms and definitions, etc." (Ramsey 1991, 42). For, "we are driven to philosophise because we do not know clearly what we mean; the question is always 'What do I mean by *x*?' and only very occasionally can we settle that without reflecting on meaning" (ibid., 47). We can do so only in the case of very simple questions, but "in order to settle more complicated questions" we "obviously need a logical structure, a system of logic," and specifically, an axiomatic system, "into which to bring them" (ibid., 46). Indeed, "all philosophy is asking whether '*p*' follows from '*q*'" where, "if '*p*', '*q*' are sufficiently analysed this is a problem of formal logic or mathematics, otherwise of philosophy" (ibid., 37).

This method, however, is inadequate. In general, the premisses are not self-evident in the sense that they are analytic, a priori, clear and distinct, or merely obvious. Moreover, the premisses are not indisputably true. Indeed, by Gödel's second incompleteness theorem, even the premisses of mathematics are not indisputably true.

6.8 Formalization

Another method of philosophy is formalization.

Thus, Hansson says that formalization consists in translating ordinary propositions into propositions of a formal language, which "by isolating important aspects," effectively "helps to bring them to light" (Hansson 2000, 166). Furthermore, "formalized treatments tend to be helpful in at least four additional ways" (ibid.). First, "formalization incites definitional and deductive economy. It brings forth questions about the interdefinability of concepts and about minimizing the set of primitive principles of inference" (ibid.). Secondly, "formalization serves to make implicit assumptions visible" (ibid.). Thirdly, "formal theories can support delicate structures that would be much more difficult to uphold and handle in the less unambiguous setting of an informal language" (ibid., 166–167). Fourthly, "formalization stimulates strivings for completeness" (ibid., 167).

This method, however, is inadequate. Formalization does not produce new ideas, it merely puts existing ideas into fair copy. Thinking that it can produce new ideas depends on the assumption that basing philosophy on mathematical logic gives not only rigour, but also substance to philosophy. This, however, is an illusion, no new substantial philosophical ideas can be obtained through formalization.

A parody of this illusion is made by Halmos: "If you think that your paper is vacuous, | Use the first-order functional calculus. | It then becomes logic, | And, as if by magic, | The obvious is hailed as miraculous" (Halmos 1985, 216).

This confirm that formalization does not give substance to philosophy.

6.9 Thought Experiments

Another method of philosophy is thought experiments.

Thus, McGinn says that the method of thought experiments consists in conducting "an 'experiment' in our mind to test" some hypothesis about the analysis of a concept, hypothesis that "we have conjectured to be true. It" involves "conceiving possible cases in which the conditions are met and then asking ourselves whether our understanding of the concept indicates that the concept applies in these cases" (McGinn 2015, 93). The hypothesis "makes certain predictions about our intuitions in possible cases, and then we check to see if our intuitions confirm or disconfirm the hypothesis. The intuitions function as evidence for or against the hypothesis, and they are grounded in our actual grasp of the concept in question" (ibid.). Thought experiments "are indeed experiments, just as we would naively suppose from the phrase. We have here a procedure that arrives at truth in a non-question-begging way by consulting a body of data ('intuitions') generated especially for the purpose" (ibid.). The "intuitions play an evidential role, so we can say that they constitute empirical ('experiential') evidence in this broad sense" (ibid., 94).

This method, however, is inadequate. Admittedly, thought experiments have contributed to science. For example, in order to reject Aristotle's assumption that the heavier body moves more rapidly than the light one, Galileo imagines that a large stone and a smaller stone are connected to each other by a string and are dropped from the top of a tower. Now, "on connecting the two" by a string, "the faster one will by partly retarded by the slower" (Galilei 1968, VIII, 107). So, if the "large stone moves with a speed of, say, eight, while" the "smaller stone moves with a speed of four, then when they are connected" by a string "the system will move with a speed less than eight" (ibid.). On the other hand, "the two stones, when tied together, make a stone larger than that which before moved with a speed of eight" (ibid.). Therefore, the system will move with a speed greater than eight. Contradiction.

But, while thought experiments have contributed to science, they have been unreliable in philosophy. Thus, to prove that the mind is distinct from the body and can exist without it, Descartes argues that "we can easily suppose" that we have "no body at all," but "we cannot suppose that we, who think these things, do not exist. For, it is a contradiction to suppose that what thinks does not, at the very time when it is thinking, exist" (Descartes 1996, VIII–1, 7). Therefore, "it is certain that I, that is, my mind," is "entirely and really distinct from my body and can be or exist without it" (ibid., IX, 62).

There is a fundamental difference between Galileo and Descartes. Galileo also conducted real experiments with inclined planes which confirmed the conclusion of his thought experiment. On the contrary, the conclusion of Descartes' thought experiments has not been confirmed by any real experiment, and is unlikely to be confirmed. In fact, in the past few decades, there has been much evidence for the identification of mental processes with bodily processes.

That thought experiments have been unreliable in philosophy depends on the fact that a philosopher using this method starts with a view he holds and concocts a thought experiment to support it, without testing the compatibility of his view with experience.

6.10 Inference to the Best Explanation

Another method of philosophy is inference to the best explanation.

Thus, Williamson says that "the methods philosophy needs for choosing between rival theories need not be so different from the more theoretical methods of natural science. We want the theory that best explains whatever evidence we can get. The method of choosing between theories on that basis is called 'inference to the best explanation'. It is widely used in both natural science and philosophy" (Williamson 2018, 78), Inference to the best explanation is "obviously fallible," but "fallibility does not justify skepticism. Inference to the best explanation is a vital means of knowledge" (Williamson 2022, 354). For, it encourages "both the accumulation of more evidence of various kinds and the development of better explanations of that evidence" (ibid., 365). Therefore, philosophy should use inference to the best explanation in "the construction and selection of philosophical theories" (ibid., 356).

This method, however, is inadequate. For, it does not offer any criterion for deciding when a given hypothesis is a better explanation than another hypothesis.

Williamson proposes simplicity as a criterion. He says that, "when two theories make the same observable predictions, inference to the best explanation may still be able to select one over the other because the former is simpler" (ibid., 355). And he also says that "philosophy is a systematic truth-directed theoretical inquiry" (ibid., 364).

But why should the fact that a theory is simpler than another one be a reason to conclude that it is true? This would be justified only on the assumption that the universe is simple, so only on the assumption that the universe is organized for our convenience. But why should the universe be organized for our convenience?

In fact, Galileo argues that natural matters do not "accommodate themselves to our comprehension," on the contrary, nature "first made things in her own way, and then made human reason skilful enough to be able to understand (but only by great labour) some of her secrets" (Galilei 1968, VII, 289).

Kant argues that it is "not nature" that adjusts "to the conditions according to which we strive to obtain a concept of it," on the contrary, it is our "reflection on the laws of nature" that "adjusts itself to nature" (Kant 2007, 15).

Therefore, the simplicity criterion is inadequate.

Moreover, inference to the best explanation involves that ultimately there is a single best explanation of a thing. But this contrasts with the fact that there are often several explanations of a thing which are useful to different purposes, so there is no single best explanation.

6.11 Intuition

Another method of philosophy is intuition.

Thus, Bealer says that "intuitions are used as evidence (or reasons) in our standard justificatory practices" and, "in philosophy, the use of intuitions as evidence is equally ubiquitous" (Bealer 1999, 30). Intuitions are "seemings; for you to have an intuition that A is just for it to seem to you that A" (Bealer 2002, 73). For example, "when you first consider one of de Morgan's laws," it does not seem to be true, "after a moment's reflection, however, something new happens: suddenly it just seems true. Of course, this kind of seeming is intellectual, not sensory or introspective (or imaginative)" (Bealer 1999, 30). Thus, intuitions are intellectual seemings whereby a proposition A suddenly seems true to you without the need to consider arguments for A. Clearly, "intuition is different from belief: you can believe things that you do not intuit (e.g., that Rome is the capital of Italy), and you can intuit things that you do not believe (e.g., the axioms of naïve set theory)" (Bealer 2002, 73). Intuitions are a basic source of evidence because they have a reliable tie to the truth, for "something is a basic source of evidence iff it has an appropriate kind of a reliable tie to the truth" (Bealer 1999, 34).

This method, however, is inadequate. Considering a proposition to be true simply because it seems true to you is completely unreliable because it is not based on data about the world but only on your impressions. The proposition may also seem true to you out of wishful thinking, namely, only because it is what you wish for.

6.12 Reflecting Equilibrium

Another method of philosophy is reflecting equilibrium.

Thus, Goodman says that reflecting equilibrium is a method by which "deductive inferences are justified by their conformity to valid general rules" (Goodman 1983, 64). And general rules "are justified by their conformity with accepted deductive practice" (ibid., 63). In the same way, "an inductive inference" is "justified by conformity to general rules, and a general rule" is justified "by conformity to accepted inductive inferences" (ibid., 64). This "looks flagrantly circular," but "this circle is a virtuous one. The point is that" inferences and general rules "are justified by being brought into agreement with each other" (ibid.). The process ends when an agreement is achieved, "and in the agreement achieved lies the only justification needed for either" (ibid.).

This method, however, is inadequate. For, it makes the justification of deductive inferences ultimately depend on conformity to accepted deductive practice, and the justification of inductive inferences ultimately depend on conformity to accepted inductive practice. But we do not know that accepted deductive or inductive practice is correct. Therefore, the alleged justification by the method of reflective equilibrium begs the question.

6.13 Modelling

Another method of philosophy is modelling.

Thus, Weisberg says that modelling is a method by which "one constructs and analyzes a model in order to learn, indirectly, about something else" (Weisberg 2016, 264). Modelling "can be contrasted with direct representation and analysis," because "in direct representation, we analyze the system itself," while in modelling, "we study a constructed intermediary" (ibid.). Modelling involves two steps, "one first constructs" a "model, and specifies the intended target of that model," then "one subjects it to analysis" (ibid.). Most commonly, models are simpler than the "target systems they represent and they are idealized relative to these targets" (ibid.). In philosophical contexts, "models are rarely used to study a single target. Instead, they are most often used to answer" questions of how it could be that a certain thing happened, or questions of what would happen if something occurred, "and hence usually have generalized systems as their targets" (ibid., 265). Moreover, "sometimes, the targets of philosophical models are themselves hypothetical systems which do not exist, such as a perfectly just society, or a universe with only two particles" (ibid.).

This method, however, is inadequate. For, even when, by developing models and testing them, it is possible to approximate what the 'real thing' is like, there is no way of knowing whether a model really matches the real thing. In particular, the link between model and reality is feeble, the model creates a toy world, but ultimately the real world is not given up for the toy world.

6.14 Transcendental Method

Another method of philosophy is the transcendental method.

Thus, Smith and Sullivan say that the transcendental method is the method based on transcendental arguments, where "a transcendental argument is an argument in the form of modus ponens in which the transcendental claim figures as the conditional premise. On such a view, a transcendental argument would state that *A* is the case, and that *B* is a condition of the possibility of *A*, and go on to conclude that *B* is the case" (Smith and Sullivan 2011, 5–6).

This method, however, is inadequate. For, from the fact that *A* is the case, and that *B* is a condition of the possibility of *A*, it does not follow that *B* is the case. For, *B* can be false, and *B* is a condition of the possibility of *A* just because *B* is false.

6.15 Experimental Philosophy

Another method of philosophy is experimental philosophy.

Thus, Stich says that "experimental philosophy" consists in "the empirical investigation of philosophical intuitions, the factors that affect them, and the psychological and neurological mechanisms that underlie them" (Stich 2015, 21). According to the method of experimental philosophy, "rather than challenging, or defending, the epistemic status of most or all philosophical intuitions, we should be exploring the epistemic status of different sorts of philosophical intuitions that, according to our best psychology and neuroscience, are produced by different mental mechanisms" (ibid., 24). Several experimental philosophers "are motivated to explore intuitions experimentally primarily because they think that by doing so they can do a better job of conceptual analysis" (ibid., 25). For, by "collecting the intuitions of a substantial number of non-philosophers," we "may discover important facts about ordinary concepts that have gone unnoticed by philosophers using more traditional methods of conceptual analysis" (ibid., 26). Other experimental philosophers are motivated to explore intuitions experimentally because they think that intuitions vary across different demographic groups, and this is an "empirical claim which is susceptible to empirical exploration" (ibid., 28). If intuitions really vary, "this is bad news for the standard philosophical assumption that the contents of people's intuitions are very likely to be true" (ibid., 29).

This method, however, is inadequate. When philosophers disagree about, say, what knowledge is, there is no empirical state of affairs whose observation will tell us who is right and who is wrong. For, no empirical state of affairs directly expresses the intuitions of the disagreeing philosophers about what knowledge is. On the other hand, for many philosophical topics, the intuitions of non-philosophers explored by experimental philosophers are unreliable because non-philosophers have no familiarity with the questions themselves and the problems they pose.

6.16 The Need for a Method to Acquire Knowledge

From what we have seen above it seems fair to conclude that the methods used in contemporary philosophy are inadequate. This does not mean, however, that philosophy does not have a method, it only means that philosophy needs a different method.

Williamson claims that "surely the toolbox of methods" of contemporary philosophy "can be improved," however this "will happen not by some melodramatic break with the past, but by" a "process of self-refinement" (Williamson 2018, 142).

But this is unjustified. The methods of contemporary philosophy require a clear break with the past, because they are the methods of analytic philosophy, which assumes that philosophy is not acquisition of knowledge.

Instead, as will be argued in Chap. 8, philosophy makes progress if and only if it can open new ways to knowledge and lead to the birth of new sciences. Therefore, philosophy makes progress if and only if it is acquisition of knowledge.

Philosophy as acquisition of knowledge needs a method for acquiring knowledge, and the methods of analytic philosophy do not serve that purpose.

Since philosophy, like science, cannot give knowledge of the essence of things, the method of philosophy cannot be essentially different from that of science. In fact, as will be argued in Chap. 18, like the method of science, the method of philosophy is the analytic method.

References

Ayer, Alfred. 1952. *Language, truth and logic*. New York: Dover.

Ayer, Alfred. 1956. *The problem of knowledge*. London: Penguin.

Bealer, George. 1999. A theory of the a priori. *Philosophical Perspectives* 13:29–55.

Bealer, George. 2002. Modal epistemology and the rationalist renaissance. In *Conceivability and possibility*, ed. Tamar Genler and John Hawthorne, 71–125. Oxford: Oxford University Press.

Carnap, Rudolf. 1962. *Logical foundations of probability*. Chicago: University of Chicago Press.

Descartes, René. 1996. *Oeuvres*. Paris: Vrin.

Galilei, Galileo. 1968. *Opere*. Firenze: Barbera.

Goodman, Nelson. 1983. *Fact, fiction, and forecast*. Cambridge: Harvard University Press.

Halmos, Paul. 1985. *I want to be a mathematician: An automathography*. New York: Springer.

Hansson, Sven. 2000. Formalization in philosophy. *The Bulletin of Symbolic Logic* 6:162–175.

Kant, Immanuel. 2007. *Critique of judgment*. Oxford: Oxford University Press.

Lewis, David. 1986a. *On the plurality of worlds*. Oxford: Blackwell.

Lewis, David. 1986b. *Philosophical papers*. Vol. 2. Oxford: Oxford University Press.

Lewis, David. 2020. Letter to Steve Pyke, 27 July 1990. In David Lewis, *Philosophical letters, Volume 1: Causation, modality, ontology*, 615–617. Oxford: Oxford University Press.

Lycan, William. 1988. *Judgment and justification*. Cambridge: Cambridge University Press.

McGinn, Colin. 2015. The science of philosophy. *Metaphilosophy* 46:84–103.

Ramsey, Frank. 1991. *Notes on philosophy, probability and mathematics*. Naples: Bibliopolis.

Schukraft, Jason. 2017. *An analysis of Edmund Gettier's: Is justified true belief knowledge?* London: Macat.

Shaffer, Michael. 2015. The problem of necessary and sufficient conditions and conceptual analysis. *Metaphilosophy* 46:555–563.

Smith, Joel, and Peter Sullivan. 2011. Introduction. In *Transcendental philosophy and naturalism*, ed. Joel Smith and Peter Sullivan, 1–25. Oxford: Oxford University Press.

Stich, Stephen. 2015. What is experimental philosophy? *Annals of the Japan Association for Philosophy of Science* 23:21–31.

Weisberg, Michael. 2016. Modeling. In *The Oxford handbook of philosophical methodology*, ed. Herman Cappelen, Tamar Szabó, and John Hawthorne, 262–286. Oxford: Oxford University Press.

Williamson, Timothy. 2018. *Doing philosophy: From common curiosity to logical reasoning*. Oxford: Oxford University Press.

Williamson, Timothy. 2022. *The philosophy of philosophy*. Hoboken: Wiley-Blackwell.

Wittgenstein, Ludwig. 1974. *Philosophical grammar*. Oxford: Blackwell.

Wittgenstein, Ludwig. 1975. *Philosophical remarks*. Oxford: Blackwell.

Wittgenstein, Ludwig. 1979. *Notebooks 1914–1916*. Oxford: Blackwell.

Wittgenstein, Ludwig. 2001. *Lectures, Cambridge, 1932–1935*. Prometheus Books.

Wittgenstein, Ludwig. 2009. *Philosophische Untersuchungen—Philosophical investigations*. Oxford: Blackwell.

Chapter 7
Utility of Philosophy

Abstract The present irrelevance of philosophy raises the question of utility of philosophy. There are several views of utility of philosophy, but this chapter considers them and argues that they are not valid. Alternatively, the chapter argues that utility of philosophy consists in the fact that philosophy can open new ways to knowledge and lead to the birth of new sciences.

Keywords The question of utility of philosophy · Views of utility of philosophy · Utility of philosophy as opening new ways to knowledge and leading to the birth of new sciences · An example of philosophy leading to the birth of a new science

7.1 Views of Utility of Philosophy

The present irrelevance of philosophy raises the question of utility of philosophy. There are several views of utility of philosophy, but this chapter considers them and argues that they are not valid.

Alternatively, the chapter argues that utility of philosophy consists in the fact that philosophy can open new ways to knowledge and lead to the birth of new sciences.

7.2 Philosophy as Pursued for the Sake of Knowledge

As already said, there are several views of utility of philosophy.

A view of utility of philosophy is that philosophy is pursued for the sake of knowledge and not for any utility to life, because knowledge is an end in itself.

Thus, Aristotle says that philosophy is pursued "for the sake of knowledge and not for any utility" (Aristotle, *Metaphysica*, A 2, 982 b 20–21). Knowledge is an end in itself, from knowledge "nothing results apart from contemplation" (Aristotle, *Ethica Nicomachea*, K 7, 1177 b 2).

C. Cellucci, *A Philosophy for the Future*, Synthese Library 508,
https://doi.org/10.1007/978-3-031-98443-3_7

The view that philosophy is pursued for the sake of knowledge and not for any utility to life, because knowledge is an end in itself, was natural in ancient Greece, where there was a sharp division between doing and knowing that was a reflection of a social organization in which useful work was done mainly by slaves. This relieved free men from labour, allowing them to devote themselves to "those sciences which do not aim" at the "necessities of life" (Aristotle, *Metaphysica*, A 1, 981 b 21–22).

Conversely, this view is incongruous in the modern world, in which scientific knowledge is not pursued only for the sake of knowing, but also to improve mankind's life and well-being.

Aristotle claims that the fact that philosophy is pursued for the sake of knowledge and not for any utility because knowledge is an end in itself, "is confirmed by the facts: when almost all the necessities of life and the things that make for comfort and recreation had been secured, only then knowledge began to be sought. So, it is clear that we pursue it for no other use" (ibid., A 2, 982 b 22–25).

But this is not valid. As will be argued in Chap. 10, knowledge did not begin to be sought only when almost all the necessities of life had been secured, it was pursued from the very beginning, because knowledge is essential to life. In fact, it is only through knowledge that life has been able to exist and preserve itself.

7.3 Philosophy as Providing Foundations for Science

Another view of utility of philosophy is that philosophy provides foundations for science.

Thus, Husserl says that "no science can have fully evident foundations," only philosophy "can provide them" and "bring forth rigorous science" (Husserl 2019, 464). Providing foundations for science is important because all sciences contain "a radical defect owing to the naiveté in which they simply accept what is pregiven" and "in which they fashion concepts, principles, and indeed ideal sciences, ideal realms of objects. These sciences can only become absolutely grounded when one descends from their beginnings and grounds to the primal grounds and primal beginnings, to the true 'archai'" (ibid., 174). This can be done only by philosophy, "in the original sense of a universal ultimately grounding science" (Husserl 1970, 207).

But this is not valid. All attempts by philosophy to provide foundations for science have failed. By Gödel's incompleteness theorems, philosophy cannot even provide foundations for mathematics.

7.4 Philosophy as Searching Out Inconsistencies in Principles

Another view of utility of philosophy is that philosophy examines critically the principles employed in science and searches out any inconsistencies there can be in them.

Thus, Russell says that "the essential characteristic of philosophy, which makes it a study distinct from science, is criticism. It examines critically the principles employed in science," it "searches out any inconsistencies there may be in these principles, and it only accepts them when, as the result of a critical inquiry, no reason for rejecting them has appeared" (Russell 1998, 87). It wants "to see whether they are mutually consistent and whether the inferences employed are such as seem valid to a careful scrutiny" (Russell 1995, 239).

But this is not valid. Philosophy cannot examine critically the principles employed in science and search out any inconsistencies there may be in them principles. For, by Gödel's second incompleteness theorem, philosophy cannot do this even in the case of mathematics.

7.5 Philosophy as Dissolving Philosophical Problems

Another view of utility of philosophy is that philosophy dissolves philosophical problems, which are only a result of the mystifying use of our language.

Thus, Wittgenstein says that a philosophical problem is "a muddle felt as a problem," a "puzzlement caused by the mystifying use of our language" (Wittgenstein 1969, 6). In fact, "philosophical problems arise when language goes on holiday" (Wittgenstein 2009, 38). Therefore, "philosophy is a struggle against the bewitchment of our understanding by means of language" (ibid., 109). In philosophy we are destroying "houses of cards, and we are clearing up the ground of language on which they stood" (ibid., 118). Philosophical problems must not be solved, they must be dissolved. By exposing the mystifying use of our language, "the philosophical problems should completely disappear" (ibid., 133)

But this is not valid, philosophical problems are not merely a result of the mystifying use of our language.

Even Hacker, who says that his conception of philosophy is wholly inspired by Wittgenstein, admits that "Wittgenstein's suggestion that philosophical problems and conceptual confusions arise only when language 'is idling'," is "incorrect and shown to be so in the sciences, in public affairs, and in the thoughts and reflections of Everyman" (Hacker 2013, xi).

In particular, several scientific results are answers to what were previously considered genuine philosophical problems.

7.6 Philosophy as Enlarging Our Conception of What Is Possible

Another view of utility of philosophy is that, although the philosophical questions that are of the profoundest interest, such as the question of whether the universe has any unity of plan, must remain insoluble to the human intellect, nevertheless philosophy must be studied because these questions enlarge our conception of what is possible.

Thus, Russell says that in philosophy "there are many questions," but "among them those that are of the profoundest interest to our spiritual life" such as the question "Has the universe any unity of plan or purpose, or is it a fortuitous concourse of atoms?" must "remain insoluble to the human intellect" (Russell 1998, 89). These questions "are asked by philosophy, and variously answered by various philosophers," but the answers "are none of them demonstrably true" (ibid.). They are "such as our capacities are essentially incapable of solving" (Russell 1994, 124). So, we cannot "include as part of the value of philosophy any definite set of answers to such questions" (Russell 1998, 91). Indeed, "the value of philosophy must not depend upon any supposed body of definitely ascertainable knowledge to be acquired by those who study it" (ibid.). Philosophy "is to be studied, not for the sake of any definite answers to its questions, since no definite answers can, as a rule, be known to be true, but rather for the sake of the questions themselves; because these questions enlarge our conception of what is possible" (ibid., 93).

But this is not valid. It assumes that the philosophical questions of the profoundest interest, such as the question of whether the universe has any unity of plan, must remain insoluble to the human intellect. This conflicts with the fact that, a few years after Russell claimed that the question of whether the universe has any unity of plan must remain unsolvable to the human intellect, Lemaître solved he question through his Big Bang theory.

Something similar had occurred to Comte who, as Hilbert recalls, "said—with the intention of mentioning a problem that was certainly unsolvable—that science would never be able to fathom the secret of the chemical composition of the heavenly bodies. A few years later this problem was solved by the spectral analysis of Bunsen and Kirchhoff" (Hilbert 1996, 1165).

The solvability of a question depends on the resources available at the time. A question unsolvable at one time can become solvable at a later time, when adequate resources become available.

7.7 Philosophy as Dealing with the Meaning of Propositions

Another view of utility of philosophy is that, while science deals with the truth of propositions, philosophy deals with their meaning, and knowing whether a proposition is true requires first knowing the meaning of the proposition.

Thus, Schlick says that, while science is the "pursuit of truth," philosophy is the "pursuit of meaning" (Schlick 1979, 217). The pursuit of meaning is essential because, "before we can know or decide whether" a proposition "is true or false we must know what this proposition says. We must know the meaning of the proposition at first. After we know" its meaning "we may be able to find out whether it is true or not" (ibid.). Once the meaning has been made clear, "we can proceed to look for the actual circumstances in the world and decide whether they make our proposition true or false" (ibid., 218). On the contrary, "in traditional philosophy," instead of pursuing meaning, most often philosophers "have tried to find out whether a certain proposition was true or false before being clear about the meaning of it" (ibid.). This has led philosophy to failure.

The pursuit of meaning is an activity that "is at work in every single science continually, because before the sciences can discover the truth or falsity of a proposition they have to get at the meaning first," and sometimes they "find, by the contradictory results at which they arrive, that they have been using words without a perfectly clear meaning" (ibid., 220). The sciences "cannot go on with the pursuit of truth before the pursuit of meaning has been successful. In this way philosophy is an extremely important factor within science" (ibid., 221). When philosophy is successful in the pursuit of meaning, it makes progress.

Meaning "is found by mere reflection about possible circumstances in the world," while truth "is decided by really discovering the existence or nonexistence of those circumstances" (ibid., 218). The reflection through which meaning is found is "the philosophic method of which Socrates' dialectical proceeding has afforded us the simplest example" (ibid., 219). Socrates asks 'What is X', "he receives various answers to his question, and in turn he asks what was meant by these answers, why a particular word was used in this way or that way, and it usually turns out that his disciple or opponent is not at all clear about his own opinion," therefore Socrates tries "to clarify thought by analyzing the meaning of our expressions and the real sense of our propositions" (ibid., 217).

But this is not valid. The pursuit of meaning is an activity that is a constituent part of science and is inseparable of it. It is not by chance that the only examples of successful pursuit of meaning mentioned by Schlick are by two scientists, Newton and Einstein. Schlick tries to avoid contradicting himself by saying that, "when Newton discovered the concept of mass he was at that time really a philosopher," and "Einstein's analysis of the meaning of the word 'simultaneity' as it is used in physics" was "a real philosophical achievement" (ibid., 219). But this is specious. Newton discovered the concept of mass as part of his quantitative theory for mass and motion, and Einstein's analysis of the concept of simultaneity was part of his discovery of the relativity of simultaneity. So, when Newton discovered the concept of mass he was really a scientist, and Einstein's analysis of the meaning of simultaneity was a real scientific achievement.

7.8 Philosophy as Clarifying Concepts by Conceptual Analysis

Another view of utility of philosophy is that philosophy clarifies concepts by conceptual analysis. Relying on the clarifying power of conceptual analysis is justified if one considers how many wonderful things mathematics does, although mathematics is ‘only’ the derivation of logical consequences from evident principles.

Thus, Dummett says that “the task of philosophy is to clarify our concepts” (Dummett 2010, 13). And “the philosopher’s only resource is the analysis of concepts we already possess, but about which we are confused; he seeks to remove that confusion” (ibid., 11). It might be objected that “I have an extremely optimistic view that clarification” through the analysis of concepts “will take us very far,” but “look at how many wonderful things mathematics does: yet” mathematics “is ‘only’ the logical derivation of consequences starting from evident principles” (Dummett 1997).

But this is not valid. As already argued in Chap. 3, it is not valid to say that philosophy clarifies concepts through conceptual analysis. Conceptual analysis cannot clarify the concepts of common use because many confusions about them depend on prejudices or lack of knowledge. Dealing with them generally requires new knowledge of the world. This contrasts with the fact that conceptual analysis permits only to assemble and marshal what we already know.

It is also not valid to compare the clarifying power of conceptual analysis with the power of mathematics, on the grounds that mathematics is ‘only’ the derivation of logical consequences from evident principles. For, mathematics is not ‘only’ the derivation of logical consequences from evident principles (see Cellucci 2022).

7.9 Philosophy as Aiming at Truth Like Science

Another view of utility of philosophy is that the aim of philosophy is truth, like the aim of science.

Thus, Popper says that, contrary to analytic philosophy, philosophy is not “an attempt either to clarify or to analyse or to ‘explicate’ concepts, or words, or languages” (Popper 1994, 178). Concepts or words “are mere tools for formulating propositions, conjectures and theories” (ibid.). The aim of philosophy is not “to analyse meanings, but to seek for interesting and important truths; that is, for true theories” (ibid.). So, the aim of philosophy is like the aim of science, because “truth” is “the aim of science: science is the search for truth” (ibid., 39). Science is “the search for objectively true, explanatory theories” (ibid., 4). In the same way, philosophy is the search for truth, each of “the truly great philosophers of the past” would “have discarded his system” had he “become convinced that” it “was not a step on the way to truth” (ibid., 178). Thus, “the history of philosophy essentially” is “part of the history of the search for truth” (ibid.).

But this is not valid. As will be argued in Chap. 11, the aim of science cannot be truth, in particular because there is no criterion truth. For the same reason, the aim of philosophy cannot be truth.

7.10 Philosophy as Consciousness of Assumptions of Science

Another view of utility of philosophy is that philosophy makes us conscious of the assumptions of science.

Thus, Laurenti, Lopes, and Abid say that, "without philosophy, science has no consciousness," it "loses its intellectual, moral, and social consciousness" (Laurenti et al. 2020, 92). Philosophy "arises as a second-order discourse—a discussion, debate, and questioning about scientific statements. Science's discourses are affirmative; philosophy's discourses are critical and negative" (ibid., 93). Parodying Kierkegaard's statement that life can only be understood backwards but must be lived forwards, "we could say that while science knows forward, philosophy criticizes backward" (ibid.). This "tension between science" which knows forward, and "philosophy" which criticizes backward, "is essential for building enlightened scientific knowledge" (ibid.). Philosophy gives "consciousness of the theoretical assumptions of science" (ibid.). Without "philosophical consciousness, science incurs epistemological naiveties" (ibid., 91).

But this is not valid. Acquiring consciousness of the assumptions of science is an integral part of the work of scientists, who are surely better qualified than philosophers to do this work. Moreover, it is not valid to say that science knows forward, philosophy criticizes backward. For, as will be argued in Chap. 18, science and philosophy proceed with the same method, the analytic method, which knows and criticizes both forward and backward.

7.11 Philosophy as Providing an Integrated Picture of Nature

Another view of utility of philosophy is that philosophy integrates the results of science into an overall picture of nature.

Thus, Kitcher says that, after the birth of modern science, "the division of labor between philosophy and natural science was more firmly instituted in the eighteenth and nineteenth centuries" (Kitcher 2023, 16). So, today the aim of philosophy is not to produce new knowledge, but to provide "an integrated picture of nature that combines the contributions of different areas of inquiry" (ibid., 12). Indeed, "different fields of investigation can be assisted by thinkers whose more synthetic perspective can alert them to missed opportunities and provide them with needed clarification" (ibid., 12–13). Then "philosophy, so understood, is a synthetic discipline, one that reflects on and responds to the state of inquiry," and "to the felt needs of individual people to make sense of the world and their place in it" (ibid., 13).

The view that the aim of philosophy is not to produce new knowledge, but to provide an integrated picture of nature that combines the contributions of different areas of inquiry, has been widely held over the last two centuries.

Thus, Spencer says that philosophy is "knowledge of the highest generality. Science means merely the family of the sciences—stands for nothing more than the sum of knowledge formed of their contributions," it "consists of truth existing more or less separated; and does not recognize these truths as entirely integrated" (Spencer 1896, 134). In fact, "knowledge of the lowest kind is un-unified knowledge; science is partially-unified knowledge; philosophy is completely-unified knowledge" (ibid., 136).

Wundt says that philosophy is "the general science which has to unite the general knowledge furnished by the individual sciences into a consistent system" (Wundt 1889, 21).

Neurath says that, "after the removal of traditional metaphysics, in constant struggle against metaphysical leanings," philosophy aims at "the creation of an encyclopedic synthesis of the sciences on uniform logical foundations" (Neurath 1983, 115).

But this is not valid. If philosophy systematizes and integrates the results of sciences into an overall picture of nature, it does what, as made clear by Galileo's revolution, no science is capable of doing, and therefore is a superscience because it has capabilities superior to those of any science. But there is no evidence that philosophy is a superscience, just think of the attacks by scientists and mathematicians on contemporary philosophy in Chap. 2.

7.12 Philosophy as Investigating Common Ideas Taken for Granted

Another view of utility of philosophy is that philosophy investigates common ideas taken for granted. For example, a mathematician takes the idea of number for granted, he may investigate the relations among numbers, but a philosopher will ask: What is a number?

Thus, Nagel says that "the main concern of philosophy is to question and understand very common ideas that all of us use every day without thinking about them" (Nagel 1987, 5). We "couldn't get along in life without taking" these ideas "for granted most of the time; but in philosophy we investigate those things themselves" (ibid.). For example, a mathematician takes the idea of number for granted, he "may investigate the relations among numbers, but a philosopher will ask, 'What is a number?'" (ibid.).

But this is not valid. In fact, a mathematician asks the question 'What is a number?' and gives an answer to it, on the basis of which he develops a theory of numbers.

7.13 Philosophy as Organ for Dealing with Social Conflicts

Another view of utility of philosophy is that philosophy can be an organ for dealing with social conflicts.

Thus, Dewey says that philosophers "have seen themselves, and have represented themselves to the public, as dealing with something which has variously been termed being, nature or the universe, the cosmos at large, reality, the truth" (Dewey 2004, vii). But in fact, "under disguise of dealing with ultimate reality, philosophy has been occupied with precious values embedded in social traditions, that it has sprung from a conflict of inherited institutions with incompatible contemporary tendencies" (ibid., 15–16). Then "the task" of "philosophy is to clarify men's ideas as to the social and moral strifes of their day. Its aim is to become so far as is humanly possible an organ for dealing with these conflicts" (ibid., 16).

But this is not valid. In fact, philosophy has never really become an organ for dealing with social conflicts.

7.14 Philosophy as Making Sense of Ourselves and Our Activities

Another view of utility of philosophy is that philosophy is part of the humanistic enterprise of making sense of ourselves and of our activities. The conclusions reached by scientific reasoning are irrelevant to it.

Thus, Williams argues that philosophy is "part of a wider humanistic enterprise of making sense of ourselves and of our activities" (Williams 2006, 197). The conclusions reached by scientific reasoning are irrelevant to it, philosophy needs "concepts and explanations which are rooted in our more local practices, our culture, and our history" (ibid., 186–187).

But this argument is not valid. For, in order to make sense of ourselves and of our activities, we need knowledge of the world and our place in it.

Even in order to acquire concepts and explanations rooted in our more local practices, our culture, and our history, we need knowledge.

As Zeki says, "the path to acquiring knowledge—whether grounded in scientific experimentation or through philosophical or humanistic speculation—must use similar mental processes" (Zeki 2014, 13). And, if "similar brain processes are involved in humanistic and scientific inference, we are led ineluctably to the view that conclusions reached by one are relevant to those reached by the other" (ibid., 14).

Therefore, Williams' attempt to oppose humanistic disciplines to scientific disciplines fails.

7.15 Philosophy as Making Us Free

Another view of utility of philosophy is that philosophy makes us free, because it is contemplation of the universe.

Thus, Russell says that "the man who has no tincture of philosophy goes through life in the prejudices derived from common sense" (Russell 1998, 91). But, "if our life is to be great and free, we must escape this prison" (ibid., 92). And "one way of escape is by philosophic contemplation" of the universe, because such contemplation, enclosing everything in a single vision, achieves "an enlargement of the Self" (ibid.). In fact, "through the infinity of the universe the mind which contemplates it achieves some share in infinity" (ibid.). The contemplation of the universe "makes us citizens of the universe," and "in this citizenship of the universe consists man's true freedom, and his liberation from the thraldom of narrow hopes and fears" (ibid., 93).

But this is not valid. For, far from making us free, the contemplation of the universe can make us even more aware of the constraints to which we are subjected. Thus, Pascal says: "When I consider the brief duration of my life, absorbed into the preceding and subsequent eternity, the small space I occupy and which I can see, engulfed in the infinite immensity of spaces that I do not know and that do not know me, I am scared" (Pascal 1904–1914, XIII, 126).

7.16 Philosophy as Demonstrating the Existence of God

Another view of utility of philosophy is that philosophy can demonstrate the existence of God.

Thus, Dummett says that "the 'worlds' of the various creatures are merely the partial and distorted projections of the one world, the world as it is in itself, upon the consciousness of those creatures" (Dummett 2006, 97). But "it makes no sense to speak of a world, or the world, independently of how it is apprehended" (ibid., 101). Therefore, the world as it is in itself "must be the world as it is apprehended by some mind" (ibid., 101–102). Also, being the world as it is in itself, it must be apprehended by some mind "not in any particular way, or from any one perspective rather than any other, but simply as it is" (ibid., 102). And this mind can only be God, because only "God has no particular point of view, no location in the world, no perspective contrasted with other perspectives" (ibid., 96). Therefore, God exists.

But this is not valid. It depends on the assumption that it makes no sense to speak of a world, or the world, independently of how it is apprehended. But scientists commonly speak of the unknown universe, namely the 95 percent of the universe which is still a mystery, despite the fact that the unknown universe is not apprehended. Therefore, it makes sense to speak of a world, or the world, independently of how it is apprehended.

7.17 An Alternative View of Utility of Philosophy

Since the above views of utility of philosophy are inadequate, an alternative view is required.

An alternative view of utility of philosophy is obtained by recognizing that the present sciences are what we already know about the world, philosophy is concerned with what we do not yet know about it, so philosophy is concerned with questions that are beyond the present sciences.

The utility of philosophy then consists in the fact that philosophy can open new ways to knowledge and lead to the birth of new sciences.

Since the things we still do not yet know about the world are many, there is much room for philosophy.

7.18 An Example

In order to open new ways to knowledge and lead to the birth of a new science, a philosopher does not have to be a particularly cultured scholar. Turing is an example of this.

Turing is "considered to be the father of computer science" (Daylight 2015, 205). But he "was not a particularly cultured scholar," in fact, "when carrying out a research, he did not like to devote himself too much to reading articles on the subject" (Numerico 2005, 31).

Turing wanted to solve Hilbert's Entscheidungsproblem: Is there an algorithm by which, given any statement of a first-order system, it can be determined whether or not the statement is provable in the system?

At the time there were the lambda calculus and the theory of recursive functions that might serve this purpose, but Turing did not start from them.

Instead, Turing posed to himself the problem of creating a theory of computation. To solve this problem, Turing analyzed the operations of a human being who makes computations, and from this he arrived at the hypothesis: "These operations include all those which are used in the computation of a number" (Turing 1936–1937, 232). He argued that the hypothesis was plausible, and on this basis he stated: "I therefore proceed with the development of the theory" (ibid.). Then, through his theory of computation, he showed that there is no algorithm of the required kind.

Gandy says: "Turing succeeded in his analysis because he was not familiar with the work of others" (Gandy 1995, 78). Admittedly, "the bare-hands, do-it-yourself approach does lead to clumsiness and error. But the way in which" Turing "uses concrete objects such as exercise books and printer's ink to illustrate and control the argument is typical of his insight and originality. Let us praise the uncluttered mind" (ibid.).

But this is misleading. Turing succeeded in his analysis not because he was not familiar with the work of others, but because he proceeded by the analytic method.

The philosophical character of Turing's work is clear from "the enthusiastic philosophical reception of Turing's approach" (Mosconi 2014, 38).

The example of Turing's creation of a theory of computation provides general indications of how philosophy can open new ways to knowledge and lead to the birth of new sciences. On this, see Chap. 18.

References

Cellucci, Carlo. 2022. *The making of mathematics: Heuristic philosophy of mathematics*. Cham: Springer.
Daylight, Edgar. 2015. Towards a historical notion of 'Turing—the father of computer science'. *History and Philosophy of Logic* 36:205–228.
Dewey, John. 2004. *Reconstruction in philosophy*. Mineola: Dover.
Dummett, Michael. 1997. Prof. Vattimo, non dimentichi la ragione. *Il Sole 24 Ore*, 3 August 1997.
Dummett, Michael. 2006. *Thought and reality*. Oxford: Oxford University Press.
Dummett, Michael. 2010. *The nature and future of philosophy*. New York: Columbia University Press.
Gandy, Robin. 1995. The confluence of ideas in 1936. In *The universal Turing machine: A half-century survey*, ed. Rolf Herken, 51–102. Wien: Springer.
Hacker, Peter. 2013. *Wittgenstein: Comparisons and context*. Oxford: Oxford University Press.
Hilbert, David. 1996. Logic and the knowledge of nature. In *From Kant to Hilbert: A source book in the foundations of mathematics*, ed. William Ewald, vol. 2, 1157–1165. Oxford: Oxford University Press.
Husserl, Edmund. 1970. *The crisis of European sciences and transcendental phenomenology*. Evanston: Northwestern University Press.
Husserl, Edmund. 2019. *First philosophy: Lectures 1923/24 and related texts from the manuscripts (1920–1925)*. Cham: Springer.
Kitcher, Philip. 2023. *What's the use of philosophy?* Oxford: Oxford University Press.
Laurenti, Carolina, Carlos Lopes, and José Abid. 2020. On usefulness of the useless: Philosophy as the consciousness of scientific knowledge. *Behavior and Philosophy* 48:91–108.
Mosconi, Jean. 2014. The development of the concept of machine computability from 1936 to the 1960s. In *Constructivity and computability in historical and philosophical perspective*, ed. Jacques Dubucs and Michel Bourdeau, 37–56. Dordrecht: Springer.
Nagel, Thomas. 1987. *What does it all mean? A very short introduction to philosophy*. Oxford: Oxford University Press.
Neurath, Otto. 1983. *Philosophical papers 1913–1946*. Dordrecht: Springer.
Numerico, Teresa. 2005. *Alan Turing e l'intelligenza delle macchine*. Milan: Franco Angeli.
Pascal, Blaise. 1904–1914. *Oeuvres*. Paris: Hachette.
Popper, Karl. 1994. How I see philosophy. In Karl Popper, *In search of a better world: Lectures and essays from thirty years*, 173–187. London: Routledge.
Russell, Bertrand. 1994. *Mysticism and logic*. London: Routledge.
Russell, Bertrand. 1995. *An outline of philosophy*. London: Routledge.
Russell, Bertrand. 1998. *The problems of philosophy*. Oxford: Oxford University Press.

Schlick, Moritz. 1979. The future of philosophy. In Moritz Schlick, *Philosophical papers*, vol. 2, 210–224. Dordrecht: Springer.
Spencer, Herbert. 1896. *First principles*. New York: Appleton.
Turing, Alan. 1936–1937. On computable numbers, with an application to the Entscheidungsproblem. *Proceedings of the London Mathematical Society* 42:230–265; 43:544–546.
Williams, Bernard. 2006. *Ethics and the limits of philosophy*. Cambridge: Harvard University Press.
Wittgenstein, Ludwig. 1969. *The blue and brown books*. Oxford: Blackwell.
Wittgenstein, Ludwig. 2009. *Philosophische Untersuchungen—Philosophical investigations*. Oxford: Blackwell.
Wundt, Wilhelm. 1889. *System der Philosophie*. Leipzig: Wilhelm Engelmann.
Zeki, Semir. 2014. Neurobiology and the humanities. *Neuron* 84:12–14.

Chapter 8
Progress in Philosophy

Abstract Closely related to the question of utility of philosophy is that of progress in philosophy. Several analytic philosophers claim that philosophy makes no progress, but this chapter argues that their arguments are not valid. On the other hand, some analytic philosophers claim that philosophy makes progress, but in the wrong sense of progress. Alternatively, the chapter argues that philosophy makes progress for the same reason for which it is useful, that is, because it can open new ways to knowledge and lead to the birth of new sciences.

Keywords The claim that philosophy makes no progress · A concept of progress in philosophy · Standard philosophy and special philosophy · Alternative solution to multiplication of philosophers

8.1 Views of Progress in Philosophy

Closely related to the question of utility of philosophy is that of progress in philosophy.

Several analytic philosophers claim that philosophy makes no progress, but this chapter argues that their arguments are not valid.

On the other hand, some analytic philosophers claim that philosophy makes progress, but in the wrong sense of progress.

Alternatively, the chapter argues that philosophy makes progress for the same reason for which it is useful, that is, because it can open new ways to knowledge and lead to the birth of new sciences.

C. Cellucci, *A Philosophy for the Future*, Synthese Library 508,
https://doi.org/10.1007/978-3-031-98443-3_8

8.2 The Argument of Disagreement

As already said, several analytic philosophers claim that philosophy makes no progress.

An argument for the claim that philosophy makes no progress is that there is inescapable disagreement among philosophers, therefore no philosophical problem ever gets solved.

Thus, Dietrich says that philosophy "has made no progress whatsoever" (Dietrich 2011, 329). In fact, "very unlike any science, no part of philosophy advances. Philosophy is, except for some modernizing, exactly the same now as it has ever been. It has not progressed one iota" (ibid., 334). For, in philosophy, there there is inescapable disagreement among philosophers, "views that are flatly contradictory are equally plausible," so "clashing points of view are ineluctable, and their existence is the only truth. Thus philosophy cannot progress" (ibid., 342).

But this argument is not valid. As will be argued later in this chapter, the opposition of philosophy to science is unjustified, disagreement among philosophers is not inescapable, some philosophical problems do get solved.

8.3 The Argument of Constancy of Language

Another argument for the claim that philosophy makes no progress is that our language has remained constant since antiquity and keeps seducing us into asking the same questions.

Thus, Wittgenstein argues that "one keeps hearing the remark that philosophy really doesn't make any progress, that the same philosophical problems that occupied the Greeks keep occupying us. But those who say that don't understand the reason it must be so. That reason is that our language has remained constant and keeps seducing us into asking the same questions" (Wittgenstein 2005, 312).

But this argument is not valid. For, what drives us to ask philosophical questions is not language but the world, in which we always encounter new situations that puzzle, intrigue, or challenge us. Language is only a means of expressing questions about the world, and the existing language is modified or extended when it is inadequate for this purpose, therefore it has not remained constant.

8.4 The Argument of Non-Finality of Solutions

Another argument for the claim that philosophy makes no progress is that solutions to philosophical problems must be final, but no solution to a philosophical problem is final.

Thus, Wittgenstein argues that "philosophical problems must be solvable really completely, in contrast to all others" (Wittgenstein 2005, 310). But no philosophical problem can be solved really completely, because "no philosophical problem can be solved until all philosophical problems are solved; which means that as long as they aren't all solved every new difficulty renders all our previous results questionable" (Wittgenstein 1969, 44).

But this argument is not valid. For, it depends on the assumption that solutions to philosophical problems must be final. But this assumption is unwarranted, because a solution to a philosophical problem is based on a hypothesis, and a hypothesis is necessarily provisional, since new data can always emerge with which the hypothesis is incompatible.

Even Wittgenstein recognizes that "in this work more than any other it is rewarding to keep on looking at questions, which one considered solved, from another quarter, as if they were unsolved" (Wittgenstein 1979, 30, 13.11.14).

Dennett says that, if philosophers were offered the two options of either (A) solving a major problem "so conclusively that there is nothing left to say," or (B) writing "a book of such tantalizing perplexity and controversy that it stays on the required reading list for centuries to come," some philosophers "admit that they would have to go for option (B)" (Dennett 2013, 429). On the contrary, scientists "tend to opt for (A) without any hesitation" (ibid.).

But scientists tend to opt for (A) without any hesitation for the wrong reason, For, not only no major philosophical problem, but also no major scientific problem, can be solved so conclusively that there is nothing left to say.

Even major mathematical problems cannot be solved so conclusively that there is nothing left to say. As Poincaré says, in mathematics there are not "solved problems and others which are not; there are only problems more or less solved" (Poincaré 2015, 377–378).

In fact, a solution to a philosophical, scientific, or mathematical problem is based on a hypothesis, and, as already said, a hypothesis is necessarily provisional, because new data can always emerge with which the hypothesis is incompatible.

8.5 The Argument of Lack of Final Goal

Another argument for the claim that philosophy makes no progress is that in philosophy a final goal cannot be formulated, so progress is paradigmatically undecidable.

Thus, Moody argues that the "the immediate impetus for continuing philosophical research is not consensus at all but perpetual and vociferous disagreement" (Moody 1986, 37). If there is progress, it "is paradigmatically undecidable progress, the sort where a final goal cannot be formulated" (ibid., 45).

But this argument is not valid. It is based on the assumption that, to decide whether there is progress, it is necessary to formulate a final goal. But this assumption is unfounded. Even for sciences it is not possible to formulate a final goal. For, sciences are journeys that have no final goal, their goal is to extend our knowledge of the world but there will always be more to know, so there is no goal for sciences other than trying to obtain ever greater knowledge of world. And yet, undeniably sciences make progress.

8.6 The Argument of Abandonment of Problems

Another argument for the claim that philosophy makes no progress is that we abandon old philosophical problems because we have different philosophical problems to solve than those that perplexed our ancestors, which therefore remain unsolved. And the same will happen to our philosophical problems.

Thus, Rorty argues that, if "philosophers, now as in Cicero's day," are "still arguing inconclusively, tramping round and round the same dialectical circles, never convincing each other," this is because "the vocabulary in which the traditional problems of Western philosophy were formulated was useful at one time but is no longer so useful" (Rorty 1999, xxi). Thus, "the vocabulary of Greek metaphysics and Christian theology" was "a useful one for our ancestors' purposes," but "we have different purposes, which will be better served by employing a different vocabulary" (ibid., xxii). Therefore, "our ancestors climbed up a ladder which we are now in a position to throw away. We can throw it away not because we have reached a final resting place, but because we have different problems to solve than those which perplexed our ancestors" (ibid.)

But this argument is not valid. It is unjustified to say that we abandon old philosophical problems because we have different philosophical problems to solve than those that perplexed our ancestors, which therefore remain unsolved. Several old philosophical problems have been solved by the sciences to which philosophy gave birth. Other philosophical problems have not been abandoned but have evolved, for example the question of the nature of knowledge evolved with the birth of modern science. And the same will happen to our philosophical problems.

8.7 The Argument of Transcendental Naturalism

Another argument for the claim that philosophy makes no progress is transcendental naturalism, that is, the view that we are unable to solve central philosophical problems because we have gaps in our cognitive skills as we have gaps in our motor skills. So, we are unable to solve those problems for the same reason for which we are incapable of unassisted flying.

Thus, McGinn argues that "the absence of progress that is so markedly characteristic of philosophy" has "a special source" (McGinn 1993, 151). That is, the "inherent limitations on our epistemic faculties" (ibid., 2). If we cannot solve central philosophical problems, it is because "our minds are not cognitively tuned to these problems" (ibid.). Philosophy makes no progress "for the same reason that we make so little progress in unassisted flying—that is, we lack the requisite equipment. We have gaps in our cognitive skills as we have gaps in our motor skills" (ibid., 13). This thesis can be called "transcendental naturalism" (ibid., 2–3).

But this argument is not valid. While we are incapable of unassisted flying, we are capable of flying with airplanes. The same goes for central philosophical problems. While our minds are incapable of solving them with our present epistemic tools, our minds might be capable of solving them with new epistemic tools. Just as Aristotle could not conceive of the epistemic tools that led to the theory of relativity, we cannot conceive of the epistemic tools that might be developed in the future and might allow human beings to solve those problems.

8.8 The Argument of Multiplication of Senses

Another argument for the claim that philosophy makes no progress is that problems that were raised thousands of years ago are still alive because the development of philosophy consists in finding new senses of problems. Different senses of problems are in permanent competition with each other, so philosophy develops through the conflicts of senses. No senses are thrown away, philosophy throws away only the things that have become generally accepted, which however constitute a trivial progress in philosophy.

Thus, Mironov argues that philosophy makes no progress because "problems that were raised thousands of years ago" are "still alive" (Mironov 2013, 12). The development of philosophy consists in finding new senses of problems, so in philosophy "there is development only "in the form of an increase in the number" of "senses" (ibid., 10). Different senses are in permanent competition with each other, so the development of philosophy is "a permanent struggle of contrary conceptions. Philosophy develops through the conflicts of senses" (ibid., 13). No senses are thrown away, "philosophy throws away only the things that have become generally accepted. Concerning such things, it is of course possible to speak about progress, but it is too trivial to do so" (ibid., 13–14).

But this argument is not valid. The things that have become generally accepted that philosophy throws away are the sciences to which philosophy gives rise which, far from constituting a trivial progress, are a great success for philosophy. For, the aim of philosophy is to acquire knowledge of the world, and when philosophy gives rise to new sciences, it fully achieves its aim.

8.9 Progress in Philosophy in the Wrong Sense

From what has been said above it is clear that the claim of several analytic philosophers that philosophy makes no progress is not valid.

On the other hand, some analytic philosophers claim that philosophy makes progress, but in the wrong sense of progress. For, they argue that philosophy makes progress because mathematical logic is part of it.

Thus, Dummett says: "Philosophy does make progress, and even achieves assured results" (Dummett 2010, 14). This "is shown by the fact that what perplexed medieval logicians is now a five-finger exercise for beginners" (ibid.). In fact, "Scholastic logicians had great trouble explaining the difference" of the words "'any' and 'every'," while "nowadays any beginning student in formal logic would be able to explain the distinction" (ibid.).

Williamson says: "Logic makes obvious progress" (Williamson 2018, 97). In fact, "far more is known in 2007 about truth than was known in 1957, as a result of technical work by philosophical and mathematical logicians" (Williamson 2022, 282).

This amounts to claiming that philosophy makes progress in the wrong sense of progress, because the progress to which Dummett and Williamson refer is not progress in philosophy but progress in mathematical logic.

Williamson claims that much of progress in mathematical logic "is progress in philosophy" (Williamson 2018, 97).

But this is not valid. For, as already stated in Chap. 2, mathematical logic is a conventional, although somewhat marginal, branch of mathematics, with no special relevance to philosophy.

Even Russell recognizes that "mathematical logic" is "not directly of philosophical importance except in its beginnings. After the beginnings, it belongs rather to mathematics than to philosophy" (Russell 2009, 33).

Therefore, contrary to Williamson's claim, progress in mathematical logic is not progress in philosophy.

8.10 An Alternative View of Progress in Philosophy

The arguments of analytic philosophers about progress in philosophy depend on their view of progress in philosophy. Since their arguments are not valid, it is necessary to formulate an alternative view of progress in philosophy.

An alternative view of progress in philosophy is implicit in the view of utility of philosophy stated in Chap. 7. It suggests that philosophy makes progress if it can open new ways to knowledge and lead to the birth of new sciences.

In terms of this view of progress in philosophy, the claim that philosophy makes no progress is unjustified.

For, as argued in Chap. 4, philosophy has played an essential role in the birth of modern science itself, it has played a part in the creation of new sciences, and has influenced developments in science and mathematics until recently.

In particular, in the past century, philosophy has led to the birth of computer science, cognitive science, artificial intelligence, and linguistics. So, philosophy has opened new ways to knowledge and led to the birth of new sciences until recently.

Therefore, in terms of the above view of progress of philosophy, philosophy makes progress.

8.11 Why the Argument of Disagreement Is Not Valid

In terms of the above view of progress in philosophy, it can be seen why the argument of disagreement, considered earlier in this chapter, is not valid.

The argument is based on the assumption that there is inescapable disagreement among philosophers, therefore no philosophical problem ever gets solved.

This assumption is a consequence of the view that philosophy is not acquisition of knowledge. According to this view, philosophical problems are completely different from scientific problems, and the fact that scientific problems can be solved does not mean that philosophical problems can also be solved.

On the contrary, according to the view that philosophy is acquisition of knowledge, philosophy in its development is a process in which different approaches to philosophy compete to address questions that are beyond the existing sciences, in the sense that they are not open questions of any of the existing sciences.

Sometimes one of these approaches to philosophy succeeds in dealing with a question that is beyond the existing sciences. Then, the question is turned into a new scientific field and the approach to philosophy gives rise to a new science.

This is not, as Dummett claims (see Chap. 5), a liberation of philosophy from an alien element, nor philosophy is ultimately what is left when philosophy is liberated from these alien elements. On the contrary, it is a progress in philosophy. For, the aim of philosophy since its birth as a discipline has been to acquire knowledge of the world, and the turning of a philosophical question into a new scientific field and the rise of a new science are a step forward in the realization of this aim.

8.12 Standard and Special Philosophy

It has been said above that philosophy in its development is a process in which different approaches to philosophy compete to address questions that are beyond the existing sciences.

At the origin of each of these approaches to philosophy there is the fact that a philosopher gives rise to it by introducing new fundamental assumptions, different from the fundamental assumptions of the existing approaches to philosophy.

A new approach to philosophy does not necessarily supplant the existing ones, it can coexist with them, albeit in competition. Even when, over time, an approach to philosophy dies out, the work of the philosopher who gave rise to it does not die out, it remains as an essential, if only polemical, reference point in all subsequent philosophical work.

But in philosophy, in addition to the introduction of new fundamental assumptions that give rise to a new approach to philosophy, there is the task of developing an existing approach to philosophy by working out the implications of its fundamental assumptions.

Marconi says that a philosopher who gives rise to a new approach to philosophy "can do without studying his antecedents, not because he is aware of all of them as logical possibilities in his titanic mind, but because, to some extent, he has no antecedents" (Marconi 2014, 133).

But this is unjustified. A new approach to philosophy does not come out of the blue. The task of developing an existing approach by working out the implications of its fundamental assumptions could serve as preparation for a new approach.

We may call 'standard philosophy' philosophy that develops an existing approach to philosophy by working out the implications of its fundamental assumptions; 'special philosophy' philosophy that introduces a new approach to philosophy.

Instead of 'standard philosophy' and 'special philosophy', one might think of using 'normal philosophy' and 'revolutionary philosophy', after Kuhn's 'normal science' and 'revolutionary science'.

Thus, Williamson says that "much contemporary philosophy is 'normal philosophy', building on assumptions without questioning them," but philosophy also includes "revolutionary activity" questioning "basic assumptions" (Williamson 2018, 109).

But this is problematic because, according to Kuhn, a new paradigm eventually supplants the existing ones and the members of a "scientific community work from a single paradigm" (Kuhn 1996, 162). On the contrary, as said above, a new approach to philosophy does not necessarily supplant the existing ones, it can coexist with them, albeit in competition.

8.13 Another Solution for the Multiplicatiom of Philosophers

The distinction between standard philosophy and special philosophy provides another solution for the multiplication of philosophers.

As we have seen in Chap. 2, Marconi argues that analytic philosophy is the solution best suited to philosophers in a situation in which professional philosophers are some tens of thousands. But this solution has led to the irrelevance of philosophy.

An alternative solution is that what most philosophers will do in a situation in which professional philosophers are some tens of thousands is standard philosophy. In this way, instead of being irrelevant, their work will contribute to the development of some existing approach to philosophy and could serve as preparation for some new approach to philosophy.

References

Dennett, Daniel. 2013. *Intuition pumps and other tools for thinking*. New York: Norton.
Dietrich, Eric. 2011. There is no progress in philosophy. *Essays in Philosophy* 12:329–344.
Dummett, Michael. 2010. *The nature and future of philosophy*. New York: Columbia University Press.
Kuhn, Thomas. 1996. *The structure of scientific revolutions*. Chicago: The University of Chicago Press.
Marconi, Diego. 2014. *Il mestiere di pensare: La filosofia nell'epoca del professionismo*. Turin: Einaudi.
McGinn, Colin. 1993. *Problems in philosophy: The limits of inquiry*. Oxford: Blackwell.
Mironov, Vladimir. 2013. On progress in philosophy. *Metaphilosophy* 44:10–14.
Moody, Todd. 1986. Progress in philosophy. *American Philosophical Quarterly* 23:35–46.
Poincaré, Henri. 2015. *The foundations of science: Science and hypothesis—The value of science—Science and method*. Cambridge: Cambridge University Press.
Rorty, Richard. 1999. *Philosophy and social hope*. London: Penguin.
Russell, Bertrand. 2009. *Our knowledge of the external world*. London: Routledge.
Williamson, Timothy. 2018. *Doing philosophy: From common curiosity to logical reasoning*. Oxford: Oxford University Press.
Williamson, Timothy. 2022. *The philosophy of philosophy*. Hoboken: Wiley-Blackwell.
Wittgenstein, Ludwig. 1969. *The blue and brown books*. Oxford: Blackwell.
Wittgenstein, Ludwig. 1979. *Notebooks 1914–1916*. Oxford: Blackwell.
Wittgenstein, Ludwig. 2005. *The big typescript TS 213*. Oxford: Blackwell.

References

Chapter 9
Other Philosophies Aimed at Knowledge

Abstract In addition to philosophy as acquisition of knowledge, there are other views of philosophy as aimed at acquiring knowledge of the world. But this chapter considers them and argues that they are not valid.

Keywords Philosophy in science · Philosophy as complementary science · Philosophy as part of science · Philosophy as non-natural science · Philosophy as ontical science

9.1 Philosophy in Science

In addition to philosophy as acquisition of knowledge, there are other views of philosophy as aimed at acquiring knowledge of the world. But this chapter considers them and argues that they are not valid.

A view of philosophy as aimed at acquiring knowledge of the world is philosophy in science, a subfield of philosophy of science which provides solutions to scientific problems.

Thus, Pradeu, Lemoine, Khelfaoui and Gingras say that "philosophy in science" is a "trend in philosophy of science, in which philosophers use philosophical tools to address scientific problems and provide scientifically useful proposals" (Pradeu et al. 2024, 376). While "most philosophers of science do philosophy 'on' science, that is, they contribute to our knowledge of the methods, concepts, objects, and problems of science," instead "some philosophers of science do philosophy 'in' science, that is, they "try to participate in resolving problems that scientists raise or encounter in their work" (ibid., 375–376).

So, philosophers in science address "a scientific problem, that is, a problem that scientists of a given field themselves can address with their usual methods" (ibid., 389). They use "philosophical tools to address the scientific problem" (ibid.). These tools are also used by scientists, so the contributions by philosophers in science "are as philosophical as they are scientific" (ibid., 392). But they "do not cease to be philosophical because they are also scientific" (ibid.).

C. Cellucci, *A Philosophy for the Future*, Synthese Library 508,
https://doi.org/10.1007/978-3-031-98443-3_9

Philosophy in science has "a significant number of instances" (ibid., 376). Thus, Malament 1977 "formulated a theorem, widely used in quantum gravity" (ibid.). Clauser and Shimony 1978 "showed that "Bell's results could be extended to actual systems" (ibid.). Therefore, philosophy in science "is a distinctive, significant trend within philosophy of science" (Pradeu et al. 2024, 378).

9.2 Limitations of Philosophy in Science

Against philosophy in science, however, it can be argued that it is not really a distinctive, significant trend within philosophy of science, because it uses tools that are also used by scientists. In fact, the contributions of Malament and of Clauser and Shimonly cited as examples of philosophy in science have been published in physics journals.

In order to claim that philosophy in science is a distinctive, significant trend within philosophy of science, philosophers in science should use tools that belong exclusively to philosophy and are not also used by scientists. Otherwise, philosophers in science are simply scientists who make contributions to the present sciences. In fact, if their articles are accepted and published in a physics journal, this means that the articles are considered standard contributions to physics.

Philosophy as acquisition of knowledge essentially differs from philosophy in science because all the questions that philosophy in science is concerned with belong to the present sciences. In contrast, according to philosophy as acquisition of knowledge, philosophy is concerned with questions that are beyond the present sciences and, if successful, it can open new ways to knowledge and lead to the birth of new sciences.

9.3 Philosophy as Complementary Science

Another view of philosophy as aimed at acquiring knowledge of the world is philosophy as complementary science, an integration of history and philosophy of science that generates scientific knowledge by doing what specialist science is presently unable to do.

Thus, Chang says that history and philosophy of science can constitute an integrated mode of study, called philosophy as "complementary science, which contributes to scientific knowledge through historical and philosophical investigations" and "asks scientific questions that are excluded from current specialist science" (Chang 2004, 3).

The need for philosophy as complementary science "arises from the fact that specialist science cannot afford to be completely open" (ibid., 237). For, "in specialist science many elements of knowledge must be taken for granted, since they are used as foundations," and "certain ideas and questions must be suppressed if they are heterodox enough to contradict or destabilize those items of knowledge that need to be taken for granted" (ibid.). Also, "not all worthwhile questions can be addressed in specialist science, simply because there are limits to the number of questions that a given community can afford to deal with at a given time," therefore "those problems that are considered either unimportant or unsolvable will be neglected" (ibid.). This is inevitable, but all the same, "suppressed and neglected questions represent a loss of knowledge, actual and potential," and the function of philosophy as complementary science is "to recover and even create such questions anew and, hopefully, some answers to them as well" (ibid.). Thus, philosophy as complementary science "can generate scientific knowledge" (ibid.).

Philosophy as complementary science can generate scientific knowledge by "two obvious methods," obvious "because they are rooted in very standard customs in philosophy and history of science" (ibid., 239). The first method is "to reconsider things that are taken for granted in current science" (ibid.). Philosophy "can raise doubts" about them, history can show how "past scientists in fact addressed the same doubts" (ibid., 239–240). Then, "after the historical record is established, philosophy can take its turn again to reassess the past arguments that have been unearthed" (ibid., 240). The second method is "to look out for apparently unusual and puzzling elements in past science" that "have not survived into modern science," to investigate "the historical reasons for their rejection and assess the philosophical cogency of those reasons" (ibid.).

9.4 Limitations of Philosophy as Complementary Science

Against philosophy as complementary science, however, it can be argued that, contrary to its claim, it cannot generate scientific knowledge by doing what specialist science is presently unable to do.

Indeed, the type of scientific knowledge that philosophy as complementary science is supposed to generate might very well be generated by specialist science but, as philosophy as complementary science admits, specialist science considers such knowledge a marginal one, inessential to the development of science. There might be some gains in opening up past questions, but this is unlikely to yield important results, much less to lead to breakthroughs in science, therefore it may not be worth the effort.

Also, it is illusory to think that philosophy as complementary science can deal with elements of knowledge that are taken for granted since they are used as foundations or are suppressed because they contradict foundations. Simply, historians and philosophers of science do not have the necessary tools to deal with these elements of knowledge.

Philosophy as acquisition of knowledge essentially differs from philosophy as complementary science because all supposed contributions to philosophy as complementary science concern questions relating to the present sciences. For, they deal with the foundations of the present sciences, or with questions belonging to the present sciences not considered by the latter. On the contrary, according to philosophy as acquisition of knowledge, philosophy is concerned with questions that are beyond the present sciences and, if successful, it can open new ways to knowledge and lead to the birth of new sciences.

An illustration of this is Turing's creation of a theory of computation, considered in Chap. 7. Turing's creation did not deal with the foundations of the then present sciences or with questions belonging to the then present sciences not considered by the latter, so it had none of the characters of contributions to philosophy as complementary science. It gave birth to a new science, and yet had a philosophical character.

9.5 Philosophy as Part of Science

Another view of philosophy as aimed at acquiring knowledge of the world is philosophy as part of science, which deals with general questions such as what sorts of things there are, what it means to exist, and what sorts of things can meaningfully be asked about what there is.

Thus, Quine says that philosophy is "concerned with our knowledge of the world and the nature of the world," it attempts "to round out 'the system of the world', as Newton put it" (Quine 2008, 5–6). While "there have been philosophers who thought of philosophy as somehow separate from science," on the contrary philosophy is "continuous with science," and indeed "a part of science" (ibid., 6).

Philosophy as part of science differs from the rest of science because it "lies at the abstract and theoretical end of science," it "is abstract through being very general," indeed, philosophy "seeks the broad outlines of the whole system of the world" (ibid.).

The most important questions philosophers have to deal with can be grouped under "two headings," that is, "the ontological questions, as they might be called: general questions as to what sorts of things there are, as well as what it means to exist, for there to be something," and "the predicative questions: questions as to what sorts of things can meaningfully be asked about what there is" (ibid., 6–7).

To deal with ontological and predicative questions, philosophy as part of science uses the method of science, which "is summed up in a word, albeit a double-barrelled word: hypothetico-deductive. First we think up a theory, a set of hypotheses," and "from this theory we then deduce what observations to expect under various observable conditions. If such an expectation is disappointed, we look to the theory for possible revisions. If not, we go on believing it" (ibid., 14). In addition to the use of the method of science, philosophy as part of science is "permitted free use of scientific findings" (Quine 1981, 85).

9.6 Limitations of Philosophy as Part of Science

Against philosophy as part of science, however, it can be argued that ontological and predicative questions are just questions of traditional metaphysics. The only difference is that, to deal with them, philosophy as part of science uses the method of science and is permitted free use of scientific findings.

But this contrasts with the fact that conflicting answers to ontological and predicative questions are consistent with current scientific findings. Therefore, philosophy as part of science does not provide any justified answer to these questions, and hence is not better off than traditional metaphysics.

Philosophy as acquisition of knowledge essentially differs from philosophy as part of science, because the latter is concerned with questions of traditional metaphysics. On the contrary, philosophy as acquisition of knowledge aims at knowledge, a type of knowledge that is not essentially different from that of the present sciences, but differs from the present sciences because it is concerned with questions that are beyond them. Dealing with such questions philosophy as acquisition of knowledge, if successful, can open new ways to knowledge and lead to the birth of new sciences.

9.7 Philosophy as Non-Natural Science

Another view of philosophy as aimed at acquiring knowledge of the world is philosophy as non-natural science, like mathematics and history.

Thus, Williamson says that "philosophy is a non-natural science" (Williamson 2018, 5). It wants "to understand the nature of everything," the "world of common sense, the world of natural science, the world of mathematics," and "mind and matter" (ibid., 3).

The term 'science' is often restricted to "the natural sciences: physics, chemistry, biology, and so on. But there are the social sciences as well," moreover "mathematics is a very salient example of science," and "history, when pursued in a scholarly way, deserves to be in that company" (Williamson 2017, 28). A science is "any kind of systematic inquiry," and "philosophy is a form of systematic inquiry in the same sense as all those other subjects" (ibid.). Philosophy is indeed "an armchair discipline" (ibid., 33). But this does not mean that it is not a science, because mathematics is a science and yet it is "an armchair activity" (ibid., 30).

Against this it might be argued: "If philosophy is a science, how come it has made so little progress?" (Williamson and Sunday-Grève 2022, 34). But this overlooks that "very little progress in natural science consists in the definitive discovery of new laws of nature. Instead, much scientific progress consists in the development of better models of natural phenomena," models which "grossly over-simplify those phenomena" but "nevertheless capture some of their key features better than previous models did" (ibid.). Now, "this applies most vividly" also "to philosophy as it concerns the human world, whose messy complexity we cannot hope to map in full detail" (ibid., 34–35). Still, in areas such as epistemology or philosophies of mind, action, language, and art, "philosophy has made the kind" of "progress typical of progress in the theoretical study of the human world" (ibid., 35). Also, in those areas philosophy overlaps scientific disciplines, for example, "the philosophy of mind overlaps theoretical psychology and cognitive science" (ibid.). Thus "philosophy appears quite naturally as a non-natural science" (ibid.).

9.8 Limitations of Philosophy as Non-Natural Science

Against philosophy as non-natural science, however, it can be argued that the fact that philosophy is a form of systematic inquiry does not make it a science, otherwise astrology should be considered a science. Indeed, Raman says that, "if by science you mean systematised knowledge, then astrology is a science. Astrology is so vast and comprehensive in its scope that it embraces all phenomena in nature," including "psychological aspect of human nature, events in human life" (Raman 1998, 142).

Moreover, it is unjustified to say that the fact that philosophy is an armchair discipline does not imply that philosophy is not a science because mathematics is a science and yet mathematics is an armchair activity. Mathematics is not an armchair activity. For, several mathematical problems have an extra-mathematical origin, several mathematical concepts are formulated to deal with extra-mathematical questions, and several mathematical theories are developed to meet extra-mathematical needs and are evaluated in terms of their ability to meet those needs. Then, mathematics necessarily involves interactions with the world beyond the armchair, therefore it needs inputs from experience (see Cellucci 2025).

A fortiori, philosophy is not an armchair discipline. Philosophy as non-natural science wants to understand the nature of everything, the world of common sense, the world of natural science, the world of mathematics, and mind and matter. But this necessarily involves interactions with the world beyond the armchair, so philosophy needs inputs from experience.

Philosophy as acquisition of knowledge essentially differs from philosophy as non-natural science because the latter assumes that philosophy can be a science without depending in any way on experience. On the contrary, according to philosophy as acquisition of knowledge, philosophy aims at extending our knowledge of the world, and therefore essentially depends on experience.

9.9 Philosophy as Ontical Science

Another view of philosophy as aimed at acquiring knowledge of the world is philosophy as ontical science, that is, science of the essence of things.

Thus, McGinn says that philosophy requires an "extension of the word 'science'" (McGinn 2021, 92). For, philosophy is a science but "is quite unlike empirical science, both in methodology and subject matter: it is an a priori discipline, removed from observation and experiment" (McGinn 2015, 84). In fact, philosophy is the "ontical science," that is, "the general science of being. It is the science of what things essentially are, what their constitutive nature is; this is why definition looms so large in philosophy" (McGinn 2021, 92). Therefore, philosophy is the science of the essence of things, which is captured by their definition.

That philosophy is not an empirical science does not mean that it is any "less a science for that," nor that it is "any less 'scientific'" (McGinn 2015, 86). Even "what are traditionally called the 'formal sciences'—arithmetic, geometry, mathematical logic, abstract computer science, game theory, information theory, general systems theory, and the like," are "not empirical" (ibid., 85). So, "it is not the status of a discipline as empirical that makes it a science" (ibid., 85–86). Thus, "it would be arbitrary to deny that mathematics is a science," only "misplaced ideology would insist that physics is a science but mathematics is not—being nonempirical is not to the point" (ibid., 87). What makes a discipline a science is instead "whether the subject achieves the right level of internal rigor and systematic organization" (ibid., 86).

Also, that philosophy is not an empirical science does not mean that it is not an experimental science. Philosophy consists "of conceptual analysis aimed at establishing a priori necessary truths" (ibid., 84). Now, "conceptual analysis is experimental science" (ibid., 93). For, it "involves the production of a hypothesis about the analysis of a concept—specifically a set of (putative) necessary and sufficient conditions—which is then tested by means of 'thought experiments'" (ibid., 92). The hypothesis "makes certain predictions about our intuitions in possible cases, and then we check to see if our intuitions confirm or disconfirm the hypothesis. The intuitions function as evidence for or against the hypothesis, and they are grounded in our actual grasp of the concept in question" (ibid., 93). Thus "thought experiments are indeed experiments," because "we have here a procedure that arrives at truth in a non-question-begging way by consulting a body of data ('intuitions') generated especially for the purpose" (ibid.).

In fact, philosophy is "a natural science" (ibid., 96). Its "subject matter consists of things that exist in the natural world, not the 'abstract world'. It is therefore entirely appropriate to say that philosophy is a natural science" (ibid.).

9.10 Limitations of Philosophy as Ontical Science

Against philosophy as ontical science, however, it can be argued that philosophy cannot be the science of the essence of things. For, this would imply that philosophy can know what science cannot know because, as made clear by Galileo's revolution, the object of science cannot be to know the essence of things. But, that philosophy can be the science of the essence of things contrasts with the fact that there is no evidence that there exists a superior source of knowledge that provides knowledge of the essence of things, let alone that philosophy is such a superior source of knowledge.

Also, what makes a discipline a science is not whether the subject achieves the right level of internal rigor and systematic organization, otherwise theology should be considered a science. Indeed, Jinkins says that "theology is a science in that it is a critical and analytical attempt to study a subject, an attempt" that is "carried out with methods appropriate to its subject matter" (Jinkins 2001, 50).

Also, it is unwarranted to say that philosophy is an experimental science because it involves the production of a hypothesis which is confirmed or disconfirmed by our intuitions, so our intuitions function as evidence for or against the hypothesis. To say this would imply that we count as evidence beliefs for which no justification is given.

Indeed, as van Inwagen says, "our 'intuitions' are simply our beliefs," philosophers "call their philosophical beliefs intuitions because 'intuition' sounds more authoritative than 'belief'," but this contrasts with the fact that our beliefs "can very easily be wrong" (van Inwagen 1997, 309).

Therefore, intuitions cannot function as evidence for or against a hypothesis.

Also, it is unwarranted to say that philosophy is a natural science because its subject matter consists of things that exist in the natural world, not the abstract world. To say that philosophy is a natural science is a contradiction in terms because the essence of things that exist in the natural world is supposed to be captured by their definition. For example, the essence of the electron is supposed to be captured by its present definition in terms of properties such as mass, electric charge, total spin. But nothing excludes that new properties of electrons can be discovered, not included in the present definition.

Philosophy as acquisition of knowledge essentially differs from philosophy as ontical science because, unlike the latter, it does not assume that philosophy has a special cognitive ability which can provide knowledge of the essence of things. It does not assume that what makes a discipline a science is whether the subject achieves the right level of internal rigor and systematic organization. It does not assume that intuition can function as evidence for or against a hypothesis.

References

Cellucci, Carlo. 2025. Mathematics and experience. *Foundations of Science* 30:473–487.

Chang, Hasok. 2004. *Inventing temperature: Measurement and scientific progress*. Oxford: Oxford University Press.

Clauser, John, and Abner Shimony. 1978. Bell's theorem: Experimental tests and implications. *Reports on Progress in Physics* 41:1881–1927.

Jinkins, Michael. 2001. *Invitation to theology*. Downers Grove: InterVarsity Press.

Malament, David. 1977. The class of continuous timelike curves determines the topology of spacetime. *Journal of Mathematical Physics* 18:1399–1404.

McGinn, Colin. 2015. The science of philosophy. *Metaphilosophy* 46:84–103.

McGinn, Colin. 2021. Philosophical philosophy. *Syzetesis* 8:89–98.

Pradeu, Thomas, Maël Lemoine, Mahdi Khelfaoui, and Yves Gingras. 2024. Philosophy in science: Can philosophers of science permeate through science and produce scientific knowledge? *The British Journal for the Philosophy of Science* 75:375–416.

Quine, Willard. 1981. *Theories and things*. Cambridge: Harvard University Press.

Quine, Willard. 2008. The ideas of Quine: Interview by Brian Macgee. In Willard Quine, *Quine in dialogue*, 5–17. Cambridge: Harvard University Press.

Raman, Bangalore. 1998. Intervention to: Validity of astrology—A radio discussion. In *Astrology and the hoax of 'scientific temper'*, ed. Gayatri Vasudev, 141–159. Delhi: Motilal Banardsidass.

van Inwagen, Peter. 1997. Materialism and the psychological-continuity account of personal identity. *Philosophical Perspectives* 11:305–319.

Williamson, Timothy. 2017. Philosophy as a scientific discipline: An interview by Hans Solberg and Max Kippersund. *Filosofisk Supplement* 12:28–34.

Williamson, Timothy. 2018. *Doing philosophy: From common curiosity to logical reasoning*. Oxford: Oxford University Press.

Williamson, Timothy, and Sebastian Sunday-Grève. 2022. Philosophy and common sense 3: Philosophy as a science. *The Philosophers' Magazine* 97:31–35.

Part II
The Ways of Knowledge

Chapter 10
The Nature of Knowledge

Abstract Developing the view that philosophy is acquisition of knowledge requires answering several questions about knowledge. The first question, of course, is: What is knowledge? A widely held view is that this question is a demand to give an analysis of the concept of knowledge. But this conflict with the fact that every analysis of the concept of knowledge is subject to counterexamples. Alternatively, the chapter argues that the question 'What is knowledge?' is a demand to give an analysis of the function of knowledge in human life, and in general in the life of all organisms. And the answer to this demand is that knowledge is a life function that serves to solve problems of biological life and, in the case of human beings, also problems of cultural life.

Keywords Concept-first approach to knowledge · Function-first approach to knowledge · Knowledge and natural selection · Biological role of knowledge · Cultural role of knowledge · Cultural evolution · Knowledge as a life function · Common-sense knowledge

10.1 Questions About Knowledge

Developing the view that philosophy is acquisition of knowledge requires answering questions about knowledge.

Answering them is important because knowledge is involved in almost all human activities.

Then, as Wang says, "if humanity is to be characterized" by "the system of human activities, one is tempted to equate the philosophy of knowledge with the philosophy of man" (Wang 1974, 388).

The questions about knowledge to be answered include at least the following: What is knowledge? What is the relation of knowledge to reality? What is the relation of knowledge, in particular scientific knowledge, to objectivity? What is the relation of knowledge to mind? Is acquisition of knowledge, in particular discovery, a rational process? What is the method of discovery? What is the role of

C. Cellucci, *A Philosophy for the Future*, Synthese Library 508,
https://doi.org/10.1007/978-3-031-98443-3_10

non-deductive rules and deductive rules in knowledge? What is the role of logic in knowledge? Is knowledge problem solving by the analytic method? What is perception? What is an explanation? What is the relation of knowledge to emotions? What is the relation of knowledge to naturalization?

These questions are the object of this Part II of the book.

10.2 Assumptions of Analytic Philosophy About Knowledge

Developing the view that philosophy is acquisition of knowledge first requires answering the question: What is knowledge?

Analytic philosophy makes the following assumptions about knowledge.

(i) The question "What is knowledge?" is a demand to give an "analysis of the concept of knowledge" (Williams 2001, 1).

(ii) In the philosophical tradition many philosophers have held "the standard analysis" of the concept of knowledge, according to which knowledge is justified true belief, so knowledge is "a special kind of belief: belief that is both true and justified" (ibid., 17). In the standard analysis, "the belief condition excludes ignorance, the truth condition excludes error, and the justification condition excludes mere opinion" (ibid., 16–17).

(iii) The standard analysis of the concept of knowledge goes "back a long way: arguably to Plato's *Theaetetus*" (ibid., 26, footnote 2).

(iv) Gettier 1963 first provided a counterexample to the standard analysis of the concept of knowledge, thus showing that the standard analysis "fails to give sufficient conditions for knowledge" (ibid., 28).

10.3 Failure of the Assumptions of Analytic Philosophy

The assumptions (i)–(iv) of analytic philosophy about knowledge, however, are not valid.

Assumption (i), that the question 'What is knowledge?' is a demand to give an analysis of the concept of knowledge, is not valid.

For, as Williamson says, "a succession of increasingly complex analyses have been overturned by increasingly complex counterexamples" (Williamson 2000, 30). The upshot of this is that the concept of knowledge "cannot be analysed into more basic concepts" (ibid., 33).

Assumption (ii), that in the philosophical tradition many philosophers have held the standard analysis of the concept of knowledge, is not valid.

For, as Antognazza says, if one examines "main figures and traditions in the history of philosophy, the 'standard' analysis begins to look surprisingly non-standard. The 'traditional' view, it appears, is not a tradition stretching back to the infancy of philosophy," rather, "it begins to look like a twentieth-century view, prominent especially in the Anglo-American world, which after the middle of the century, has repeatedly tried" and "failed to correct itself" (Antognazza 2015, 169). In fact, "the analysis of knowledge as belief plus the addition of some conditions does not work. Perhaps this is why, historically, the traditional accounts of knowledge central to Western philosophy did not take knowledge to be a kind of belief which meets certain criteria. The project of finding what should be added to belief in order to turn it into knowledge would have been regarded by much pre-twentieth century epistemology as absurd" (ibid.).

Assumption (iii), that the standard analysis of the concept of knowledge goes back to Plato's *Theaetetus*, as well as by Williams, is made by several other analytic philosophers.

Thus, Gettier says: "Plato seems to be considering some such definition at *Theaetetus* 201, and perhaps accepting one at *Meno* 98" (Gettier 1963, 121, footnote 1).

Chisholm says: "The traditional or classic" definition of knowledge, "and the one proposed in Plato's dialogue, the *Theaetetus*, is that knowledge is justified true belief" (Chisholm 1989, 90).

Pritchard says: "In this work," the *Theaetetus*, "Plato was offering us an early version of the classical tripartite account of knowledge, whereby knowledge is understood as justified true belief" (Pritchard 2016, 127).

But assumption (iii) is not valid because, on the contrary, in the *Theaetetus* Plato criticizes the analysis of the concept of knowledge as justified true belief.

In fact, Socrates says: "Are you satisfied, and do you set it down, that knowledge is true belief accompanied by justification?" (Plato, *Theaetetus*, 202 c 7–8). No, "knowledge cannot" be "true belief accompanied by justification" (ibid., 210 b 1–2). For, justification is itself knowledge. Then, to say that knowledge is true belief accompanied by justification would be to say that knowledge is true belief accompanied by knowledge. But "it would be utterly silly to say, when you are trying to find out what knowledge is," that knowledge "is true belief accompanied by knowledge" (ibid., 210 a 7–8).

Rather than going back to Plato's *Theaetetus*, the standard analysis of the concept of knowledge was formulated by Lewis.

Indeed, Lewis says: "Knowledge is belief which not only is true but also is justified" (Lewis 1946, 9).

This confirms Antognazza's statement that the standard analysis is not a tradition stretching back to the infancy of philosophy but a twentieth-century view.

Assumption (iv), that Gettier first provided a counterexample to the standard analysis of the concept of knowledge, is not valid.

For, even without counting Plato's argument against it, counterexamples were implicit in Meinong and Russell.

For example, Russell says: "If a newspaper, by an intelligent anticipation, announces the result of a battle before any" official confirmation "has been received, it may by good fortune announce what afterwards turns out to be the right result, and it may produce belief in some of its less experienced readers" (Russell 1998, 76). Their belief is true, and is also justified because the newspaper has a reputation of being reliable. But they "cannot be said to have knowledge" (ibid.). For, it is only by chance that the announcement of the newspaper is true.

Even mathematical knowledge is not true justified belief. For, by Gödel's second incompleteness theorem, mathematical knowledge cannot be said to be true and justified.

10.4 Failure of Other Assumptions of Analytic Philosophy

In addition to the assumptions (i)–(iv) considered above, analytic philosophy makes other assumptions about knowledge, but these are also not valid.

(1) Analytic philosophy assumes that knowledge is propositional knowledge, that is, knowledge expressed in propositions, because "knowledge requires judgements—thoughts or beliefs that can be true or false. Such thoughts demand propositional content: the sort of content that is expressed by complete sentences" (Williams 2001, 140).

This assumption is not valid, because there is non-propositional knowledge, for example, perception.

According to analytic philosophy, perception is not knowledge, because "knowledge must involve propositional content and so cannot consist in a mere relation to a particular. Sensing a sense-datum is no more knowing anything than is standing next to a lamp-post" (Williams 2001, 97).

But this is based on the assumption that perception is a passive process, and so consists in a mere relation to a particular, it is merely sensing a sense-datum. This assumption is not valid because, as will be argued in Chap. 19, perception is not a passive process.

(2) Analytic philosophy assumes that knowledge must be conscious, because having knowledge involves "the sort of awareness that is involved in knowledge properly so-called: the sort of knowledge that involves the ability to make propositionally articulable claims" (ibid., 99). Then animals do not have knowledge because, "although animals can be trained to respond differentially to environmental stimuli, this does not entitle us to see a pigeon pecking a red disc to get food as saying that there is a red disc present" (ibid., 174). Responses become knowledge only if they can be treated "as reasons for further judgements. This sort of thing comes with learning a language. Animals do not go in for it" (ibid.).

This assumption is not valid because, as will be argued in Chap. 19, perception is unconscious. For, it is based on processes that take place in us at too low level and too high speed to be accessible to direct introspection.

Moreover, as will be argued below, even unicellular organisms such as the prokaryotes must have knowledge, otherwise they could not survive.

(3) Analytic philosophy assumes that an analysis of the concept of knowledge must not describe "just a factual state or condition," it must have a "normative status," that is, it must fix the characters of knowledge once and for all, because "this

normative dimension distinguishes philosophical theories of knowledge from straightforwardly factual inquiries" (ibid., 11).

This assumption is not valid, because knowledge comes in forms that change over time, thus there is no guarantee that an analysis of the concept of knowledge, adequate for the present form of knowledge, will remain adequate for future forms of knowledge. Therefore, an analysis of the concept of knowledge cannot have a normative status.

10.5 The Function-First Approach to Knowledge

Rather than as a demand to give an analysis of the concept of knowledge, the question 'What is knowledge?' must be seen as a demand to give an analysis of the function of knowledge in human life, and in general in the life of all organisms. This involves moving from a concept-first approach to knowledge, that is, an approach which aims to give an analysis of the concept of knowledge, to a function-first approach, that is, an approach which aims to give an analysis of the function of knowledge in human life, and in general in the life of all organisms.

According to a widespread view, knowledge is an end in itself. In Chap. 7 we have seen the origin of this view with Aristotle. The view has been repeatedly reaffirmed until today.

Thus, Ebbersmeyer says: "Philosophy" is "the pursuit of knowledge for its own sake," indeed philosophy is "the realm of disinterested knowledge. Pursuing knowledge for its own sake, stands in opposition to instrumental research that serves concrete and practical goals. By disinterested I" mean "dissociated from immediate utilitarian goals, material interests, and short-term benefits" (Ebbersmeyer 2024, 114–115). Therefore, I can use "the expression passion for knowledge to characterise philosophy, rather than the more literal translation" of the term 'philosophia' as "love for wisdom" (ibid., 116)

But this view is not valid. Knowledge is not an end in itself but has a vital role. The resources whose lack hinders the rise, preservation and reproduction of life are not only food and sexual partners, but also knowledge. Human beings acquire knowledge about the environment and, on this basis, adopt behaviours that ensure their survival. Therefore, knowledge serves to solve the problem of survival. Human beings seek knowledge not merely because they wish but, more importantly, because they must.

Knowledge serves to solve the problem of survival not only of human beings but of all organisms. All organisms seek knowledge not as an end in itself, but as a means to solve the problem of survival, recognizing those aspects of the environment which make the difference between preserving life and destroying it, and implementing appropriate responses to them. Solving the problem of survival is a difficult and risky task if over ninety per cent of the species that have developed on Earth have disappeared.

All organisms survive using energy sources present in the environment, and avoiding dangers that could destroy them. They can do so only if they acquire knowledge about the environment. Like human beings, all organisms in general acquire knowledge about the environment and, on this basis, adopt behaviours that ensure their survival. Thus, knowledge is essential to survival of all organisms.

Then, Aristotle is right when he says that "by nature, all men tend to knowledge" (Aristotle, *Metaphysica*, A 1, 980 a 21). But, contrary to Aristotle's view, they do not tend to knowledge only for the sake of knowledge because from knowledge nothing results apart from contemplation. They tend to knowledge because knowledge plays a vital role, it is essential to their survival.

In fact not only all men, but all organisms tend to knowledge because knowledge is essential to their survival.

Since knowledge is essential to survival of all organisms, all organisms are cognitive systems, and life itself owes its existence and preservation to a cognitive process.

10.6 Generality of the Role of Knowledge in Survival

That knowledge is essential for the survival of all organisms also holds for the most elementary ones, such as the prokaryotes, the unicellular organisms that were the first form of life to appear on Earth.

The prokaryotes have molecules on their plasma membrane that act as sensors and send data to the DNA.

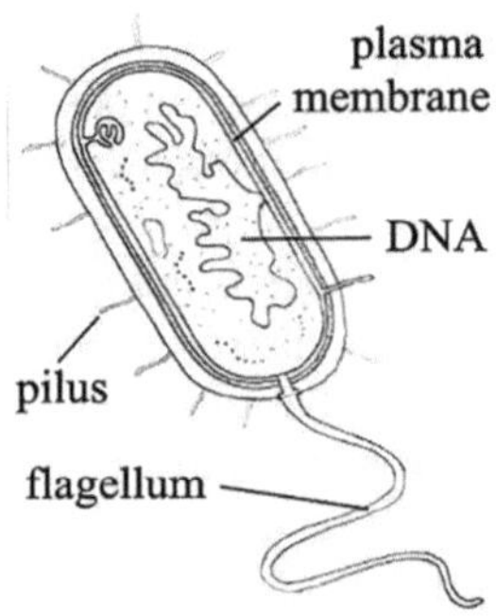

Through the information encoded in the DNA, the prokaryotes interpret the data and thus acquire some knowledge about the environment. Based on this, they move in a liquid environment by means of the flagellum and pili, and so they ensure their survival.

10.7 Knowledge and Consciousness

That knowledge is essential for the survival of all organisms implies that all organisms have knowledge.

As we have seen above, this conflicts with the assumption of analytic philosophy that knowledge requires consciousness, therefore animals do not have knowledge because animals do not have consciousness.

But the assumption of analytic philosophy that knowledge requires consciousness is unjustified. Consciousness is neither a necessary nor a sufficient condition even for human knowledge.

Indeed, on the one hand, consciousness is not a necessary condition for human knowledge, because a significant part of human knowledge is arrived at through processes that are unconscious. They are unconscious because they occur too fast and at too low a level in mind to be accessible to direct inspection. Such is perceptual knowledge because, as will be argued in Chap. 19, perception is unconscious inference.

On the other hand, consciousness is not a sufficient condition for human knowledge, because beliefs arrived at consciously and considered to be knowledge, may turn out not to be knowledge. Thus, Frege arrived consciously at his axiom system and considered it to be knowledge, but it turned out not to be knowledge, because it led to Russell's paradox.

10.8 Biological Role of Knowledge

That knowledge is essential for the survival of all organisms also implies that knowledge is a life function.

Life functions are usually described saying that they are "a series of highly organized and related activities that allow living organisms to live, grow, and maintain themselves," and consist of "movement, ingestion, digestion, transport, respiration, synthesis, assimilation, growth, secretion, excretion, regulation (sensitivity), and reproduction" (Scott Senisi and Fong 2016, 6).

But this description of life functions is incomplete: it omits knowledge. Knowledge is a life function because it serves to solve the problem of survival of human beings, and in general of all organisms.

Moreover, knowledge serves to solve the problem of survival not only of individual organisms, but also of entire species. For, it plays an essential part in natural selection.

Natural selection is usually described by saying that it “is a two-step process” (Mayr 2002, 131). At the first step, “consisting of all the processes leading to the production of a new zygote,” a “new variation is produced” (ibid.). At the second step, “that of selection (elimination), the ‘goodness’ of the new individual is constantly tested, from the larval (or embryonic) stage until adulthood and its period of reproduction” (ibid., 132). Those “individuals who are most efficient in coping with the challenges of the environment and in competing with other members of their population and with those of other species, will have the best chance to survive until the age of reproduction and to reproduce successfully” (ibid.). They “are the ones that are ‘fittest to survive’” (ibid.). Such individuals are “an adaptation,” namely have traits “the possession of which favors the individual in the struggle for existence” (ibid., 165). These traits “were acquired by natural selection or, if they arose by chance, their maintenance was favored by selection” (ibid.).

But again, this description of natural selection is incomplete: it omits knowledge. This omission makes it inexplicable how our earliest human ancestors were able to survive, and therefore how we, their descendants, can be here.

For, compared to large mammals, our earliest human ancestors were weak and vulnerable creatures, so, in the midst of stronger competing or threatening species, they could not survive by sheer physical strength.

They could survive only by outsmarting those species, showing greater ingenuity in gathering and using knowledge about the environment. Thus, that our earliest human ancestors could survive, becomes explicable only if knowledge plays an essential role in natural selection in human beings.

In general, knowledge plays an essential role in natural selection in all species. Therefore, knowledge is necessary not only to avoid short-term threats to the survival of individual organisms. The latter can only be ensured for a limited period of time, because all organisms sooner or later die.

Other is the case of genes, which contain the instructions of the project of components of organisms. The characters of the project are propagated by promoting the reproduction of genes. With respect to a species, knowledge serves to propagate the characters of the project by promoting the reproduction of genes. Thus, knowledge serves to solve the problem of survival of entire species.

The present theories of natural selection take no account of the essential role of knowledge, so they cannot explain how our earliest human ancestors could survive in the midst of stronger competing or threatening species. Thus, there is need for a new science, integrating the present theories of natural selection evolution with a theory of knowledge.

Since knowledge serves to solve the problem of survival of individual organisms and entire species, knowledge has a biological role.

10.9 Cultural Role of Knowledge

In addition to the biological role, knowledge has a cultural role. For, in human beings, knowledge serves not only to solve the problem of survival, but also to improve the quality of life, from the routine workings of everyday life to global issues.

The cultural role of knowledge is different from the biological role, because it is based on non-genetic transmission of information. Indeed, a culture is a body of knowledge that human beings transmit from generation to generation non-genetically, that is, not through DNA, but through what they do and communicate.

On the other hand, however, the cultural role of knowledge is not opposed to the biological role but is continuous with it, because a culture essentially depends on the biological makeup of organisms.

Moreover, even in its cultural role, knowledge can influence natural selection. For, the body of knowledge that forms a culture gives human beings the ability to design their environment to some extent, modifying it so as to make it more suitable for them. If several generations repeatedly modify their environment in the same way, this can lead to changes in the process of natural selection.

An example of this is the culture of pastoralism, which led to the domestication of livestock and the activities associated with the production of milk and its derivatives. This modified the environment of certain human populations for enough generations to select those genes that today make many adults lactose tolerant.

In general, the body of cognitions that forms a culture greatly enhances the capabilities of human beings, enabling them to achieve goals unattainable by natural selection alone.

10.10 Cultural Evolution

The cultural role of knowledge goes together with cultural evolution, namely the fact that a body of knowledge, not only is transmitted from generation to generation non-genetically, but can undergo changes or expansions in the succession of generations.

The origin of cultural evolution can be traced back to the agricultural revolution of some 10,000 years ago, which marked the transition from hunter-gatherer communities to settled agricultural communities. This led to a change in lifestyle, an increase in population, and in a growth of innovation which is known as the "10,000 year explosion" (Cochran and Herpending 2009, 226).

With the agricultural revolution, the crucial factor for survival, rather than the nomadic movement to new locations, became the ability to solve problems concerning the control of the environment.

These problems prompted the search for hypotheses to solve them, the hypotheses gave rise to new problems and new hypotheses, and so on. This led to a continuous growth of knowledge.

Once this stage was reached, the biological role of knowledge became only part of the story, cultural evolution took on a primary role. Human beings were no longer only a result of natural selection, but also a result of cultural evolution.

Cultural evolution does not reduce to natural selection. Natural selection is slow, it takes thousands of unfavourable mutations before a favourable mutation emerges. Cultural evolution is much faster, being a result of non-genetic interactions between billions of human beings.

Moreover, the world changes continually and irregularly, presenting situations that are dissimilar to those that occurred in the evolutionary past of human beings. If their problem-solving resources were always strained to the limit, human beings could easily fail when certain critical situations arise, and if such failures had occurred frequently in our evolutionary past, we would not be here. To be able to cope with critical situations during times of peak demand, human beings must have excess capacity to spare for other issues at slack times. Thanks to it, in normal times they may engage in activities not directly useful for survival. This excess capacity is a result of cultural evolution.

Then it is unjustified to say, as Popper does, that "from the amoeba to Einstein is just one step (Popper 1979, 246). Cultural evolution represents an important leap forward with respect to biological evolution. Therefore, from the amoeba to Einstein is not just one step, but several substantial steps.

Nevertheless, although cultural evolution does not reduce to natural selection, cultural evolution depends on the biological makeup of human beings, so it develops on the basis of natural selection. Therefore, cultural evolution is continuous with natural selection.

Moreover although, thanks to cultural evolution, human beings can devote only part of their efforts to survival, in order to survive they must exercise control over their environment and improve it, and this requires knowledge. Thus, survival remains a primary purpose of knowledge, and knowledge plays an essential role in it.

As Sagan says, "knowledge is prerequisite to survival," indeed "our future depends on how well we know this Cosmos in which we float like a mote of dust in the morning sky" (Sagan 1980, 4).

10.11 Science and Survival

It has been said above that cultural evolution is continuous with natural selection. This also holds of science, which is both a product of cultural evolution and a form of cultural evolution in its own right. In fact, science plays an important part in survival.

Mach says that, although "science apparently grew out of biological and cultural development as its most superfluous offshoot," today "we can hardly doubt that it has developed into the factor that is biologically and culturally the most beneficial. Science has taken over the task of replacing tentative and unconscious adaptation by a faster variety that is fully conscious and methodical" (Mach 1976, 361).

In fact, science provides an essential contribution to solving the survival problem, it is an extension of the activities through which our remotest ancestors solved their survival problem.

Such activities, and those underlying science, involve similar cognitive processes. Our earliest human ancestors solved the problem of survival making hypotheses about the location of predators or prey on the basis of clues they found in the environment—such as footprints, crushed or bent grass and vegetation, bent or broken branches or twigs, mud displaced from streams, excrements, and so on. Scientists solve the problem of survival making hypotheses on the basis of clues they find in the world.

The role of science is particularly relevant because the Earth's human population is experiencing an exponential growth. It was one billion in 1804, two billion in 1927, eight billion in 2022. This puts severe strain on resources such as food, water, energy, healthcare, and increases pollution levels and deforestation. Since the Earth is not naturally adapt to handle this exponential growth, science is indispensable to address this condition. Therefore, science is important to human survival.

10.12 Mathematics and Survival

Even mathematics plays an important part in survival, because of its role in science. But mathematics played an important part in survival already for our earliest human ancestors. For, to solve the problem of survival, they had to make hypotheses not only about the location, but also about the shape and number of predators or prey. These hypotheses were mathematical in kind, so they required some mathematical capacities.

These mathematical capacities were a result of natural selection and are present not only in human beings but also in some animal species.

The mathematical capacities present in both human beings and some animal species can even be quite sophisticated.

For example, if you toss a frisbee and observe how a dog runs to try to catch it on the fly when it comes down, you will see that the dog does not run in a straight line, but along an arc that ends at the point where the frisbee falls. This is because the dog moves so that the trajectory of the frisbee appears to it as a straight line (see Shaffer et al. 2004). The dog is able to move in this way thanks to a complex mathematics that natural selection has built into its visual and motor system, which leads the dog to move in a way that allows it to keep the frisbee fixed in its visual field.

The same mathematical capacity is present in human beings. Thus, baseball players, when they run to catch the ball, do not run in a straight line but along an arc, and they do so for the same reason as dogs when they run to catch a frisbee.

10.13 Knowledge and Ultimate Aim

Since knowledge is essential for survival, knowledge has an immediate aim, that is, survival. On the other hand, however, knowledge does not have an ultimate aim. Survival is only a relative aim, it is an aim only for the majority of non-human animals and human beings, but there is no reason why survival should be absolutely preferable.

Hume even says: "It is not contrary to reason to prefer the destruction of the whole world to the scratching of my finger" (Hume 2007, I, 267).

In fact, natural selection does not have an ultimate aim.

Already Aristotle said: "What forbids that nature should work with no end, and not for the sake of the better?" (Aristotle, *Physica*, B 8, 198 b 17).

Since natural selection does not have an ultimate aim, human life and knowledge do not have an ultimate aim.

10.14 Knowledge as a Life Function

What has been said above provides an analysis of the function of knowledge in human life and, in general, in the life of all organisms. The analysis complements the claim made earlier that knowledge is a life function.

According to this analysis, knowledge is a life function that serves to solve problems of biological life and, in the case of human beings, also problems of cultural life.

This view of knowledge is different in kind from the view of knowledge as justified true belief.

While the latter is an instance of the concept-first approach to knowledge, the view of knowledge as a life function is an instance of the function-first approach to knowledge. For, it provides an analysis of the function of knowledge in human life, and in general in the life of all organisms.

10.15 Role of Language in Human Knowledge as a Life Function

What is the role of language in human knowledge as a life function? This question consists of two components. What is the role of language in an individual's discovery of new knowledge? What is the role of knowledge in the process by which newly discovered knowledge becomes part of a culture? These two components have different answers.

On the one hand, language has no necessary role in an individual's discovery of new knowledge.

In fact, several mathematicians and scientists say that language plays no part in their research work.

Thus, Hadamard says: “Words are totally absent from my mind when I really think,” and “do not reappear in my consciousness before I have accomplished or given up the research” (Hadamard 1954, 75).

Einstein says: “The words or the language, as they are written or spoken, do not seem to play any role in my mechanism of thought. The psychical entities which seem to serve as elements in thought” are, “in my case, of visual and some of muscular type” (Einstein 1954, 142–143).

This is confirmed by evidence from neuroscience, which “demonstrates that all regions of the language network are largely ‘silent’ during all tested forms of thought, including mathematical reasoning” (Fedorenko et al. 2024, 579). In fact, “all types of thought tested to date are possible without language” (ibid.). Therefore, “language is not necessary or sufficient for thought” (ibid., 576).

On the other hand, language has an important role in the process through which newly discovered knowledge becomes part of a culture, because language is essential for knowledge transmission.

Again, this is confirmed by evidence from neuroscience, which demonstrates that language is “a tremendously useful tool for knowledge transmission. The cumulative effect of this transmission—knowledge building on knowledge—along with increased sophistication of our social and problem-solving abilities is plausibly what enabled us to create human civilizations” (Fedorenko et al. 2024, 583). Indeed, “it is unlikely that our species’ success would have been possible without the cumulative culture that was enabled by the external usage of language” (ibid.). Therefore, “instead of providing the key substrate for thinking and reasoning, language likely transformed our species by enabling cross-generational transmission of acquired knowledge” (ibid.).

10.16 Common-Sense Knowledge and Natural Selection

A particular type of knowledge is common-sense knowledge.

Common-sense knowledge is usually described as “what everyone knows and therefore what everyone has” (Ledwig 2007, 134).

Analytic philosophers interpret this as meaning that common-sense knowledge is what most people, belonging to a particular culture and in a particular period, know and therefore have. So, common-sense knowledge varies from one culture and period to another.

Thus, Dummett says that “we may properly speak of certain views as forming, at a particular period and in a particular culture, a part of common-sense” knowledge, that is, knowledge “on which most of those who belong to that culture at that time habitually rely in their everyday thinking” (Dummett 1993, 390). Therefore, common-sense knowledge is knowledge that is “culturally conditioned and subject to historical evolution” (ibid., 392).

Williamson says: "By 'common-sense knowledge' in a society I mean whatever most of its members know. Thus common-sense knowledge varies from one society to another" (Williamson 2018, 8).

But this is not valid. For, if common-sense knowledge is what everyone knows and therefore what everyone has, then common-sense knowledge cannot be knowledge that only people belonging to a given culture at one time have, it must be knowledge that all human beings have.

In fact, there is evidence that there exists knowledge that all human beings have.

As Spelke says, all human beings have "a set of cognitive systems that are phylogenetically ancient, innate, and universal across humans: systems of core knowledge" (Spelke 2011, 287). This set of cognitive systems is also possessed by animals, it is a result of natural selection and "evolved over hundreds of millions of years. At least six distinct" cognitive "systems serve to represent highly abstract properties of the unchanging navigable environment, of movable objects, of number, and of the living, animate, and social beings who populate our world" (Spelke 2022, xvii).

If common-sense knowledge is what everyone knows and therefore what everyone has, then it can only be a result of natural selection. So, common-sense knowledge must be knowledge that all human beings have as a result of natural selection. This type of knowledge actually exists, as it is clear from the fact that all human beings have a set of systems of core knowledge.

References

Antognazza, Maria Rosa. 2015. The benefit to philosophy of the study of its history. *British Journal for the History of Philosophy* 23:161–184.

Chisholm, Roderick. 1989. *Theory of knowledge*. London: Prentice-Hall.

Cochran, Gregory, and Henry Herpending. 2009. *The 10,000 explosion: How civilization accelerated human evolution*. New York: Basic Books.

Dummett, Michael. 1993. *The seas of language*. Oxford: Oxford University Press.

Ebbersmeyer, Sabrina. 2024. Philosophy as passion for knowledge: What kind of history of philosophy for the 21st century? *SATS—Northern European Journal of Philosophy* 25 (2): 113–131.

Einstein, Albert. 1954. A testimonial. In *The psychology of invention in the mathematical field*, ed. Jacques Hadamard, 142–143. Mineola: Dover.

Fedorenko, Evelina, Steven Piantadosi, and Edward Gibson. 2024. Language is primarily a tool for communication rather than thought. *Nature* 630:575–586.

Gettier, Edmund. 1963. Is justified true belief knowledge? *Analysis* 23 (6): 121–123.

Hadamard, Jacques. 1954. *The psychology of invention in the mathematical field*. Mineola: Dover.

Hume, David. 2007. *A treatise of human nature*. Oxford: Oxford University Press.

Ledwig, Marion. 2007. *Common sense: Its history, method, and applicability*. New York: Peter Lang.

Lewis, Clarence. 1946. *An analysis of knowledge and valuation*. La Salle: Open Court.

Mach, Ernst. 1976. *Knowledge and error: Sketches on the psychology of enquiry*. Dordrecht: Springer.

Mayr, Ernst. 2002. *What evolution is*. London: Phoenix.

Popper, Karl. 1979. *Objective knowledge: An evolutionary approach*. Oxford: Oxford University Press.

Pritchard, Duncan. 2016. Epistemology. In Duncan Pritchard, *What is this thing called philosophy?* 121–158. London: Routledge.
Russell, Bertrand. 1998. *The problems of philosophy*. Oxford: Oxford University Press.
Sagan, Carl. 1980. *Cosmos*. New York: Random House.
Scott Senisi, Ann, and Elizabeth Fong. 2016. *Body structures and functions*. Boston: Cengage Learning.
Shaffer, Dennis, Scott Krauchunas, Marianna Eddy, and Michael McBeath. 2004. How Dogs Navigate to Catch Frisbees. *Psychological Science* 15:437–441.
Spelke, Elizabeth. 2011. Natural number and natural geometry. In *Space, time and number in the brain*, ed. Stanislas Dehaene and Elizabeth Brannon, 287–317. London: Elsevier.
Spelke, Elizabeth. 2022. *What babies know: Core knowledge and composition*. Vol. 1. Oxford: Oxford University Press.
Wang, Hao. 1974. *From mathematics to philosophy*. London: Routledge.
Williams, Michael. 2001. *Problems of knowledge: A critical introduction to epistemology*. Oxford: Oxford University Press.
Williamson, Timothy. 2000. *Knowledge and its limits*. Oxford: Oxford University Press.
Williamson, Timothy. 2018. *Doing philosophy: From common curiosity to logical reasoning*. Oxford: Oxford University Press.

Chapter 11
Knowledge and Reality

Abstract Developing the view that philosophy is acquisition of knowledge also requires answering the question: What is the relation of knowledge to reality? There are different types of knowledge, each of which involves a different relation to reality, but a type of knowledge whose relation to reality is particularly relevant is scientific knowledge. The relation of scientific knowledge to reality is the primary theme of this chapter. This involves considering what the aim of science is. Since antiquity, a widely held view has been that the aim of science is truth, but this chapter argues that this view is not valid. Alternatively, the chapter argues that the aim of science is plausibility.

Keywords Truth as the aim of science · Impossibility of a criterion of truth · Plausibility as the aim of science · Science and certainty · Science and common-sense knowledge

11.1 Truth as the Aim of Science

Developing the view that philosophy is acquisition of knowledge also requires answering the question: What is the relation of knowledge to reality?

There are different types of knowledge, each of which involves a different relation to reality, but a type of knowledge whose relation to reality is particularly relevant is scientific knowledge. The relation of scientific knowledge to reality is the primary theme of this chapter.

This involves considering what the aim of science is.

From antiquity it has been affirmed that the aim of science is truth.

Thus, Aristotle says that "the aim of theoretical science is truth" (Aristotle, *Metaphysica*, α 1, 993 b 20–21).

That the aim of science is truth has also been reaffirmed in the past century.

Thus, Husserl says that "science seeks to be a means towards the greatest possible conquest of the realm of truth" (Husserl 2001, I, 18).

C. Cellucci, *A Philosophy for the Future*, Synthese Library 508,
https://doi.org/10.1007/978-3-031-98443-3_11

Popper says that "the task of science is the search for truth," in particular, for "truth which has a high degree of explanatory power" (Popper 2002, 311).

Sankey says that "the aim of science is to discover the truth about the natural world," therefore "progress in science must consist in progress toward the truth. This may be understood as a cumulative build-up of truths or as convergence on the truth" (Sankey 2020, 69).

From a recent survey it appears that "on average, physicists tend towards" the view that the aim of science is truth, "and slightly more so than philosophers of science" (Henne et al. 2024, 3).

However, this chapter argues that the view that the aim of science is truth is not valid. Alternatively, the chapter argues that the aim of science is plausibility.

11.2 Impossibility of a Criterion of Truth

The view that the aim of science is truth presupposes that there is a criterion of truth, that is, a standard by which true propositions can be distinguished from false propositions. But, a criterion of truth is impossible. This can be seen by slightly modifying Frege's argument that a definition of truth is impossible. The resulting argument is as follows.

In order to have a criterion of truth, "certain characteristics" of truth "would have to be specified. And, in application to any particular case, the question would always arise whether it were true that the characteristics were present" (Frege 1984, 353). To answer this question we would need a criterion of truth. So "we should be going round in a circle" (ibid.). Therefore, a criterion of truth is impossible.

11.3 Impossibility That Truth Is the Aim of Science

Despite its wide support, the view that the aim of science is truth is untenable. Truth cannot be the aim of science for the following reasons.

(1) Saying that the aim of science is truth conflicts with the fact that a criterion of truth is impossible, so true propositions cannot be distinguished from false propositions. But distinguishing them is essential if the aim of science is truth. Therefore, if the aim of science is truth, the aim of science is unachievable.

(2) Saying that the aim of science is truth conflicts with the fact that the history of science offers us many examples of theories that were once thought to be true but were later found to be in conflict with reality. This is impossible, because what is true is true forever.

(3) Saying that the aim of science is truth conflicts with the fact that there are theories that are confirmable but mutually incompatible, and hence cannot all be true. An example is provided by quantum mechanics and general relativity. As Abdeen and de Silva observe, "quantum fluctuations arising from the uncertainty

principle" lead "to violent distortions of the space-time fabric, in sharp contrast to the smooth world view of general relativity" (Abdeen and de Silva 2013, 58).

11.4 Truth and Mythology

The concept of truth is a remnant of Greek mythology. In fact, Aesop says that Prometheus "decided one day to sculpt a statue of Truth," then he put the statue "in the kiln," and when it "had been thoroughly baked, he infused" it "with life: sacred Truth walked with measured steps" (Aesop, *Fables* 530: *Prometheus and Truth*).

Originally Greek philosophy had several elements in common with Greek mythology, but gradually got rid of them. Yet, it retained some of them, and truth was one of the remaining elements. Indeed, as we have seen, Aristotle says that the aim of theoretical science is truth.

Aristotle also says that "he who loves myth is, in a way, a philosopher" (Aristotle, *Metaphysica*, A 2, 982 b 19).

But, like all concepts of a mythical origin, truth is useless for modern science. As Hempel says, "the conception of science as a search for truth is strictly untenable" (Hempel 2000, 77).

11.5 Plausibility as the Aim of Science

Since truth cannot be the aim of science, an alternative view of the aim of science is needed.

An alternative view is that the aim of science is plausibility. In particular, the aim of science is to discover plausible hypotheses about the world.

A hypothesis is said to be plausible if the arguments for it are stronger than the arguments against it, on the basis of experience.

When a hypothesis has been advanced, we check its plausibility. If the hypothesis is plausible, then for the moment the hypothesis is approved. In fact, the term 'plausible' comes from the Latin 'plaudere', which means 'to applaud', 'to approve'.

The hypothesis, however, is approved only for the moment, because new data can always emerge with which the hypothesis is incompatible.

If this occurs, the hypothesis is not rejected outright. Rather, it is put on a waiting list, subject to further investigation.

Plausibility is a relative concept. A hypothesis is plausible if the arguments for it are stronger than those against it on the basis of experience. But experience does not constitute a definitive basis for evaluation, because experience can change over time. So, a hypothesis that is plausible at one time can become implausible at a later time, and conversely, a hypothesis that is implausible at one time can become plausible at a later time.

For this reason it has been said above that, if new data emerge with which the hypothesis is incompatible, the hypothesis is not rejected outright, it is put on a waiting list, subject to further investigation.

11.6 Insufficiency of a Single Case for Refutation

That, if new data emerge with which the hypothesis is incompatible, the hypothesis is not rejected outright, contrasts with the widespread opinion that, in this case, the hypothesis should be immediately rejected.

Thus, Einstein says: "No amount of experimentation can ever prove me right," but "a single experiment can prove me wrong" (Calaprice and Einstein 2011, 476).

Feynman says: "In general," when "we look for a new law," first "we guess it," then "we compute the consequences of the guess" and compare them "with experiment;" if some consequence "disagrees with experiment it is wrong. That is all there is to it" (Feynman 1965, 156).

Popper says that "so long as a theory stands up to several tests we can design, it is accepted, if it does not" stand a single observation or experiment, "it is rejected" (Popper 2002, 72).

But this is not valid. Already Galileo made it clear that it is wrong to say: "Ad destruendum sufficit unum [A single case is sufficient for refutation]" (Galilei 1968, VII, 222). To say this is to be like "the man who wanted to tear down his house to the foundations, saying that it was uninhabitable and architecturally flawed, only because the chimney made too much smoke; and he would have done it if a friend had not pointed out that it was enough to fix the chimney without tearing down the rest" (ibid., VI, 533). Indeed if, "because of any particular new accident which is discovered in some part of the heavens, one had to change the whole structure of the world, one would never get anywhere" (ibid., VI, 354). In fact, "the motions, sizes, distances, and arrangements of spheres and stars are never observed to be so correct that they do not need constant corrections" (ibid.).

Therefore, as already said, if new data emerge with which a hypothesis is incompatible, the hypothesis is not rejected outright, it is put on a waiting list, subject to further investigation.

An example of this is provided by the hypotheses of Newton's theory of gravity. When a seventh planet, Uranus, was discovered, its orbit was observed to deviate considerably from the orbit calculated from the hypotheses of Newton's theory, but these hypotheses were not rejected outright, they were put on a waiting list.

It was hypothesized that there was an eighth planet, Neptune, and Adams and Leverrier calculated the mass and position the new planet would have to cause the observed deviation from the calculated orbit. Eventually, Neptune was observed less than one degree away from the calculated position.

11.7 Adequacy of Plausibility as the Aim of Science

Contrary to the view that the aim of science is truth, the view that the aim of science is plausibility is adequate.

(1) While a criterion of truth is impossible, a criterion of plausibility is possible. In fact, to determine whether a hypothesis is plausible, we see if the arguments for the hypothesis are stronger than those against it, on the basis of experience.

(2) While a theory that is true at one time cannot become not true at a later time because what is true is true forever, a theory that is plausible at one time can become implausible at a later time because the arguments against it can become stronger than those for it, on the basis of the experience available at that time.

(3) While two theories that are both confirmable but mutually incompatible cannot both be true, they can both be plausible, because the arguments for and against one of them can be different from the arguments for and against the other one, on the basis of experience.

11.8 Plausibility and New Information

If a hypothesis is plausible, then the hypothesis agrees with experience, because the arguments for the hypothesis are stronger than those against it on the basis of experience. But this does not mean that the hypothesis does not provide new information with respect to experience.

For example, let us consider hyperbolic geometry, the non-Euclidean geometry obtained by replacing Euclid's fifth postulate with the hyperbolic postulate: Given any straight line and any point not on it, there are at least two distinct straight lines through that point which do not intersect that straight line.

A model of hyperbolic geometry is the Beltrami-Klein model, which is obtained by taking a circle in the Euclidean plane and interpreting points as the points inside the circle, and lines as chords of the circle (excluding their endpoints).

All Euclid's postulates except the fifth hold in the Beltrami-Klein model, and the hyperbolic postulate also holds. For, as the following figure suggests, given any chord c of the circle and any point A inside the circle not on c, there are at least two distinct chords of the circle through A which do not intersect c.

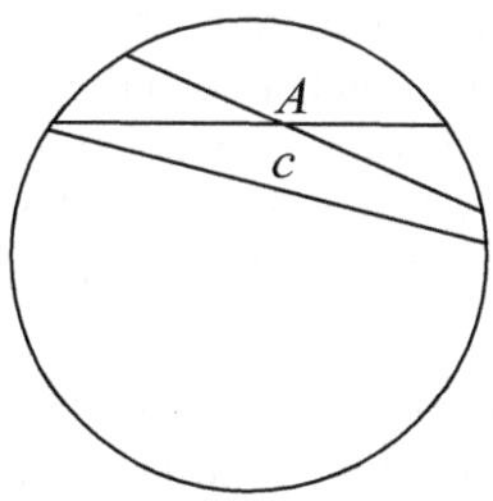

Since the Beltrami-Klein model is a circle in the Euclidean plane, it follows that, if Euclidean geometry is consistent, then so is hyperbolic geometry.

Therefore, the hypotheses of hyperbolic geometry agree with Euclidean experience. But this does not mean that they do not provide new information with respect to Euclidean experience. For, the hyperbolic postulate cannot be deduced from Euclid's postulates.

11.9 Plausibility and Truth

Plausibility is different from several other concepts to which it might seem similar, in particular, truth, probability, warranted assertibility, non-falsification, and persuasiveness.

Plausibility is different from truth.

Peirce says that "what we mean by the truth" is "the opinion which is fated to be ultimately agreed to by all who investigate" (Peirce 1931–1958, 5.407).

This suggests that plausibility is similar to truth. But this is not the case, between plausibility and truth there are important differences.

A criterion of truth is impossible. On the contrary, a criterion of plausibility is possible.

Moreover, truth is an absolute concept, because a true proposition is true forever. Instead, plausibility is a relative concept.

In fact, plausibility is alternative to truth. For, by Gödel's incompleteness theorems, even the aim of mathematics cannot be truth but only plausibility. (For more on this, see Cellucci 2022).

11.10 Plausibility and Probability

Plausibility is different from probability.

Pólya says that we can "use the calculus of probability to render more precise our views on plausible reasoning" (Pólya 1954, II, 116). For, "the calculus of plausibilities obeys the same rules as the calculus of probabilities" (Pólya 1941, 457).

This suggests that plausibility is similar to probability. But this is not the case, between plausibility and probability there is an important difference.

Plausibility involves a comparison between the arguments for and the arguments against a hypothesis, on the basis of experience, so plausibility is not a mathematical concept. Instead, probability is a mathematical concept.

This is made clear already by Kant. He says that "plausibility is concerned with whether, in the cognition, there are more grounds for the thing than against it" (Kant 1992, 331). Then plausibility is not a mathematical concept. Conversely, "probability is a fraction, whose denominator is the number of all the possible cases, and whose numerator contains the number of winning cases" (ibid., 328). Then probability is a mathematical concept, and in fact "there is a mathematics of probability" (ibid., 331). (For more on Kant's distinction between plausibility and probability, see Capozzi 2013, 2019, 2020).

11.11 Plausibility and Warranted Assertibility

Plausibility is different from warranted assertibility.

Rescher says that "plausible (sub-demonstrative) reasoning" establishes "what later philosophers called 'warranted assertability'" (Rescher 2007, 128).

This suggests that plausibility is similar to warranted assertibility. But this is not the case, between plausibility and warranted assertibility there are important differences.

In fact, Dewey says that warranted assertibility is the status a proposition acquires in an inquiry that is part of a scientific enterprise that ultimately converges to the ideal limit of truth because, as Peirce says, truth is "the ideal limit towards which endless investigation would tend to bring scientific belief" (Dewey 1938, 345, footnote 6). Thus, the scientific enterprise leads to "the institution of conditions which remove need for doubt" (ibid., 7).

Instead, plausibility is not the status a proposition acquires in an inquiry that is part of a scientific enterprise that ultimately converges to the ideal limit of truth. For, not only a hypothesis that is implausible at one time can become plausible at a later time, but a hypothesis that is plausible at one time can become implausible at a later time.

Moreover, if the aim of science is plausibility, then the scientific enterprise does not lead to the institution of conditions which remove need for doubt. For, a hypothesis that is plausible is not certain, because there is no guarantee that a counterexample will never be found, so there is always need for doubt.

11.12 Plausibility and Non-Falsification

Plausibility is different from non-falsification.

Popper says that, in order to decide whether a hypothesis or theory is acceptable, "conclusions are drawn" from the hypothesis "by means of logical deduction" (Popper 2005, 9). The conclusions are then compared with "the results" of "experiments" (ibid., 10). If the conclusions have not been falsified by these results, the theory is, for the time being. non-falsified because "we have found no reason to discard it. But if" the "conclusions have been falsified, then their falsification also falsifies the theory from which they were logically deduced," therefore we "may always overthrow it" (ibid.).

This suggests that plausibility is similar to non-falsification. But this is not the case, between plausibility and non-falsification there is an important difference.

Unlike plausibility, non-falsification does not involve any comparison between the arguments for and against a hypothesis or theory. If the results of experiments are not in conflict with the conclusions deduced from the hypothesis, the latter, for the time being, is considered to be acceptable, on the other hand, if they are in conflict with the hypothesis, the hypothesis is discarded.

This contrasts with the fact that, on the one hand, for a hypothesis to be acceptable, it is not enough that the results of experiments are not in conflict with the hypothesis, otherwise a myriad of hypotheses without interest would be acceptable. On the other hand, for a hypothesis to be discarded, it is not enough that the results of experiments are incompatible with the hypothesis, because the experiments are based on other hypotheses, and the latter could turn out to be implausible. Therefore, it would be unjustified to discard the hypothesis without further investigation.

11.13 Plausibility and Persuasiveness

Plausibility is different from persuasiveness.

Walton says that "the very best definition of plausibility was given by Carneades," whose "most important legacy to philosophy was his famous theory of plausibility" (Walton 2001, 152). According to it, "something is plausible if it seems, or appears to be true, or if it fits in with other things we accept as true, or if it tested, and passes the test" (ibid., 153).

But this is not valid. For, Carneades' theory is not a theory of plausibility but a theory of persuasiveness, and does not concern knowledge but the conduct of life.

In fact, Carneades says that, while a criterion of truth is impossible, there is "a criterion for the conduct of life," and this criterion is "the persuasive appearance" (Sextus Empiricus, *Adversus Logicos*, I.166, ed. Mutschmann). Namely "the appearance that seems to be true" (ibid., I.169).

Walton himself admits that Carneades' theory of "plausibility is not a theory of knowledge or belief. It is a guide to rational acceptance or commitment, a guide to action" (Walton 2001, 153).

11.14 Plausibility and Endoxa

While plausibility is different from truth, probability, warranted assertibility, non-falsification, and persuasiveness, on the contrary plausibility is somewhat similar to Aristotle's endoxa.

Aristotle says that endoxa are opinions accepted "by everyone, or by the great majority, or by the wise, and among them either by all of them, or by the great majority, or by the most famous and esteemed" (Aristotle, *Topica*, A 1, 100 b 21–23).

This is often taken to be Aristotle's definition of endoxa. But this is not the case, otherwise the superficial or biased opinions of the great majority would be endoxa.

Rather, Aristotle's definition of endoxa is implicit in his assertion that, for each opinion concerning some given subject, we must "examine the arguments for it and the arguments against it" (ibid., Θ 14, 163 a 37–b 1). For, "going through the difficulties on either side, we shall more readily discern the true as well as the false in any subject" (ibid., A 2, 101 a 35–36). In fact, "if the objections are answered and the endoxa remain, we shall have proved the case sufficiently" (Aristotle, *Ethica Nicomachea*, Z 1, 1145 b 6–7).

Then, endoxa are opinions on which we can agree on the basis of a careful examination of the arguments for and against them, from which the former turn out to be stronger than the latter. This justifies the claim that plausibility is somewhat similar to Aristotle's endoxa.

Striker even uses 'plausible' "to translate the Greek endoxon" (Striker 2009, 77).

However, that plausibility is somewhat similar to Aristotle's endoxa does not mean that it is the same as endoxa. For, according to Aristotle, endoxa are continuous with truth. Indeed, he says that "to aim at endoxa and to aim at truth belong to the same disposition" (Aristotle, *Rhetorica*, 1355 a 17–18). On the contrary, plausibility is alternative to truth.

11.15 The Plausibility Test Procedure

To determine whether a hypothesis is plausible, the following plausibility test procedure can be used.

(1) Deduce conclusions from the hypothesis.

(2) Compare the conclusions with each other, to check that the hypothesis does not lead to contradictions.

(3) Compare the conclusions with other hypotheses already known to be plausible, to check that the hypothesis is compatible with them.

(4) Compare the conclusions with experience, to check that the hypothesis is compatible with it.

If the hypothesis passes the plausibility test procedure, the hypothesis is plausible and so it is approved.

However, as already said, the hypothesis is approved only for the moment, because new data can always emerge with which the hypothesis is incompatible. Then the hypothesis is put on a waiting list, subject to further investigation.

11.16 Plausibility and the Adversarial System

The process through which a hypothesis is shown to be plausible can be compared to the adversarial system for settling legal disputes between two parties.

The adversarial system is designed to provide adequate motivation for each party of the dispute. It has two key features: advocates who put forward arguments in support of each party's case, and a judge and jury who make a decision. The advocates of each party argue their case as strongly and effectively as they can. The judge and jury listen to what each party has to say and decide between the two parties.

In the case of establishing the plausibility of a hypothesis, the advocates are the scientists who present arguments for and against the hypothesis, the judge and jury is the wider community of scientists. But the process can also take place within a single scientist.

11.17 Science and Certainty

The view that the aim of science is plausibility implies that science is not certain. This conflicts with the widespread opinion that science is certain.

Thus, Husserl says that science aims at truths that "must carry with them an absolute certainty," namely an "absolute indubitability" (Husserl 1960, 14). Therefore, they must satisfy the condition of "excluding in advance every doubt as 'objectless', empty" (ibid., 16).

But this is not valid. For, a theory that is plausible at one time can become implausible at a later time, and conversely, a theory that is implausible at one time can become plausible at a later time.

As Poincaré says, "in our relative world all certainty is a lie" (Poincaré 1910, VII).

Even mathematics is not certain because, by Gödel's second incompleteness theorem, it is impossible to demonstrate by absolutely reliable means that mathematics is true and certain.

From this it seems fair to conclude that nothing is certain.

Against the conclusion that nothing is certain it might be argued that it is self-defeating, because it implies that the conclusion itself is not certain. But this argument is not valid. For, the fact that the conclusion is not certain confirms that nothing is certain.

In this connection it is interesting to cite Peirce who says: "I am a man of whom critics have never found anything good to say" (Peirce 1931–1958, 1.10). Only once "in all my lifetime have I experienced the pleasure of praise—not for what it might bring but in itself. That pleasure was beatific; and the praise that conferred it was meant for blame. It was that a critic said of me that I did not seem to be absolutely sure of my own conclusions. Never, if I can help it, shall that critic's eye ever rest on what I am now writing; for I owe a great pleasure to him; and, such was his evident animus, that should he find that out, I fear the fires of hell would be fed with new fuel in his breast" (ibid.). For, by saying so, that critic had implicitly recognized that "we cannot in any way reach perfect certitude nor exactitude. We never can be absolutely sure of anything" (ibid., 1.147).

Of course, as Russell says, "uncertainty, in the presence of vivid hopes and fears, is painful, but must be endured if we wish to live without the support of comforting fairy tales" (Russell 2004, 17). Indeed, "to teach how to live without certainty, and yet without being paralysed by hesitation, is perhaps the chief thing that philosophy, in our age, can still do for those who study it" (ibid.).

11.18 Common-Sense Knowledge and Scientific Knowledge

In Chap. 10 it has been argued that common-sense knowledge is knowledge that all human beings have as a result of natural selection. On the other hand, scientific knowledge is a result of cultural evolution. From this it follows that scientific knowledge is different from common-sense knowledge.

However, some philosophers and scientists do not merely say that scientific knowledge is different from common-sense knowledge, they claim that scientific knowledge is opposed to common-sense knowledge.

Thus, Sellars says that "the common-sense world, the world of the 'manifest image' is, in the Kantian sense, phenomenal," that is, "the way in which things as they really are appear to minds," but this "by no means guarantees that" it "is an adequate representation of the way things are" (Sellars 2002, 306). With respect to the manifest image, "the scientific image presents itself as a rival image. From its point of view the manifest image" is "an 'inadequate'" likeness "of a reality which first finds its adequate (in principle) likeness in the scientific image" (Sellars 1991, 20).

Wolpert says that "both the ideas that science generates and the way in which science is carried out are entirely counterintuitive and against common sense," indeed "science does not fit with our natural expectations" (Wolpert 1992, 1). There are "rare exceptions to the rule that all scientific ideas are contrary to common sense" (ibid., 5). Indeed, "if something fits in with common sense it almost certainly isn't science" because "the way in which the universe works is not the way in which common sense works: the two are not congruent" (ibid., 11).

Feyerabend says that science "invites us to reject commonsense" (Feyerabend 1981–2016, I, xii). There is an "antagonism between commonsense and comprehensive theories" (ibid., I, 3). Therefore, commonsense must be completely replaced by science. If "the complete replacement of commonsense" by science "succeeds, then man will have succeeded in ridding himself of prejudices which would have prevented him from ever seeing the world as it is and from figuring out the details of its real structure" (ibid., I, 127–128). For, "science not only produces predictions, it is also about the nature of things" (ibid., I, 3).

But the claim that scientific knowledge is opposed to common-sense knowledge is not valid. Simply, scientific knowledge gives an interpretation of what the senses suggest which is more plausible than the interpretation given by common-sense knowledge.

For example, the senses suggest that the sun rises each morning, it goes across the sky and sets each evening, and in the night the moon, the stars and the planets become visible and move in the sky.

Common-sense knowledge interprets what the senses suggest as meaning that all heavenly objects, including the sun, the moon, the planets and the stars, move around the earth in a circular orbit, while the earth is at the centre of the universe and is fixed and immovable.

Conversely, Copernican astronomy interprets what the senses suggest as meaning that the planets move around the sun, and the moon moves around the earth, in a circular orbit, while the sun is at the centre of the universe and is fixed and immovable, just like the stars.

The interpretation provided by Copernican astronomy is more plausible than the interpretation provided by common-sense knowledge on the basis of experience, but it is not plausible in an absolute sense. For, the sun is not at the centre of the universe, nor it is fixed and immovable, and the planets do not move around the sun, nor does the moon move around the earth, in a circular orbit, they move in an elliptical orbit.

Although scientific knowledge is not opposed to common-sense knowledge, there is nevertheless an important difference between them with regard to errors.

Both common-sense knowledge and scientific knowledge contain errors, although scientific knowledge in much smaller proportions than common-sense knowledge. But, as will be argued in Chap. 18, scientific knowledge is based on a method, the analytic method, which is capable of self-correcting. Therefore, unlike common-sense knowledge, scientific knowledge is self-improving.

11.19 Common-Sense Knowledge and Truth

The above argument that scientific knowledge is not opposed to common-sense knowledge essentially depends on the view that the aim of science is plausibility. For, the view that the aim of science is truth is unable to account for this.

Thus, Sankey argues that "commonsense beliefs" have "survival value because they are for the most part true. Our species could not have survived if the majority of the commonsense beliefs on which we base our everyday interaction with the world were false" (Sankey 2014, 17–18). In fact, "a robust sense of reality provides us with a reasonable degree of practical certainty that things are by and large as they appear to us" (ibid., 15). Science "starts from common sense" and "does not lead to the overthrow of common sense" (ibid., 23). Given common sense, the view that the aim of science is truth "is the most natural position to adopt as an interpretation of scientific inquiry into the world around us" (ibid.). Therefore, scientific knowledge is not opposed to common-sense knowledge.

This argument, however, is not valid. The claim that commonsense beliefs have survival value because they are for the most part true, conflicts with the fact that a criterion of truth is impossible, so it is impossible to recognize that a belief is true. Then, our species could not have survived because the majority of the commonsense beliefs on which we base our everyday interaction with the world are true. For, our species would have been unable to recognize that they are true. Moreover, the claim that a robust sense of reality provides us with a reasonable degree of practical certainty that things are by and large as they appear to us, conflicts with the fact that common sense knowledge is based on our cognitive system which, as will be argued in Chap. 19, has limitations. Because of these limitations, things are not by and large as they appear to us.

References

Abdeen, Mohamed, and Nadish de Silva. 2013. Incompatibility of general relativity with quantum mechanics. *Proceedings of Technical Sessions* 29:57–63.

Calaprice, Alice, and Albert Einstein. 2011. *The ultimate quotable Einstein*. Princeton: Princeton University Press.

Capozzi, Mirella. 2013. *Kant e la logica I*. Naples: Bibliopolis.

Capozzi, Mirella. 2019. Le ipotesi secondo Kant: Requisiti, giustificazione, status epistemico e euristica. *Syzetesis* 6:153–189, 183–184.

Capozzi, Mirella. 2020. *Wahrscheinlichkeit* and *Scheinbarkeit*: A key issue in Kant's logic and philosophy. In *Kants Schriften in Übersetzungen, Sonderheft 15 of Archiv für Begriffsgeschichte*, ed. Gisela Schlüter, 513–533. Hamburg: Meiner.

Cellucci, Carlo. 2022. *The making of mathematics: Heuristic philosophy of mathematics*. Cham: Springer.

Dewey, John. 1938. *Logic: The theory of inquiry*. New York: Holt, Rinehart and Winston.

Feyerabend, Paul. 1981–2016. *Philosophical papers*. Cambridge: Cambridge University Press.

Feynman, Richard. 1965. *The character of physical law*. Cambridge: The MIT Press.

Frege, Gottlob. 1984. *Collected papers on mathematics, logic, and philosophy*. Oxford: Blackwell.

Galilei, Galileo. 1968. *Opere*. Florence: Barbera.

Hempel, Carl. 2000. *Selected philosophical essays*. Cambridge: Cambridge University Press.

Henne, Céline, Hannah Tomczyk, and Christoph Sperber. 2024. Physicists' views on scientific realism. *European Journal for Philosophy of Science* 14:10.

Husserl, Edmund. 1960. *Cartesian meditations: An introduction to phenomenology*. The Hague: Martinus Nijhoff.

Husserl, Edmund. 2001. *Logical investigations*. London: Routledge.

Kant, Immanuel. 1992. *Lectures on logic*. Cambridge: Cambridge University Press.

Peirce, Charles. 1931–1958. *Collected papers*. Cambridge: Harvard University Press.

Poincaré, Henri. 1910. *Savants et écrivains*. Paris: Flammarion.

Pólya, George. 1941. Heuristic reasoning and the theory of probability. *The American Mathematical Monthly* 48:450–465.

Pólya, George. 1954. *Mathematics and plausible reasoning*. Princeton: Princeton University Press.

Popper, Karl. 2002. *Conjectures and refutations: The growth of scientific knowledge*. London: Routledge.

Popper, Karl. 2005. *The logic of scientific discovery*. London: Routledge.

Rescher, Nicholas. 2007. *Dialectics: A classical approach to inquiry*. Frankfurt: Ontos Verlag.

Russell, Bertrand. 2004. *History of Western philosophy*. London: Routledge.

Sankey, Howard. 2014. Scientific realism and basic common sense. *Kairos* 10:11–24.

Sankey, Howard. 2020. Scientific realism and the conflict with common sense. In *New approaches to scientific realism*, ed. Wenceslao Gonzales, 68–83. Berlin: de Gruyter.

Sellars, Wilfrid. 1991. *Science, perception and reality*. Atascadero: Ridgeview.

Sellars, Wilfrid. 2002. *Kant's transcendental metaphysics: Sellars' Cassirer lectures notes and other essays*. Atascadero: Ridgeview Publishing Company.

Striker, Gisela. 2009. Introduction—Commentary. In Aristotle, *Prior analytics*, Book I, xi–xiii, 67–246. Oxford: Oxford University Press.

Walton, Douglas. 2001. Abductive, presumptive and plausible arguments. *Informal Logic* 21:141–169.

Wolpert, Lewis. 1992. *The unnatural nature of science*. Cambridge: Harvard University Press.

Chapter 12
Knowledge and Objectivity

Abstract Developing the view that philosophy is acquisition of knowledge also requires answering the question: What is the relation of knowledge, in particular scientific knowledge, to objectivity? It has been claimed that objectivity in science is impossible, but this chapter argues that this is unjustified. There are several views of objectivity, the chapter considers them and argues that they are not valid. Alternatively, the chapter argues that objectivity consists in plausibility.

Keywords The denial of objectivity · Different views of objectivity · Objectivity as plausibility · Objectivity in mathematics

12.1 The Denial of Objectivity

Developing the view that philosophy is acquisition of knowledge also requires answering the question: What is the relation of knowledge, in particular scientific knowledge, to objectivity?

According to Kuhn, objectivity in science is impossible because, "when scientists must choose between competing theories," they may "reach different conclusions," and "with respect to divergences of this sort, no set of choice criteria yet proposed is of any use" (Kuhn 1977, 324). For, the choice between competing theories is a choice between different paradigms, and different paradigms are "incommensurable ways of seeing the world" (Kuhn 1996, 4). So, the transition between different paradigms "is a transition between incommensurables" (ibid., 150). Being a transition between incommensurables, the transition cannot be "forced by logic and neutral experience" (ibid.). Although each side "may hope to convert the other to his way of seeing his science and its problems, neither may hope to prove his case. The competition between paradigms is not the sort of battle that can be resolved by proofs" (ibid., 148). Each side can only "try, by persuasion, to convert the other" (ibid., 198). A "lifelong resistance, particularly from those whose productive careers have committed them to an older tradition of normal science, is not a

C. Cellucci, *A Philosophy for the Future*, Synthese Library 508,
https://doi.org/10.1007/978-3-031-98443-3_12

violation of scientific standards" (ibid., 151). There is no "point at which resistance becomes illogical or unscientific" (ibid., 159).

But this is unjustified. Of course, the competition between paradigms is not the sort of battle that can be resolved by proofs. Nevertheless, each side can try to convert the other not by rhetorical means but by reason, by showing that the arguments for one paradigm are stronger than the arguments for another, on the basis of experience. Even a lifelong resistance to a paradigm can have a positive side, it can ensure that the paradigm is given every chance to overcome opposition, rather than being abandoned prematurely.

12.2 Different Views of Objectivity

Contrary to Kuhn, several people argue that objectivity in science is possible.

However, there are several views of objectivity, objectivity as: independence of human cognitive abilities, absolute conception of the world, view from nowhere, approximation to truth, knowledge without knower, shared criteria for theory choice, methods conducive to truth, invariance under all admissible transformations, knowledge of primary properties, value-freedom, avoidance of epistemic risks, and mechanization of representation.

This chapter considers them and shows that they are not valid.

Alternatively, the chapter argues that objectivity consists in plausibility. It consists in plausibility even in the case of mathematics.

12.3 Objectivity as Independence of Human Cognitive Abilities

A view of objectivity is that scientific knowledge is objective because it is independent of human cognitive abilities.

Thus, Frege says that scientific knowledge is objective because it "is independent of our sensation, intuition and imagination, and of all construction of mental pictures out of memories of earlier sensations" (Frege 1960, 36).

But this is not valid. For, science is a human product, and therefore is intrinsically dependent on human cognitive abilities. In order to have knowledge independent of human cognitive abilities, we should have knowledge of the world which is completely independent of our biological makeup. But the only knowledge of the world we can have is through our ways of apprehending it, which essentially depend on our biological makeup. Therefore, it is impossible to have knowledge independent of human cognitive abilities.

As Calvino's Mr. Palomar says, "we can know nothing about what is outside us if we overlook ourselves," because "the universe is the mirror in which we can contemplate only what we have learned to know in ourselves" (Calvino 1985, 119).

12.4 Objectivity as Absolute Conception of the World

Another view of objectivity is that scientific knowledge is objective because it is the absolute conception of the world.

Thus, Williams says that scientific knowledge is objective because it is "the absolute conception" of the world, that is, the "conception of the world that might be arrived at by any investigators" (Williams 2006, 139). The "notion of an absolute conception can serve to make effective a distinction between 'the world as it is independent of our experience' and 'the world as it seems to us'" (ibid.).

But this is not valid. For, any investigators would express their conception of the world through a theory. But, as argued in Chap. 11, a theory can be at most plausible, and a theory that is plausible at one time can become implausible at a later time because the arguments against it can become stronger than those for it, on the basis of the experience available at that time. Then the conception of the world would change.

Therefore, it is impossible to arrive at the absolute conception of the world. It is only possible to arrive at the relative conception of the world that investigators reach at one time, through the theories that are plausible at that time.

12.5 Objectivity as View from Nowhere

Another view of objectivity is that scientific knowledge is objective because it is a view from nowhere, namely, a view of the world that is from no particular point of view and apprehends the world as centerless.

Thus, Nagel says that scientific knowledge is objective because it is a view "from nowhere," namely a view of the world that is from "no particular point of view at all" and apprehends "the world as centerless" (Nagel 1986, 61).

But this is not valid. For, it is impossible to have a view from nowhere, every judgment is made from and depends on a particular perspective.

Nagel claims that I can have a view from nowhere "conceiving the world as a place that includes the person I am within it, as just another of its contents—conceiving myself from outside, in other words" (ibid., 63). By so doing, "I can step away from the unconsidered perspective of the particular person I thought I was. Next comes the step of conceiving from outside all the points of view and experiences of that person and others of his species" (ibid.).

But, to have a view from nowhere, it is not enough to conceive from outside all the points of view and experiences of a particular person and others of his species.

It would be necessary to conceive from outside all the points of view and experiences of every possible being, even very different from human beings, but this is impossible.

12.6 Objectivity as Approximation to Truth

Another view of objectivity is that scientific knowledge is objective because it provides an approximation to truth.

Thus, Popper says that scientific knowledge is objective because it provides an approximation to truth, since "the aim of science is truth in the sense of better approximation to truth" (Popper 1979, 57). This "is a clearer and a more realistic aim than the search for truth" (ibid.). For example, "Newton never believed that his theory was really the last word, and Einstein never believed that his theory was more than a good approximation to the true theory" (ibid.)

But this is not valid. For, it presupposes the ability to decide which of the two theories comes closest to the truth, and this requires observation and experiment. But observation and experiment depend on the perspectives and background assumptions the scientist brings to the observation and experiment setting, which depend on the theory. This makes it impossible to evaluate if a theory is a better approximation to truth than another one. What is more important, this is impossible because a criterion of truth is impossible.

12.7 Objectivity as Knowledge Without Knower

Another view of objectivity is that scientific knowledge is objective because it is knowledge without knower, which belongs to the world of objective contents of thought.

Thus, Popper says that scientific knowledge is objective because it is knowledge without knower, since in addition to the World 1 "of physical objects or of physical states" and the World 2 "of states of consciousness, or of mental states," there is the World 3 "of objective contents of thought" (Popper 1979, 106). World 3 contains "problems, conjectures, theories, arguments" (ibid., 107). World 3 is the world of "knowledge in the objective or impersonal sense" (ibid., 286). It is the world of "knowledge without a knower" (ibid., 109). The problems, conjectures, theories, arguments contained in World 3 are "autonomous. They are in no sense made by us; rather, they are discovered by us; and in this sense they exist, undiscovered, before their discovery by us; and in this sense they exist, undiscovered, before their discovery" (ibid., 160–161). Indeed, in World 3 "there are many theories in themselves and arguments in themselves and problem situations in themselves which have never been produced or understood and may be never produced or understood by men" (ibid., 116).

But this is not valid. Popper's description of World 3 is inconsistent. For, while saying that in World 3 there are many theories, arguments and problem situations which have never been produced by men, Popper also says that World 3 "originates as a product of human activity. One can even admit that" World 3 "is man-made" (ibid., 159). World 3 "is a natural product of the human animal, comparable to a spider's web" (ibid., 112).

Moreover, Popper says that Frege's theory belonged to World 3 and it is a "fact, belonging to Worlds 3, that there was an inconsistency in Frege's theory" (Popper 1977, 57). In general, World 3 contains all theories, both consistent and inconsistent theories. Additionally, "every scientific theory" has "a huge set of important consequences. Whether or not these have as yet been discovered" (ibid., 40). They are all contained in World 3. Then, containing inconsistent theories and their consequences, World 3 contains all sorts of statements, true and false, because "if a theory contains a contradiction, then it entails everything" (Popper 2002, 429).

But the fact that World 3 contains all sorts of statements, true and false, flatly contradict Popper's statement that World 3 is the world of knowledge in the objective or impersonal sense. False statements are not knowledge.

12.8 Objectivity as Based on Shared Criteria for Theory Choice

Another view of objectivity is that scientific knowledge is objective because there are shared criteria for theory choice, that is, accuracy, consistency, and simplicity.

Thus Kuhn, inconsistently with his claim that objectivity in science is impossible, says that there are criteria that "provide the shared basis for theory choice," that is, "accuracy, consistency," and "simplicity" (Kuhn 1977, 322). A theory "should be accurate," that is, the "consequences deducible from" it "should be in demonstrated agreement with the results of existing experiments and observations" (ibid., 321). A theory "should be consistent, not only internally or with itself, but also with other currently accepted theories" (ibid., 321–322). A theory "should be simple, bringing order to phenomena that in its absence would be individually isolated and, as a set, confused" (ibid., 322).

But this is not valid. Kuhn himself says that these criteria do not provide the shared basis for theory choice in the case of Copernicus's system and Ptolemy's system.

Indeed, as regards the accuracy criterion, Kuhn says that Copernicus's system "was not more accurate than Ptolemy's," and "if Kepler or someone else had not found other reasons to choose heliocentric astronomy," Copernicus's work "might have been forgotten" (ibid., 323).

As regards the consistency criterion, Kuhn says that "both Ptolemy's and Copernicus's" astronomical theories "were internally consistent," but "heliocentric astronomy, which required the earth's motion, was inconsistent with the existing scientific explanation of these and other terrestrial phenomena. The consistency criterion, by itself, therefore, spoke unequivocally for the geocentric tradition" (ibid.). That is, it favoured Ptolemy.

As regards the simplicity criterion, Kuhn says that, in a sense, it favoured Copernicus, because "Copernicus required only one circle per planet, Ptolemy two. In that sense the Copernican theory was the simpler" (ibid., 324). But, in another sense, if "the two systems were compared in terms of the actual computational labor required to predict the position of a planet at a particular time," then "Copernicus's system offered" astronomers "no laborsaving techniques; in that sense it was not simpler than Ptolemy's" (ibid.).

Therefore, the case of Copernicus's system and Ptolemy's system shows that the criteria of accuracy, consistency, and simplicity do not provide the shared basis for theory choice.

12.9 Objectivity as the Result of Methods Conducive to Truth

Another view of objectivity is that scientific knowledge is objective because it is the result of methods conducive to truth, as the empirical and practical success of science shows.

Thus, Sankey says that scientific knowledge is objective because "the methods of the sciences are highly reliable truth-conducive tools of inquiry" (Sankey 2024, 9). They promote "the epistemic objectivity of science by excluding subjective factors in favour of epistemically relevant factors" (ibid., 5). It is their use that "leads us to the objective truth about the objective world" (ibid., 9). Given the "empirical and practical success of the sciences," it "would be nothing short of a miracle were science to be so successful, and for our methods of scientific inquiry not to lead to the truth" (ibid.).

But this is not valid. The empirical and practical success of science does not show that the methods of science lead to truth. For, as argued in Chap. 11, the view that the aim of science is truth is inadequate. In particular, even if the methods of science led to truth, we could not recognize this because a criterion of truth is impossible. On the contrary, the aim of science is plausibility, and the empirical and practical success of science shows that the methods of science lead to plausibility.

12.10 Objectivity as Invariance Under All Admissible Transformations

Another view of objectivity is that scientific knowledge is objective because it is invariant under all admissible transformations as discovered by science in the course of its development.

Thus, Nozick says that scientific knowledge is objective because "an objective fact is one that is invariant under all admissible transformations" (Nozick 2001, 82). It "is this invariance that constitutes something as an objective truth, and it underlies and explains" other features of objective truths, such as "accessibility from different angles, intersubjectivity, and independence" (ibid., 76). However, if an objective fact is one that is invariant under all admissible transformations, "how is it to be determined which transformations are the admissible ones?" (ibid., 79). This "cannot be known a priori but only through the bootstrap process of scientific investigation," in fact "science discovers the admissible transformations for the domains it investigates and explains. Advances in each domain of knowledge will discover the admissible transformations" central "to that domain" (ibid., 84).

But this is not valid. The claim that scientific knowledge is objective because it is invariant under all admissible transformations gives rise to a circle. For, it makes which scientific knowledge is objective depend on which transformations are admissible, and which transformations are admissible depend on which scientific knowledge is objective.

12.11 Objectivity as Knowledge of Primary Properties

Another view of objectivity is that scientific knowledge is objective because it is knowledge of primary properties, as opposed to knowledge of secondary properties which is subjective.

Thus, Searle says that knowledge of primary properties is objective because its "truth or falsity can be established independently of the feelings, attitudes, prejudices, preferences, and commitments, of investigators" (Searle 2008, 8). Conversely, knowledge of secondary properties, such as "Water tastes better that wine," is "subjective. It is a matter of opinion" (ibid.). Of course, "all representations are from a perspective, from a point of view. So when I say, 'Water consists of H_2O molecules', that is a description at the level of atomic structure," while "at some other level of description, at the level of subatomic physics, for example, we might wish to say that water consists of quarks, muons, and other sundry sub-atomic particles" (ibid.). Nevertheless, the knowledge that 'Water consists of H_2O molecules' is objective. Thus, "the fact that all knowledge claims are perspectival does not preclude epistemic objectivity" (ibid.).

But this is not valid. When we say that objectivity is knowledge of primary properties, we refer to a relative notion of objectivity, namely the notion of objectivity relative to a given perspective, not to an absolute notion. For, not only 'Water consists of H_2O molecules' is relative to the atomic perspective while 'Water consists of quarks, muons, and other sundry sub-atomic particles' is relative to the sub-atomic perspective, but both are relative to human beings. To have an objective science of the world it would be necessary to have knowledge independent of the particular perspective of any possible creature, even very different from human beings, but this is impossible.

12.12 Objectivity as Value-Freedom

Another view of objectivity is that scientific knowledge is objective because it is value-free, that is, it can be gathered, assessed and accepted without depending on value judgments. Specifically, scientific knowledge has the characters of impartiality, neutrality, and autonomy.

Thus, Lacey says that scientific knowledge is objective because "science is value free," since "impartiality, neutrality and autonomy are constitutive values of scientific practices and institutions" (Lacey 2002, 522). Impartiality means that a scientific theory is accepted "if and only if it manifests the cognitive values" of science, such as truth, accuracy or explanatory power, "according to the highest standards, in the light of available empirical data" (ibid.). Neutrality means that scientific theories make no value statements about the world, so they "do not entail that any particular values should be adopted" (ibid., 523). Autonomy means that "research practices are conducted" without "interference from 'outside' interests, powers and values," but only with the aim "to extend research into" new "domains" (ibid., 524).

But this is not valid. As argued in Chap. 5, values guide scientists in selecting the scientific problems to work on and the hypotheses to solve them, and in evaluating the plausibility of hypotheses, because determining whether the arguments for a hypothesis are stronger than the arguments against it, requires value criteria. In particular, every scientific revolution leads to a change in the values that guide scientists.

12.13 Objectivity as Avoidance of Epistemic Risks

Another view of objectivity is that scientific knowledge is objective because it avoids important epistemic risks arising from our imperfections as epistemic agents.

Thus, Koskinen says that scientific knowledge is objective because it is knowledge with respect to which "important epistemic risks arising from our imperfections as epistemic agents have been effectively averted" (Koskinen 2020, 1196). In fact, "we are imperfect as epistemic agents in many ways: we are prone to wishful thinking, we can be fooled by illusions, as communities we may share ideas and values without even realizing we do so, and so on. Our imperfections lead to many different kinds of epistemic risks" (ibid., 1197).

But this is not valid. Scientific knowledge is objective not merely in the negative sense that it avoids important epistemic risks arising from our imperfections as epistemic agents. It is objective in the positive sense that it is plausible, that is, the arguments for it are stronger than the arguments against it, on the basis of experience.

12.14 Objectivity as Mechanization of Representation

Another view of objectivity is that scientific knowledge is objective because it is based on the ideal of mechanization of representation.

Thus, Daston and Galison say that scientific knowledge is objective because "the ideal of scientific representation" is "mechanical objectivity," that is, the "search to replace individual volition and discretion in depiction by the invariable routines of mechanical reproduction" (Daston and Galison 1992, 98). The "machine was fundamental to the very idea of mechanical objectivity" (ibid., 119). For, "the machine, now in the form of techniques of mechanical reproduction, held out the promise of images uncontaminated by interpretation" (ibid., 120).

But this is not valid. For, the devices by means of which scientific representations are mechanized are built and work on the basis of certain theories of scientific representation, and the plausibility of these theories has not been established mechanically, but by comparing the arguments for them and the arguments against them, on the basis of experience.

12.15 Objectivity as Plausibility

Since the above views of objectivity are not valid, an alternative view is necessary.

An alternative view is that scientific knowledge is objective because it is plausible. The aim of science is not truth but plausibility, so scientific theories are not true but plausible. Then theories are objective because they are plausible.

This view of objectivity is based on the adversarial system (see Chap. 11). That is, the objectivity of a scientific theory consists in the fact that some scientists present arguments for a theory, other scientists present arguments against it, and the wider community of scientists judges whether the arguments for the theory are stronger than those against it, on the basis of experience.

12.16 Objectivity in Mathematics

The view of objectivity as plausibility also applies to mathematical knowledge. Even mathematical knowledge is objective because it is plausible.

Against this view of objectivity, Grosholz argues: "I do not think that" the view "that we should focus on plausibility instead of truth, and construe justification as always somewhat empirical and revisable, captures the peculiar nature of mathematical reasoning" (Grosholz 2016, 49). For, "the things of mathematics are determinate and unchanging as the things of nature are not: why would we ever revise our belief that the sum of the squares of the two oppositive sides of a Euclidean right triangle is equal to the square of its hypotenuse, or that 2+3=5?" (ibid.).

This argument, however, is not valid. If we do not want to revise our belief that a Euclidean right triangle has a certain property or that 2+3=5, it is not because the things of mathematics are determinate and unchanging as the things of nature are not. It is rather because we have evolved in a world that, at our scale, is mostly made up of continuous objects such that a Euclidean right triangle has that property, and discrete objects such that 2+3=5. Therefore, our belief is a result of natural selection.

As Dehaene says, "at our scale, the world is mostly made up of separable objects that combine into sets according to the familiar equation 1+1=2. This is why" natural selection "has anchored this rule in our genes" (Dehaene 1997, 249). But "perhaps our arithmetic would have been radically different if, like cherubs, we had evolved in the heavens where one cloud plus another cloud was still one cloud" (ibid.).

The propositions of mathematics as discipline are not determinate and unchanging and their justification is not unrevisable. For, they are based on hypotheses and, by Gödel's second incompleteness theorem, hypotheses cannot be said to be true but only plausible. And plausible hypotheses are not determinate and unchanging, and their justification is not unrevisable.

References

Calvino, Italo. 1985. *Mr. Palomar*. New York: Harvester Books.

Daston, Lorraine, and Peter Galison. 1992. The image of objectivity. *Representations* 40:81–128.

Dehaene, Stanislas. 1997. *The number sense: How the mind creates mathematics*. Oxford: Oxford University Press.

Frege, Gottlob. 1960. *The foundations of arithmetic: A logico-mathematical enquiry into the concept of number*. New York: Harper.

Grosholz, Emily. 2016. *Starry reckoning: Reference and analysis in mathematics and cosmology*. Cham: Springer.

Koskinen, Inkeri. 2020. Defending a risk account of scientific objectivity. *The British Journal for the Philosophy of Science* 71:1187–1207.

Kuhn, Thomas. 1977. *The essential tension: Selected studies in scientific tradition and change*. Chicago: The University of Chicago Press.

Kuhn, Thomas. 1996. *The structure of scientific revolutions*. Chicago: The University of Chicago Press.

Lacey, Hugh. 2002. The ways in which the sciences are and are not value free. In *In the scope of logic, methodology and philosophy of Science*, ed. Peter Gärdenfors, Jan Woleński, and Katarzyna Kijania-Placek, 519–532. Dordrecht: Springer.

Nagel, Thomas. 1986. *The view from nowhere*. Oxford: Oxford University Press.

Nozick, Robert. 2001. *Invariances: The structure of the objective world*. Cambridge: Harvard University Press.

Popper, Karl. 1977. Materialism criticized. In Karl Popper and John Eccles, *The Self and its brain*, 51–68. Berlin: Springer.

Popper, Karl. 1979. *Objective knowledge: An evolutionary approach*. Oxford: Oxford University Press.

Popper, Karl. 2002. *Unended quest: An intellectual autobiography*. London: Routledge.

Sankey, Howard. 2024. The objectivity of science. *Journal of Philosophical Investigations* 17 (45): 1–10.

Searle, John. 2008. *Philosophy in a new century: Selected essays*. Cambridge: Cambridge University Press.

Williams, Bernard. 2006. *Ethics and the limits of philosophy*. Cambridge: Harvard University Press.

Chapter 13
Knowledge and Mind

Abstract Developing the view that philosophy is acquisition of knowledge also requires answering the question: What is the relation of knowledge to mind? A widely held view is that the mind is different from the body and knowledge belongs to the mind alone, but this chapter argues that this view is not valid. Alternatively, the chapter argues that the mind consists of certain faculties of the body and knowledge is based on these faculties, which are implemented by processes internal and external to the body.

Keywords Disembodied knowledge · Embodied knowledge · Processes internal to the body · Processes external to the body

13.1 Two Views of the Relation of Knowledge to Mind

Developing the view that philosophy is acquisition of knowledge also requires answering the question: What is the relation of knowledge to mind?

This question is complementary to the question 'What is the relation of knowledge to reality?' considered in Chap. 11, because the latter concerns the relation of knowledge to the object of knowledge, and the question 'What is the relation of knowledge to mind?' concerns the relation of knowledge to the subject of knowledge.

With respect to the relation of knowledge to mind, two views can be distinguished, the view of disembodied knowledge and the view of embodied knowledge.

13.2 The View of Disembodied Knowledge

According to the view of disembodied knowledge, the mind is distinct from the body, knowledge belongs to the mind alone, and not to the composite of mind and body, it is based only on representations of the mind, traditionally called ideas, and

C. Cellucci, *A Philosophy for the Future*, Synthese Library 508,
https://doi.org/10.1007/978-3-031-98443-3_13

its object is not the world but these representations. The body plays no essential role in it.

The view of disembodied knowledge is primarily due to Descartes.

Indeed, Descartes says that "the mind is really distinct from the body" (Descartes 1996, VII, 226). Knowledge "belongs to the mind alone, and not to the composite of mind and body" (ibid., VII, 83). Knowledge does not arise because external objects "conveyed the ideas into our mind through sense organs," but because they "conveyed something which gave the mind occasion to form the ideas" (ibid., VIII-2, 359). Knowledge is based only on ideas, because "we cannot have any knowledge of things except by the ideas we conceive of them" (ibid., III, 476). The object of knowledge are ideas, because ideas are "all that is immediately perceived by the mind" (ibid., VII, 181). In knowledge, the mind "turns in some way toward itself, and regards one of the ideas that are within itself" (ibid., VII, 73).

The view of disembodied knowledge has contemporary exponents.

Thus, Fodor says that knowledge belongs to the mind alone because "there can be no serious hope that" a being "whose psychology is effectively identical to that of some organism can be described by physical natural kind predicates" (Fodor 1974, 105). Our knowledge of things is based only on "mental representations (often called 'ideas' in the older literature)" (Fodor 1981, 26). For "we have access to the world only via the ways in which we represent it" (ibid., 241). The object of knowledge are mental representations. The view that our knowledge is based on mental representations is "a Good Old Theory—one to which both Locke and Descartes (among many others) would certainly have subscribed" (ibid., 26).

13.3 Inadequacy of the View of Disembodied Knowledge

The view of disembodied knowledge is inadequate. Its main shortcomings are the following.

(1) The view of disembodied knowledge cannot account for the fact that many features of our knowledge depend on the kind of body we have. Therefore, our knowledge essentially depends on the fact that we have a body with a certain constitution. In contrast, the view of disembodied knowledge involves an absolute separation of the mind from the body. As Auden says, with the view of disembodied knowledge "Self was the one city, | The cell where each must find his comfort and his pain, | The body nothing but a useful favourite machine | To go upon errands of love and to run the house, | While the mind in its study spoke with its private God" (Auden 2022, Volume 1, 310).

(2) The view of disembodied knowledge cannot account for the fact that all organisms have knowledge, even the most elementary ones such as the prokaryotes. Thus, Fodor says that the prokaryotes do not have knowledge because they are like thermostats, they only "respond selectively to such properties as, temperature and light intensity, and these properties are surely prime candidates for nomicness" (Fodor 1986, 12). That is, they only respond to the laws of physics. Knowledge is

based on mental representations, and "it would be preposterous to attribute mental representations" to the prokaryotes, for "where would they keep them?" (ibid., 3).

(3) The view of disembodied knowledge leads to an infinite regress because, according to it, the object of knowledge is not the world but representations of the mind. This implies that inside us there should be an homunculus that perceives these representations. But then, for the same reason, inside this homunculus there should be another homunculus that perceives the representation of this homunculus, and so on ad infinitum.

13.4 The View of Embodied Knowledge

An alternative to the view of disembodied knowledge is the view of embodied knowledge. According to it, the mind consists of certain faculties of the body that do not reduce to the brain, and knowledge is based on these faculties, therefore knowledge is a natural process.

The faculties of the body on which knowledge is based are implemented not only by processes internal to the body, but also by processes external to it, from material to symbolic ones, such as pictures, models, maps, and symbol systems. The processes external to the body cooperate with those internal to the body, forming an integrated cognitive system with them. Just as the blind man's cane is an integral part of how the blind man perceives the world, the external processes are an integral part of how we know the world.

That the processes external to the body cooperate with those internal to the body forming an integrated cognitive system with them, implies that there is no sharp separation between the subject and the external world.

As Calvino's Mr. Palomar says, "you think of the ego as one who is peering out of your own eyes as if leaning on a window sill, looking at the world stretching out before him in all its immensity" (Calvino 1985, 114). Then "the world is out there; and in here, what do we have? The world still—what else could there be?" (ibid.). The 'I', the ego, is itself "a piece of the world that is looking at another piece of world," or perhaps is "the window through which the world looks at the world" (ibid.).

The view of embodied knowledge dates back at least to Dewey.

In fact, Dewey says that "knowledge-getting" is "not an event going on exclusively within the cortex," but "hands and feet, apparatus and appliances of all kinds are as much a part of it as changes in the brain" (Dewey 2004, 9). Then, "since these physical operations (including the cerebral events) and equipments are a part" of knowledge-getting, knowledge-getting "is mental, not because of a peculiar stuff which enters into it or of the peculiar non-natural activities which constitute it, but because of what physical acts and appliances do: the distinctive purpose for which they are employed and the distinctive results which they accomplish" (ibid.).

13.5 Some Implications of the View of Embodied Knowledge

That, according to the view of embodied knowledge, knowledge is based on certain faculties of the body, has implications that highlight the differences between the view of embodied knowledge and that of disembodied knowledge.

(a) The faculties of the body on which knowledge is based are present not only in human beings but, although to a different extent, in all organisms, including the most elementary ones such as the prokaryotes, therefore all organisms have knowledge. This conflicts with the view of disembodied knowledge, according to which knowledge belongs to human beings only. Thus, Descartes says: "We refuse to allow that" animals "have any knowledge of things" (Descartes 1996, X, 415). For, "animals do not see as we do when we are aware that we see; but only as we do when our mind is elsewhere" (ibid., I, 413).

(b) The faculties of the body on which knowledge is based include unconscious processes. Thus, as will be argued in Chap. 19, perception is unconscious inference. This conflicts with the view of disembodied knowledge, according to which knowledge is always conscious. Thus, Descartes says: "There can be nothing in the mind, in so far as it is a thinking thing, of which it is not conscious" (ibid., VII, 246). In fact, "when we set about making use of some faculty, if that faculty does indeed exist in the mind, we are immediately conscious of it" (ibid., VII, 246–247).

(c) The faculties of the body on which knowledge is based are vehicles of emotions. This conflicts with the view of disembodied knowledge, according to which, since knowledge belongs to the mind alone and the body plays no essential role in it, knowledge does not depend on emotions, because emotions belong to the body. Thus, Descartes says that emotions "are all ordained by nature to relate to the body," none of them relates to the mind, and they "are not given to the mind except in so far as it is joined with the body" (ibid., XI, 430).

13.6 Adequacy of the View of Embodied Knowledge

Contrary to the view of disembodied knowledge, the view of embodied knowledge is adequate.

(1) The view of embodied knowledge can account for the fact that many features of our knowledge depend on the kind of body we have. For, according to it, the mind consists of certain faculties of the body and knowledge is based on these faculties, therefore many features of knowledge necessarily depend on the kind of body we have.

(2) The view of embodied knowledge can account for the fact that all organisms have knowledge, even the most elementary ones such as the prokaryotes. For, without knowledge, the prokaryotes would have been "unable to survive in the chaotic, dangerous environment in which life first appeared" (Reber et al. 2023, 9). The prokaryotes "routinely make decisions based on the positive and negative aspects of

objects and events," because "they have well-developed perceptual processes that are the source for making decisions and forming preferences based on what was sensed" (ibid., 53). If they "detect that the temperature has risen to potentially dangerous levels, they actively swim away into cooler waters," and "if they make contact with an acidic molecule, they experience it as a negative object and withdraw" (ibid.).

(3) The view of embodied knowledge does not lead to an infinite regress because, according to it, the object of knowledge is the world.

13.7 Processes Internal and External to the Body

As already said, according to the view of embodied knowledge, the faculties of the body on which knowledge is based are implemented not only through processes internal to the body, but also through processes external to the body. Processes external to the body cooperate with those internal to the body, forming an integrated cognitive system with them.

There is abundant evidence for this, starting with the archaeological evidence which shows that, between 130,000 to 60,000 years ago, "the Neanderthals and archaic Homo sapiens" had a mental activity that, compared to ours, "appears to have been substantially more reliant on their brains alone—brains that are unlikely to have been significantly different to those that we possess today" (Mithen 2000, 208). Therefore, "the range and scope of their mental activity was markedly constrained when compared to the way that we think today" (ibid.). But 60,000 years ago, with "the discovery of using material culture to extend the mind, humans escaped those mental constraints imposed by the brain. By extending the mind into material culture a new evolutionary trajectory was initiated" (ibid., 217). This gave rise to the cultural role of knowledge some 10,000 years ago. Thus, "in little more than 50,000 years, by having 'given up' their near-total reliance on the brain alone, humans have mentally evolved from having thoughts about no more than stone tools and acquiring food, to having thoughts about subjects no less than the origins of the universe and the nature of the human mind itself" (ibid.).

13.8 The Sapient Paradox

The rise of the cultural role of knowledge some 10,000 years ago was the beginning of our 'sapient' status. That our sapient status started emerging only some 10,000 years ago raises the "sapient paradox: that the biological basis of our species has been established" perhaps "for as much as 200,000 years," while "the novel behavioural aspects of our 'sapient' status have taken so long to emerge" (Renfrew 2008, 2041).

Towards solving the sapient paradox, it can be observed that our sapient status started emerging roughly at the same time as the beginning of the sedentary revolution, the transition from nomadic to permanent year-round settlement. This "in some areas soon led to the rise of urban life and of state societies and indeed to the rise of literacy" (ibid., 2043).

The sedentary revolution gave rise to new material conditions, cultural practices and interpersonal relationships, setting human beings new challenging tasks. Human beings could cope with them only by acquiring new capacities, and they acquired them through processes external to the body. This solves the sapient paradox, because these processes allowed human beings to overcome the limitations of internal processes.

13.9 Clearing Up a Misunderstanding

Against the claim that processes external to the body cooperate with those internal to the body forming an integrated cognitive system with them, it could be argued that between internal processes and external processes there is an essential difference. Only internal processes have a cognitive content, external processes do not have any cognitive content in themselves, they can derive a cognitive content only from internal processes.

Thus, Fodor argues that, if something "has content, then either it is mental (part of a mind) or the content is 'derived' from something that is mental. 'Underived' content" is "the mark of the mental; underived content is what minds and only minds have" (Fodor 2009). The content of an external process can only be derived from something that is mental, and "a thing's having derived intensionality thus depends on someone's thinking about it" (ibid.). The "world can't play the kind of role" that is played by "representations," because "the world doesn't represent anything; least of all itself," the world "isn't about anything; it just is" (ibid.).

This argument, however, is not valid, because it is based on a misunderstanding. Saying that processes external to the body cooperate with those internal to the body forming an integrated cognitive system with them, does not mean that external processes are cognitive by themselves. It only means that, when internal processes are integrated with certain external processes, the result of the integration satisfies a properly motivated criterion for being cognitive. Thus, what is cognitive is not the external processes themselves, but the integration of internal processes with external processes.

External processes have quite different properties from internal ones. This is why the integration of internal processes with external processes allows the cognitive subject to perform tasks that he could not perform exclusively through internal processes. If external processes had the same properties as internal processes, they could not enhance the latter in any essential way.

13.10 Strengthening the Mind with External Processes

The integration of internal processes with external processes allows the cognitive subject to perform tasks that he could not perform only through internal processes.

Preverbal babies are able to do some simple arithmetic because there is "a number module" which "is built into all our brains when we are born" (Butterworth 1999, 9). But the power of the number module is limited, so to extend it we use a number of external tools resulting from cultural evolution. This means that our brain "contains these two elements: a number module and our ability to use the mathematical tools supplied by our culture" (ibid., 7).

The Neanderthals and the archaic Homo sapiens must have had a number module, but they could not use external tools supplied by culture because these tools had not yet been developed. Therefore, their mental power was confined to what their brain was able to do on its own, without the aid of external tools, which was very little.

Even Frege recognizes that external processes greatly enhance the power of the mind. Indeed, he says that "to think we must use sense symbols" (Frege 1964, 155). Without them, we would "hardly raise ourselves to the level of conceptual thought" (ibid., 156). The "concept itself is first gained by our symbolising it, for, since the concept is of itself imperceptible to the senses, it requires a perceptible representative in order to appear to us. Thus it is that the sensuous opens up for us the world of the non-sensuous" (ibid.).

13.11 Integration and Brain Plasticity

The integration of internal processes with external processes requires new capacities of the human brain.

These new capacities depend on brain plasticity, "the ability of the nervous system to change its activity in response to intrinsic or extrinsic stimuli by reorganizing its structure, functions, or connections" (Mateos-Aparicio and Rodriguez-Moreno 2019).

The architecture of the human brain is laid down under tight genetic constraints. Nevertheless, it has a certain variability range, so the nervous system can rewire itself to a certain extent as the result of experiences that occur during an individual lifetime.

Specifically, when the human brain is faced with tasks for which it was not prepared by natural selection, it partially reconverts some cerebral circuits initially selected to support evolutionary relevant functions, but sufficiently plastic for acquiring new functions. Thus, cognitive abilities evolved for evolutionary purposes are co-opted for new tasks.

Saying that the human brain partially reconverts a limited number of cerebral circuits, means that the human brain is not a blank slate but is highly structured, with specialized cerebral circuits that carry out specific mental functions. However, the structure of the human brain is sufficiently modifiable through experience, especially in young children, so as to make it capable of functions different from those for which it was prepared by natural selection.

On the other hand, the modifications of the human brain can never totally overturn its structure. The reconversion of a limited number of cerebral circuits is possible only insofar as it remains within a certain range. Thus, the new capacities reflect the intrinsic constraints of the underlying architecture of the human brain. Therefore, the cultural inventions which invade evolutionary older braid circuits inherit many of their structural constraints.

The modifications in question are not genetically but only culturally heritable, so they must be reiterated by every child after birth. Every child can participate into the phylogenetic acquisition of culture only through a personal, ontogenetic acquisition of skills and experience.

13.12 Processes External to the Body

Several processes external to the body can cooperate with those internal to the body, forming an integrated cognitive system with them. Some significant examples are the following.

(1) *Writing*. To acquire much knowledge, writing is necessary. This has sometimes been denied. Thus, Plato challenged the importance of writing for knowledge by referring to the myth of Theuth, the god inventor of writing, who went to Thamous, the king of Egypt, and offered him his invention, saying: "O king, this invention will make the Egyptians wiser and more capable of memory, because it has been invented as an elixir of memory and wisdom" (Plato, *Phaedrus*, 274 e 4–7). But Thamous declined the offer, replying: "What you, as the father of writing, now say out of affection for it, is the opposite of what writing is capable of making. In fact, it will produce forgetfulness in the minds of those who will have learned it, because they will not practice their memory. For, putting their trust in writing, they will depend for their memory on external characters, instead of using their own memory within them" (ibid., 274 e 9–275 a 5). Therefore, "you offer your pupils the appearance of wisdom, not true wisdom" (ibid., 275 a 6–7). Only "the discourse that is written down, with knowledge, in the mind of the learner" is "the living, breathing discourse of the man who knows, of which the written discourse can be properly said to be only a semblance" (ibid., 276 a 5–6, 8–9).

Contrary to Plato's challenge to the importance of writing for knowledge, without the help of writing the processes internal to the body would not have gone very far. Only the invention of writing made possible abstract thinking and permitted the birth of philosophy and science, because writing is essential to solve abstract problems. Inventing writing must not have been easy if the earliest form of writing only appeared about 5500 years ago.

(2) *Figures in elementary geometry*. To acquire knowledge in elementary geometry, figures are generally necessary. For example, suppose we want to demonstrate that, in any triangle, the sum of the internal angles is equal to two right angles. To this purpose, we draw a triangle *ABC* and, through the vertex *A*, we draw a line *DE* parallel to *BC*.

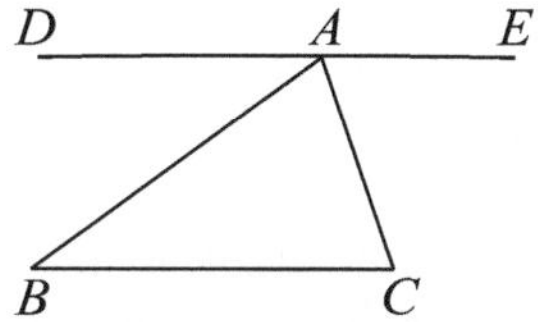

Since *DE*, *BC* are parallel, the alternate angles *ABC*, *DAB* are equal, and so are the alternate angles *ACB*, *EAC*. Hence, the sum of the angles *ABC*, *ACB* is equal to the sum of the angles *DAB*, *EAC*. Add the angle *BAC* to each. Then the sum of the angles *ABC*, *ACB*, *BAC* is equal to the sum of the angles *DAB*, *EAC*, *BAC*, and hence to the sum of the angles *DAB*, *BAC*, *EAC*, so it is equal to two right angles. Therefore, the sum of the angles *ABC*, *ACB*, *BAC* is equal to two right angles.

This demonstration uses a representation external to the body, consisting of a figure that is literally drawn by hand, and must be drawn in the right way. The demonstration involves eye vision, hand movement, and motor coordination between the eyes and the hand. The person making the demonstration must draw a figure external to the body and use properties of the figure to arrive at the result.

(3) *Elementary arithmetic operations*. To make elementary arithmetic operations, in general external representations are necessary. For example, suppose we want to calculate the product 7398 × 3587. Few people could do this without using an external representation. We common mortals break the problem into subproblems, that is, we make the partial products 7 × 8, 7 × 9, etc., based on the multiplication table which we memorized at school, then we add up the resulting numbers, again how we learned at school.

This involves an external representation, consisting of number symbols written on paper that are literally manipulated by hand, and must be manipulated in the right order. Thus, making the product involves eye vision, hand movement, and motor coordination between the eyes and the hand. And, of course, it involves the use of material means to write the symbols. What is performed in the mind are partial products and sums, which we find based on memory.

(4) *Algebraic equations*. Equations play a role in algebra similar to that of figures in elementary geometry. For example, suppose we want to solve the problem: if the area of a square of 100 square meters is equal to the area of two smaller squares the side of one of which is three quarters of the other, what is the length of the sides of the two smaller squares? Few people would be able to do this without using a representation external to the body. We common mortals write the following equations

$$\begin{cases} x^2 + y^2 = 100 \\ 4x - 3y = 0 \end{cases}$$

and through suitable substitutions we find that $x = 6$ and $y = 8$.

This involves a representation external to the mind, consisting of algebraic symbols written on paper, which are literally manipulated by hand. Thus, solving the problem involves eye vision, hand movement, and motor coordination between the eyes and the hand. And, of course, it involves the use of material means to write the algebraic symbols.

Algebraic formulas play a role comparable to that of geometric figures. As Peirce says, "the very idea" of algebra "is that it presents formulae which can be manipulated, and that by observing the effects of such manipulation we find properties not to be otherwise discerned," where formulae "are the icons par excellence of algebra" (Peirce 1931–1958, 3.363).

(5) *Logical deductions*. Let us consider a problem proposed by Carroll. Given the following six propositions as premisses, determine what conclusion can be drawn from them:

(1) I trust every animal that belongs to me;
(2) Dogs gnaw bones;
(3) I admit no animals into my study, unless they will beg when told to do so;
(4) All the animals in the yard are mine;
(5) I admit every animal, that I trust, into my study;
(6) The only animals, that are really willing to beg when told to do so, are dogs.

Few people would be able to do this without using a representation external to the body. We common mortals use the external representation provided by mathematical logic. In the universe of animals, we let $A(x)$ = 'x is admitted into my study', $B(x)$ = 'x is trusted by me', $C(x)$ = 'x is a dog', $D(x)$ = 'x gnaws bones', $E(x)$ = 'x is in the yard', $H(x)$ = 'x belongs to me', $K(x)$ = 'x will beg when told to do so'. Then (1)-(6) are represented by:

(1) $\forall x(H(x) \rightarrow B(x))$;
(2) $\forall x(C(x) \rightarrow D(x))$;
(3) $\forall x(A(x) \rightarrow K(x))$;
(4) $\forall x(E(x) \rightarrow H(x))$;
(5) $\forall x(B(x) \rightarrow A(x))$;
(6) $\forall x(K(x) \rightarrow C(x))$.

From (4), (1), (5), (3), (6), (2), in that order, we obtain $\forall x(E(x) \rightarrow D(x))$. Therefore, the conclusion that can be drawn from the premisses (1)–(6) is: "All the animals in the yard gnaw bones" (Carroll 1977, 186).

13.13 The Mind as an Incomplete Cognitive System

External processes enable the mind to perform tasks that it otherwise could not perform. In this sense, the mind is an incomplete cognitive system.

This conflicts with the view of disembodied knowledge, according to which, as Fodor says, "how the world is makes no difference to one's mental states" (Fodor 1981, 228).

On the contrary, according to the view of embodied knowledge, how the world is makes all the difference to one's mental states, because the integration of internal processes with external ones produces a major change.

The integration of internal processes with external ones is a continuing process, because new external processes are created all the time, and the integration gives rise to ever more powerful cognitive systems.

The present psychology and cognitive science do not account for this fact. Then, there is need for a new science, capable of dealing with the integration of internal processes with external ones.

13.14 The Subject of Knowledge

The fact that external processes enable the mind to perform tasks that it could not otherwise perform, implies that the cognitive system is not just the mind, but the mind plus some external processes, which are then part of the cognitive system. Just as the blind man's cane is a constituent of the way the blind man perceives the world, these external processes are constituents of the way human beings know the world.

Therefore, the subject of knowledge is not just the mind, but the cognitive system consisting of the mind plus some external processes. Which specific external processes are used may be inessential. Thus, in the case of writing, it may be inessential whether one uses pen and paper or a computer. What is essential is that the external processes perform the relevant function.

Since the subject of knowledge is the cognitive system consisting of the mind plus some external processes, knowledge depends on these two components. In order to have knowledge it is essential a coordination between them.

13.15 Knowledge and Other Minds

The external processes that can cooperate with internal processes forming an integrated cognitive system with them, include other minds.

Our hunter-gatherer ancestors relied on foraging and hunting for subsistence, and moved from place to place in response to changes in the season and climate. Climate amelioration triggered the domestication of animals and plants. Dogs were the first animals to be domesticated because they could be used for hunting wild animals, followed by goats and sheep which could provide milk, wool, and meat. Interaction with domesticated animals eventually led to selective hunting and then to herding. On the other hand, domestication of plants gave rise to agriculture, which led people to become sedentary, establishing the first villages that over time grew in their size becoming small towns.

The establishment of villages and small towns gave rise to completely new situations and changes in the environment. This led to new material conditions, cultural practices and interpersonal relationships, setting human beings new challenging tasks. Human beings could cope with them only by interacting with other human beings.

Tollfsen says that interactions with other human beings give rise to a "collective mind" (Tollefsen 2006, 140). For, "the mind is where cognition and belief are. If this is so, then, yes, there are collective minds" (ibid., 147).

But this is unjustified. Interactions with other human beings do not give rise to a collective mind encompassing the processes of several human beings. The mind remains an individual one, which however takes advantage of the external processes of other human beings.

References

Auden, Wystan Hugh. 2022. *The complete works of W. H. Auden: Poems*. Princeton: Princeton University Press.

Butterworth, Brian. 1999. *The mathematical brain*. London: Macmillan.

Calvino, Italo. 1985. *Mr. Palomar*. New York: Harvester Books.

Carroll, Lewis. 1977. *Symbolic logic*. New York: Clarkson N. Potter.

Descartes, René. 1996. *Oeuvres*. Paris: Vrin.

Dewey, John. 2004. *Essays in experimental logic*. Mineola: Dover.

Fodor, Jerry. 1974. Special sciences and the disunity of science as a working hypothesis. *Synthese* 28:77–115.

Fodor, Jerry. 1981. *Representation: Philosophical essays on the foundations of cognitive science*. Brighton: Harvester Press.

Fodor, Jerry. 1986. Why paramecia don't have mental representations. *Midwest Studies in Philosophy* 10:3–23.

Fodor, Jerry. 2009. Where is my mind? *London Review of Books* 31:13–15.

Frege, Gottlob. 1964. On the scientific justification of a concept-script. *Mind* 73:155–160.

Mateos-Aparicio, Pedro, and Antonio Rodriguez-Moreno. 2019. The impact of studying brain plasticity. *Frontiers in Cellular Neuroscience* 13:Article 66.

Mithen, Steven. 2000. Mind, brain and material culture: An archaeological perspective. In *Evolution and the human mind: Modularity, language and meta-cognition*, ed. Peter Carruthers and Andrew Chamberlain, 207–217. Cambridge: Cambridge University Press.

Peirce, Charles. 1931–1958. *Collected papers*. Cambridge: Harvard University Press.

Reber, Arthur, František Baluška, and William Miller. 2023. *The sentient cell: The cellular foundations of consciousness*. Oxford: Oxford University Press.

Renfrew, Andrew. 2008. Neuroscience, evolution and the sapient paradox: The factuality of value and of the sacred. *Philosophical Transactions of the Royal Society B* 363:2041–2047.

Tollefsen, Deborah. 2006. From extended mind to collective mind. *Cognitive Systems Research* 7:140–150.

Chapter 14
Knowledge and Rationality

Abstract Developing the view that philosophy is acquisition of knowledge also requires answering the question: Is acquisition of knowledge, in particular discovery, a rational process? A widely held view is that discovery is not a rational process, but this chapter argues that this view is not valid. Alternatively, the chapter argues that discovery is a rational process.

Keywords Rationality of discovery · Subjectivity · Intuition · Genius · Algorithmicity · Illogicalness · Chance · Serendipity · Method of discovery

14.1 The Character of Rationality

Developing the view that philosophy is acquisition of knowledge also requires answering the question: Is acquisition of knowledge, in particular discovery, a rational process?

The term 'rational' comes from the Latin 'ratio', one of whose meanings is 'relation'. In fact, to be rational, or rationality, is a matter of the relation of means to given ends. It can be defined as the capacity to choose appropriate means to given ends. It has nothing to do with the choice of ends.

That this is the proper concept of rationality has been repeatedly asserted since antiquity.

Thus, Aristotle says that "we deliberate not about ends, but about means to ends" (Aristotle, *Ethica Nicomachea*, G 3, 1112 b 11–12). Indeed, "a doctor does not deliberate about whether he shall heal, nor an orator whether he shall persuade, nor a statesman whether he shall produce good order, nor does anyone else deliberate about his end" (ibid., Γ 3, 1112 b 12–15).

Dewey says that "rationality is an affair of the relation of means and consequences," that is, of the relation "between means (methods) employed and conclusions attained as their consequence" (Dewey 1938, 9).

C. Cellucci, *A Philosophy for the Future*, Synthese Library 508,
https://doi.org/10.1007/978-3-031-98443-3_14

Russell says that rationality "signifies the choice of the right means to an end that you wish to achieve. It has nothing whatever to do with the choice of ends" (Russell 1954, viii).

Against this concept of rationality, Rescher argues that "the rationality of ends is an indispensable component of rationality at large. The rationality of our actions hinges critically both on the appropriateness of our ends and on the suitability of the means by which we pursue their cultivation. Both of these components" are "alike essential to full-fledged rationality" (Rescher 1988, 100). In particular, "the pursuit of what we want," that is, of our ends, "is rational only in so far as we have sound reasons for deeming this to be want-deserving" (ibid., 99).

This argument, however, is not valid. For, one cannot assert of any ends that they are appropriate. In fact, suppose that one asserts of certain ends that they are appropriate. Then the question arises: Why are these ends appropriate? If one answers that they are appropriate because there are good reasons for considering them appropriate, then the question arises: Why are such reasons good reasons? If one answers that they are good reasons because there are good reasons for considering them good reasons, then the question arises: Why are these reasons good reasons? And so on, ad infinitum.

Therefore, rationality cannot be concerned with the end, but only with the suitability of means to a given end.

Now, the end of discovery is to solve a problem. Then, discovery is a rational process if it provides appropriate means to solve a problem, that is, if it provides a method of discovery.

Several arguments have been advanced against the claim that discovery is a rational process. This chapter considers them and argues that they are not valid.

14.2 The Argument of Illumination

An argument against the claim that discovery is a rational process is that discovery is a sudden illumination that is the result of unconscious work.

Thus, Poincaré says that discovery is a "sudden illumination, a manifest sign of long, unconscious prior work" (Poincaré 2015, 389). In particular, discovery involves "making new combinations" of ideas "already known" and choosing "those that are useful and which are only a small minority" (ibid., 386). To this aim, the self must "form all the possible combinations" of ideas already known because, "if it produces only a small part of these combinations, and if it makes them at random, there would be small chance that" the useful ones "would be found among them" (ibid., 392–393). Now, forming all the possible combinations of ideas already known cannot be the work of "the conscious self," since the latter "is narrowly limited" (ibid., 392). Instead, it must be the work of "the unconscious" or "subliminal self" (ibid., 390). For, only the unconscious self is "able in a short time to make more different combinations than the whole life of a conscious being could encompass" (ibid., 392). Once the unconscious self has made all the possible combinations, it chooses the useful ones on the basis of the "feeling of mathematical beauty" (ibid., 391). For, "the useful combinations are precisely the most beautiful" (ibid., 392). They are those "which are capable of developing in us a sort of aesthetic emotion" (ibid., 391). Thus, with respect to combinations of ideas, the feeling of mathematical beauty plays "the role of the delicate sieve" (ibid., 392).

But this argument is not valid. Poincaré assumes that the useful combinations will be included among all the possible combinations of ideas already known, and hence among all the possible notions that can be defined in terms of ideas already known. But this conflicts with a result similar to Tarski's undefinability theorem, according to which, "for every deductive science in which arithmetic is contained it is possible to specify arithmetical notions which, so to speak, belong intuitively to this science, but which cannot be defined on the basis of this science" (Tarski 1983, 276). Therefore, there is no guarantee that the useful combinations will be included among all the possible combinations of ideas already known. In fact, rather than making combinations of ideas already known, in general discovery involves introducing new ideas.

Moreover, Poincaré assumes that, among all the possible combinations of ideas already known, the unconscious self chooses the useful ones on the basis of the feeling of mathematical beauty. But this conflicts with the fact that the feeling of mathematical beauty has often been an obstacle to choosing useful ideas. Thus, the feeling of mathematical beauty led Galileo to held fast to Copernicus' circular orbits for planets, which contrasted with observations, rejecting Kepler's elliptical orbits, which agreed with them. For, Galileo opposed mannerism in favour of Raphael's classicism, so Kepler's compressed elliptic orbits seemed to him unbearable aesthetic deformations. Therefore, he said that, "for the maintenance of perfect order among the parts of the world, it is necessary to say that movable bodies are movable only circularly" (Galilei 1968, VII, 56).

14.3 The Argument of Psychology

Another argument against the claim that discovery is a rational process is that discovery is a process that can be analyzed only psychologically, not logically.

Thus, Reichenbach says that discovery "can be analyzed only psychologically, not logically," because it "is a process" that "cannot be portrayed by a rational procedure controlled by logical rules" (Reichenbach 1949, 434). Therefore, "I refuse to answer the challenge of setting up rules of a logic of discovery. There are no such rules" (ibid.). Rationalization "belongs in the context of justification; it can be applied only when given inductive conclusions are to be judged appropriate to given facts" (ibid.). In particular, "testing the relations between given observational data and given inductive conclusions is a procedure expressible in terms of theorems of the calculus of probability" (ibid.).

This argument, however, is not valid. Psychological studies of discovery rely on self-reports by scientists and mathematicians. An example of this is Einstein's self-report of the discovery of his theory of gravity: "The breakthrough came suddenly one day. I was sitting on a chair in my patent office in Bern. Suddenly a thought struck me: If a man falls freely, he would not feel his weight. I was taken aback. This simple thought experiment made a deep impression on me. This led me to the theory of gravity" (Einstein 1982, 47).

But psychological studies of discovery do not provide a substantial analysis of the discovery process. So, in the case of Einstein's self-report, psychological studies focus on the fact that the thought of a man falling freely led Einstein to his theory of gravity, but they do not provide an analysis of the process that led Einstein to the thought of a man falling freely.

14.4 The Argument of Dark Ways

Another argument against the claim that discovery is a rational process is that discovery evades distinct analysis because productive thinking follows dark ways.

Thus, Reichenbach says that discovery "evades distinct analysis" because "productive thinking is bound to follow its own dark ways," what "we observe as its constituents are isolated crystallizations of largely subconscious currents hidden below a haze of emotional processes" (Reichenbach 1966, 1).

This argument, however, is not valid. Productive thinking does not follow dark ways. Thus, already Greek mathematicians invented a method of discovery, the analytic method, they used it as a basis for their mathematical practice, and even reported their processes of discovery by publishing their analyses (see Chap. 15).

The analytic method is a rational procedure. For, in the analytic method the hypotheses for solving mathematical problems are obtained by non-deductive rules, non-deductive rules are as objective as deductive rules and are justified much in the same sense. Moreover, the choice among alternative hypotheses obtained by non-deductive rules is made by comparing the arguments for and against them, on the basis of experience.

14.5 The Argument of Intuition

Another argument against the claim that discovery is a rational process is that discovery is achieved through intuition.

Thus, Einstein says: "The supreme task of the physicist is to arrive at those universal elementary laws from which the cosmos can be built up by pure deduction" (Einstein 1954, 226). But "there is no logical path to these laws; only intuition, resting on sympathetic understanding of experience, can reach them" (ibid.).

Rubbia says: "Many people think that a scientific discovery is a very rational process" because "mathematics and equations and so on make it look extremely formal and extremely strict" (Rubbia 1997, 197). But in fact "scientific discovery is an irrational act. It's an intuition which turns out to be reality at the end of it—and I see no difference between a scientist developing a marvellous discovery and an artist making a painting, or music or something" (ibid.).

This argument, however, is not valid. For, it does not consider that an unrestrained intuition would generate so many hypotheses that it would be impossible to test which of them is plausible. There must be some criterion for picking up some hypothesis among them, and for judging it worth testing. This criterion will act as a rule of discovery, so at least in that sense, discovery is a rational process.

Moreover, intuition is totally unreliable as a means of discovery.

For example, intuition tells us that a continuous curve is the place of successive positions of a continuously moving point. But Peano defined something that, according to this concept of continuous curve, is a continuous curve, but "goes through every point of a square" (Peano 1973, 144). This implies that the concept of continuous curve based on intuition is incorrect. For, a curve is a one-dimensional object, a square is a two-dimensional object, and Peano's curve, while being a curve, is also a square.

As another example, intuition tells us that physical objects have properties independent of whether they have been observed or not. So, based on this intuition, Einstein declined to accept quantum theory by saying: "It seems hard to see what cards the Lord God is holding. But I cannot for a moment believe that he plays dice and employs telepathic capabilities (as is expected of him by the current quantum theory)" (Kleinknecht 2017, 157).

The unreliability of intuition sharply contrasts with the success of mathematics science. It would be nothing short of a miracle if mathematics and science were successful and yet depended on intuition.

14.6 The Argument of Genius

Another argument against the claim that discovery is a rational process is that discovery is an exclusive product of genius.

Thus, Kuhn says that an essential requisite "for the beginning of an episode of discovery" is "the individual skill, wit, or genius" (Kuhn 1977, 173). The "most profound sort of genius in physical science is that displayed by men" who "enunciate a whole new theory that brings potential order to a vast number of natural phenomena" (ibid., 188). It is the kind of "genius that leaps ahead of the facts, leaving the rather different talent of the experimentalist and instrumentalist to catch up" (ibid., 194). Scientific theories are "imaginative posits, invented in one piece for application to nature" (ibid., 279). They are born through "flashes of intuition," and although "such intuitions depend upon experience," they "are not logically or piecemeal linked to particular items of that experience" (Kuhn 1996, 123).

Byers says that there are "people who find a way to transcend their limitations" and "dare to do what appears to be impossible" (Byers 2007, 16). The "impossible is rendered possible through acts of genius," in particular, "mathematics boasts genius in abundance" (ibid.). Genius is a matter of insight, and "insight often reveals itself in a flash" (ibid., 329). Therefore, "mathematics transcends logic" (ibid., 26).

This argument is a remnant of the Romantic myth of genius, according to which discovery is the result of extraordinary thought processes that are the privilege of the genius.

Thus, Novalis says that discoveries, "are leaps—(intuitions, resolutions)" that are the privilege "of the genius—of the leaper par excellence" (Novalis 2007, 28). The genius is a creator, and "whoever is able to bring forth numerous" creations "is called a genius" (ibid., 194). In fact, the genius "is en état de createur absolu" (ibid., 140). Genius does not only pertain to mathematical creations, it also "belongs to experimenting, that is to say, that wondrous ability to capture the sense of Nature—and to act in her spirit" (ibid., 219).

This argument, however, is not valid. It is at odds with facts, because a discovery is not made by a native, spontaneous genius, but by a person having a prepared mind. And a person having a prepared mind is not a person who has studied all the existing literature, but a person experienced in using the analytic method.

Moreover, a discovery is usually made by a person totally absorbed in the problem. For example Newton, when "asked how he had been able to discover his mechanical system of the universe," answered: "Nocte dieque incubando [By thinking about it day and night]" (Ortega y Gasset 1957, 42). Newton's cat took advantage of this by "munching on all the food" that Newton, when absorbed in a problem, "left untouched" (Bardi 2006, 28).

14.7 The Argument of Algorithmicity

Another argument against the claim that discovery is a rational process is that discovery would require an algorithmic method, but there exists no algorithmic method of discovery. Instead, creative ingenuity is required.

Thus, Carnap says that "one cannot simply follow a mechanical procedure based on fixed rules to devise a new system of theoretical concepts, and with its help a theory" (Carnap 1966, 33). For, there cannot be "a computer into which we can put all the relevant observational sentences and get, as an output, a neat system of laws that will explain the observed phenomena" (ibid.). Instead, "creative ingenuity is required" (ibid.).

Hempel says that there are "no generally applicable 'rules of induction', by which hypotheses or theories can be mechanically derived or inferred from empirical data. The transition from data to theory requires creative imagination" (Hempel 1966, 15). In fact, "theories are not derived from observed facts, but invented in order to account for them" (ibid.).

This argument, however, is not valid. For, it is based on the assumption that knowledge and methods of acquisition of knowledge must be infallible, and only algorithmic methods can be infallible. But this overlooks that, by Gödel's second incompleteness theorem, not even mathematical knowledge and methods of acquisition of mathematical knowledge can be said to be infallible.

The argument is also based on the assumption that discovery either is obtained by an algorithmic method or is reached through creative ingenuity. But, between the dullness of algorithmic methods and the inscrutability of creative ingenuity, there is an intermediate region inhabited by heuristic methods, which are neither algorithmic nor involve creative ingenuity. The purpose of a method of discovery is not to dispense with the need for intelligence through algorithmic methods, but to expand natural intelligence providing it with heuristic means—means capable of guiding natural intelligence, although not infallibly.

14.8 The Argument of Indeterminateness

Another argument against the claim that discovery is a rational process is that, while deductive rules can be definitively stated, the rules of discovery cannot be definitively stated, so they are an indeterminate set.

Thus, Hilbert says that deductive rules "can be discovered and definitively stated" (Hilbert 1967, 475). 'Deductive rules' comprises not only the rules for first-order logic, but also the rules for second-order logic, because only by comprising them one may hope that "the axiom systems for number theory as well for analysis are complete" (Hilbert 1998, 231). While deductive rules can be definitively stated, the rules for a logic of discovery cannot be stated, because discovery is not a rational process, it is obtained "by separating and collecting ideas in fortunate ways" (Hilbert 2000, 243).

This argument, however, is not valid. For, by the strong incompleteness theorem for second-order logic, there is no consistent set of rules for second-order logic capable of deducing all logical consequences from given premisses, so deductive rules cannot be definitively stated. Therefore, if the rules of discovery are an indeterminate set, even deductive rules are an indeterminate set.

14.9 The Argument of Illogicality

Another argument against the claim that discovery is a rational process is that discovery, in particular mathematical discovery, involves processes which appear haphazard and illogical. Logic can only explain the verificational structure of mathematics.

Thus, Feferman says that "the creative and intuitive aspects of mathematical work evade logical encapsulation" (Feferman 1998, 178). The "actual historical and individual processes of mathematical discovery appear haphazard and illogical" (ibid., 77). Logic can only "explain what constitutes the underlying content of mathematics and what is its organizational and verificational structure" (ibid., 92).

This argument, however, is not valid. For, saying that discovery is based on processes that appear haphazard and illogical conflicts with the fact that, since antiquity, mathematicians have used heuristic procedures in which hypotheses were found by non-deductive rules, so by logical means.

On the other hand, saying that logic can explain what constitutes the underlying content of mathematics and what is its organizational and verificational structure, conflicts with Gödel's incompleteness theorems.

14.10 The Argument of Irrationality

Another argument against the claim that discovery is a rational process is that discovery essentially involves irrational processes, without a frequent dismissal of reason there is no progress.

Thus, Feyerabend argues that many discoveries were possible only because "all the elements that characterize the context of discovery opposed the dictates of reason," and "these irrational elements were permitted to have their way" (Feyerabend 1993, 116). Indeed, "without a frequent dismissal of reason, no progress" (ibid., 158).

In particular, "Copernicanism and other 'rational' views exist today only because reason was overruled at some time in their past" (ibid., 116).

Against Copernicanism, the Aristotelians argued that, if the earth moves and a stone is thrown off a tower, then during the time taken by the stone in its fall the tower would advance in the direction of the earth's motion, so the stone would not hit the ground at the tower's base. But a simple experiment shows that the stone does nothing of the kind. Therefore, the earth does not move and Copernicanism is "inconsistent with facts" (ibid., 39).

But Galileo replaces the Aristotelians' argument with considerations that are "arguments in appearance only" (ibid., 65). He turns "an experience which partly contradicts the idea of the motion of the earth" into "an experience that confirms it" (ibid., 71). However, "the experience on which Galileo wants to base the Copernican view is not but the result of his own fertile imagination," it "has been invented" (ibid., 65).

This argument, however, is not valid. It is unjustified to say that, without a frequent dismissal of reason, there is no progress. For, a theory that is plausible at one time can become implausible at a later time, and conversely, a theory that is implausible at one time can become plausible at a later time. This was the case of the Ptolemaic theory and the Copernican theory. So, science's progress is a rational process.

In particular, Galileo does not overrule reason, he acts in accordance with it. For, he argues that, if the earth moves, the motion of the stone will be "a compound of two motions," one with which the stone "grazes the tower, and another one with which it follows the course of the tower" (Galilei 1968, VII, 165). Then, the stone will not "describe a simple straight and perpendicular line, but rather an inclined, and perhaps not straight, one" (ibid.). Therefore, "from just seeing the falling stone graze the tower, you cannot affirm with certainty that it describes a straight and perpendicular line, unless you first assumed that the earth is standing still" because, "if the earth were moving, the stone's motion would be inclined and not perpendicular" (ibid.). But the assumption that the earth is standing still is the conclusion the Aristotelians want to prove, so the Aristotelians' argument "is assuming as known what it is trying to prove" (ibid., VII, 166). Thus, the Aristotelians' argument begs the question, so it is not valid.

Then, Galileo does not replace the Aristotelians' argument with considerations which are arguments in appearance only, he does not turn an experience which partly contradicts the idea of the motion of the earth into an experience that confirms it. For, his argument is not meant to prove the earth's motion, but only to prove that the Aristotelians' argument against the earth's motion is not valid. As Finocchiaro says, "eventually, Galileo did formulate several positive arguments for the earth's motion" (Finocchiaro 2010, 129). But the argument against the Aristotelians' argument was not meant to be a positive argument, only a negative one.

14.11 The Argument of Good Luck

Another argument against the claim that discovery is a rational process is that discovery is just a matter of good luck.

Thus, Pólya says that "the first rule of discovery is to have brains and good luck. The second rule of discovery is to sit tight and wait till you get a bright idea" (Pólya 1971, 172).

This argument, however, is not valid. For, it is of the same kind as the argument of *Molière's* Bachelierus: "Mihi a docto doctore domandatur causam et rationem quare opium facit dormire. A quoi respondeo, Quia est in eo virtus dormitiva, cujus est natura sensus assoupire [I am asked by a learned doctor for the cause and reason why opium makes one sleep. To which I reply, Because there is in it a dormitive virtue, whose nature is to make the senses drowsy]" (Molière, *The Imaginary Invalid*, Act III, Interlude III).

Indeed, Pólya's argument amounts to saying that the reason why the brain discovers solutions to problems is that there is in it a discoveritive virtue, whose nature is to make the brain inventive.

Moreover, Pólya's argument conflicts with the fact that Pólya admits that there are "procedures (mental operations, moves, steps) which are typically useful in solving problems. Such procedures are practiced by every sane person sufficiently interested in his problems" (Pólya 1971, 172). The best of them is "the method of analysis, or method of 'working backwards'" (ibid., 225).

14.12 The Argument of Chance

Another argument against the claim that discovery is a rational process is that all discoveries are made by chance.

Thus, Crick says that, in science, "chance is the only source of true novelty" (Crick 1982, 58).

This argument, however, is not valid. For, it does not consider that the so-called 'discoveries by chance' are not made by anyone but only by a person with a prepared mind. As Pasteur says, "fortune favours only the prepared mind" (Pasteur 1922–1939, VII, 131).

Moreover, the so-called 'discoveries by chance' are made only by a person who keeps an eye open for something unexpected, and reflects on whether this unexpected occurrence might have some significance. For, as Heraclitus says, "if one does not expect the unexpected, one will not find it out" (Heraclitus 22 B 18, ed. Diels-Kranz).

14.13 The Argument of Serendipity

Another argument against the claim that discovery is a rational process is that all significant discoveries are made by serendipity.

Thus, Karen says that all "significant discoveries are made by serendipity" (Karen 1974, 25).

Serendipity is variously understood. But, as Gillies says, "in serendipity proper, someone fails to discover what he or she was looking for, but discovers something else unexpected instead" (Gillies 2014, 37).

However, that all significant discoveries are made by serendipity, contrasts with the fact that it is very difficult to find genuine examples of discoveries made looking for one thing, but discovering something else unexpected instead.

Fleming's discovery of penicillin is commonly cited as a "serendipitous discovery" (Ban 2006, 339). But it is not really so.

For, Fleming had made research on antiseptics, and at the time he discovered penicillin was making research into the staphylococcus bacterium. He had read a paper which argued that colonies of staphylococci changed colour if they were kept at room temperature for several days. Then Fleming formulated the hypothesis that the colour of colonies of staphylococci was an indicator of their virulence in causing disease and wanted to test this hypothesis. To this purpose, he prepared colonies of staphylococci in glass dishes and left the dishes on his bench, examining them every few days to see if changes in the colour could be observed. After a while, however, one of his glass dishes became contaminated with mold, and Fleming noticed that the colonies of staphylococci, which would normally have covered the glass dish, appeared to have dissolved in the vicinity of the mold.

At the time, contaminations of glass dishes frequently occurred in laboratories, and normally the contaminated dishes were simply discarded. Thus, when Fleming showed the contaminated glass dish to his laboratory colleagues, they were uninterested. They "merely glanced at it," and "promptly forgot all about it" (Hare 1970, 55). But Fleming thought that this contaminated glass dish had some unexpected characters that were worth investigating. Therefore, he abandoned his planned investigation on colour changes and started a new investigation on the contaminated glass dish, aimed to find out why the colonies of staphylococci dissolved in the vicinity of the mold. As a result of this new investigation, Fleming arrived at the conclusion that the mold must be producing an anti-bacterial agent, which he later called penicillin.

Then Fleming's discovery was not a discovery by serendipity. Of course, what Fleming found was not what he had originally looked for, namely change of colour, but another thing, namely that the mold must be producing an anti-bacterial agent. But Fleming had abandoned his planned investigation on colour changes and started a new investigation on the contaminated glass dish, to find out why the colonies of staphylococci dissolved in the vicinity of the mold. So, what he found was what he had looked for after abandoning the original investigation and starting a new investigation.

14.14 The Argument of No Scientific Method

Another argument against the claim that discovery is a rational process is that scientific method does not exist.

Thus, Popper says that "Plato, Aristotle, Bacon and Descartes" believed "that there existed a method of finding scientific truth. In a later and slightly more sceptical period" others "believed that there existed a method, if not of finding a true theory, then at least of ascertaining whether or not some given hypothesis was true" (Popper 2000, 6). But in fact "there is no method of discovering a scientific theory," and there is "no method of ascertaining the truth of a scientific hypothesis" (ibid.). Therefore, "scientific method does not exist," and "I ought to know, having been, for a time at least, the one and only professor of this non-existent subject within the British Commonwealth" (ibid., 5).

This argument, however, is not valid. For, it makes the success of science absolutely improbable. This contrasts with the fact that science has produced theories that have been successful in addressing several aspects of the world. The argument cannot explain this fact and therefore makes it appear miraculous.

14.15 The Search for a Method of Discovery

From what has been said above it is clear that the arguments against the claim that discovery is a rational process are not valid. Therefore, it seems fair to conclude that discovery is a rational process.

Now, as already said, the end of discovery is to solve a problem. Then, discovery is a rational process if it provides appropriate means to solve a problem, so, if it provides a method of discovery.

Since discovery is a rational process, this raises the question: What is the method of discovery?

An answer to this question is given in Chap. 15, in which it is argued that the method of discovery is the analytic method.

References

Ban, Thomas. 2006. The role of serendipity in drug discovery. *Dialogues in Clinical Neuroscience* 8:335–344.

Bardi, Jason. 2006. *The calculus wars: Newton, Leibniz, and the greatest mathematical clash of all time*. New York: Thunder's Mouth Press.

Byers, William. 2007. *How mathematicians think: Using ambiguity, contradiction, and paradox to create mathematics*. Princeton: Princeton University Press.

Carnap, Rudolf. 1966. *Philosophical foundations of physics: An introduction to the philosophy of science*. New York: Basic Books.

Crick, Francis. 1982. *Life itself: Its origin and nature*. New York: Simon and Schuster.

Dewey, John. 1938. *Logic: The theory of inquiry*. New York: Holt, Rinehart & Winston.

Einstein, Albert. 1954. *Ideas and opinions*. New York: Crown Publishers.

Einstein, Albert. 1982. How I created relativity. *Physics Today* 35:45–47.

Federman, Solomon. 1998. *In the light of logic*. Oxford: Oxford University Press.

Feyerabend, Paul. 1993. *Against method*. London: Verso.

Finocchiaro, Maurice. 2010. *Defending Copernicus and Galileo: Critical reasoning in the two affairs*. Dordrecht: Springer.

Galilei, Galileo. 1968. *Opere*. Firenze: Barbera.

Gillies, Donald. 2014. Serendipity and mathematical logic. In *From a heuristic point of view*, ed. Cesare Cozzo and Emiliano Ippoliti, 23–39. Newcastle upon Tyne: Cambridge Scholars Publishing.

Hare, Ronald. 1970. *The birth of penicillin and the disarming of microbes*. London: Allen and Unwin.

Hempel, Carl. 1966. *Philosophy of natural science*. Englewood Cliffs: Prentice-Hall.

Hilbert, David. 1967. The foundations of mathematics. In *From Frege to Gödel: A source book in mathematical logic, 1879–1931*, ed. Jean van Heijenoort, 464–479. Cambridge: Harvard University Press.

Hilbert, David. 1998. Problems of the grounding of mathematics. In *From Brouwer to Hilbert: The debate on the foundations of mathematics in the 1920s*, ed. Paolo Mancosu, 227–233. Oxford: Oxford University Press.

Hilbert, David. 2000. Lecture at the international congress of mathematicians, Paris 1900. In *The Hilbert challenge*, ed. Jeremy Gray, 240–282. Oxford: Oxford University Press.

Karen, Robert. 1974. *An introduction to behavior theory and its applications*. New York: Harper & Row.

Kleinknecht, Konrad. 2017. *Einstein and Heisenberg: The controversy over quantum physics*. Cham: Springer.
Kuhn, Thomas. 1977. *The essential tension: Selected studies in scientific tradition and change*. Chicago: The University of Chicago Press.
Kuhn, Thomas. 1996. *The structure of scientific revolutions*. Chicago: The University of Chicago Press.
Novalis (von Hardenberg, Friedrich). 2007. *Notes for a romantic encyclopedia: Das Allgemeine Brouillon*. Albany: State University of New York Press.
Ortega y Gasset, José. 1957. *On love: Aspects of a single theme*. New York: Greenwich Editions/ Meridian Books.
Pasteur, Luis. 1922–1939. *Oeuvres*. Paris: Masson.
Peano, Giuseppe. 1973. *Selected works*. London: Allen & Unwin.
Poincaré, Henri. 2015. *The foundations of science: Science and hypothesis—The value of science—Science and method*. Cambridge: Cambridge University Press.
Pólya, George. 1971. *How to solve it: A new aspect of mathematical method*. Princeton: Princeton University Press.
Popper, Karl. 2000. *Realism and the aim of science*. London: Routledge.
Reichenbach, Hans. 1949. *The theory of probability: An inquiry into the logical and mathematical foundations of the calculus of probability*. Berkeley: University of California Press.
Reichenbach, Hans. 1966. *Elements of symbolic logic*. New York: The Free Press.
Rescher, Nicholas. 1988. *Rationality. A philosophical inquiry into the nature and the rationale of reason*. Oxford: Oxford University Press.
Rubbia, Carlo. 1997. Asking nature. In *Passionate minds: The inner world of scientists*, ed. Lewis Wolpert and Alison Richards, 195–202. Oxford: Oxford University Press.
Russell, Bertrand. 1954. *Human society in ethics and politics*. London: Allen and Unwin.
Tarski, Alfred. 1983. *Logic, semantics, metamathematics*. Indianapolis: Hackett.

Chapter 15
Knowledge and Method

Abstract Developing the view that philosophy is acquisition of knowledge also requires answering the question: What is the method of discovery? This chapter argues that the method of discovery is the analytic method.

Keywords Analytic method · Analytic method and intuition · Analytic method and non-finality · Analytic method and abduction · Analytic method and reductio ad absurdum

15.1 Algorithmic and Heuristic Methods

Developing the view that philosophy is acquisition of knowledge also requires answering the question: What is the method of discovery?

Methods can be divided into two types, algorithmic and heuristic. An algorithmic method is a procedure for solving problems which involves only mechanical processes. A heuristic method is a procedure for solving problems which also involves non-mechanical processes.

Algorithmic methods cannot solve many problems, for example, as we have seen in Chap. 7, they cannot solve Hilbert's Entscheidungsproblem. But a method need not be algorithmic, it can be heuristic.

Lakatos humorously says that "primitive men prefer decision-procedures. With the help of a decision-procedure one can decide mechanically whether a conjecture is true or false. Primitive men worship algorithms. Their concept of rationality" is "essentially algorithmic" (Lakatos 1978, II, 72). This "primitive heritage is the main obstacle to heuristic style" (ibid., footnote 3). Contrary to primitive men, "the Greeks" found "a heuristic procedure," which is "a standard pattern of the logic of discovery" (ibid.).

C. Cellucci, *A Philosophy for the Future*, Synthese Library 508,
https://doi.org/10.1007/978-3-031-98443-3_15

The heuristic procedure that the Greeks found was the analytic method. Thus, the answer of Greek mathematicians to the question 'What is the method of discovery?' was that the method of discovery is the analytic method. Greek mathematicians not only made their discoveries by the analytic method, but also described how they arrived at them in a number of works, called the 'Analytic Corpus'.

As Pappus says, "the so-called Analytic Corpus" is "a special body of doctrine provided for the use of those who, after going through the ordinary elements, wish to acquire the power of solving problems," it "is the work of three men, Euclid, the author of the *Elements,* Apollonius of Perga, and Aristaeus the Elder, and proceeds by the method of analysis" (Pappus, *Collectio*, VII, 634.1–11, ed. Hultsch).

The answer of Greek mathematicians to the question 'What is the method of discovery?' remains valid today, the method of discovery is the analytic method.

Thus, Mach says that "the greatest and most important scientific discoveries owe their origins to the analytic method" (Mach 1976, 191). This "method became conscious already in the most ancient and simple exact natural science, namely geometry" (ibid., 192). Since "the field of natural science in general is incomparably richer and wider than geometry," we may "expect that the analytic method will yet produce fundamentally new principles" (ibid., 196).

15.2 The Analytic Method

The analytic method, or method of analysis, is the method according to which, to solve a problem, one looks for some hypothesis that is a sufficient condition for solving the problem, that is, such that a solution to the problem can be deduced from the hypothesis.

The hypothesis is obtained from the problem, and possibly other data already available, by some non-deductive rule, such as induction, analogy, generalization, specialization, metaphor, metonymy, definition, diagrams. (For a formulation of these rules, see Cellucci 2022, Chapter 7).

The hypothesis must be plausible, that is, such that the arguments for it are stronger than the arguments against it, on the basis of experience.

But the hypothesis, in turn, is a problem that must be solved, and it is solved in the same way as the original problem.

That is, one looks for some hypothesis that is a sufficient condition for solving the problem posed by the previous hypothesis, it is obtained from the latter, and possibly other data already available, by some non-deductive rule, and must be plausible. And so on.

The analytic method can be schematically represented as follows.

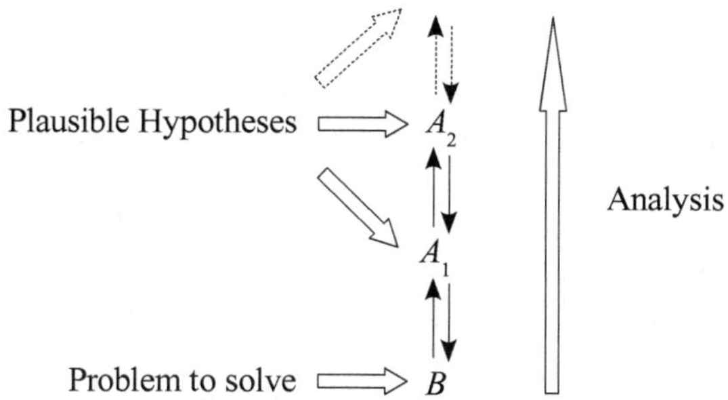

15.3 Some Remarks on the Analytic Method

The following remarks complement the above description of the analytic method.

(1) In the analytic method, when a hypothesis has been found that is a sufficient condition for solving the problem and is plausible, the hypothesis, in turn, is a problem that must be solved. For, although the hypothesis provides a solution to the problem, the hypothesis may not provide a complete understanding of all aspects of the problem. Moreover, the hypothesis may raise new questions not involved in the problem.

(2) In the analytic method, the hypothesis to solve a problem is obtained from the problem, and possibly other data already available, by some non-deductive rule because non-deductive rules are ampliative, that is, the conclusion is not contained in the premisses, so non-deductive rules can produce something new. On the contrary, deductive rules are non-ampliative, that is, the conclusion is contained in the premisses, so deductive rules cannot produce anything really new. (For more on the non-ampliativity of deductive rules, see Chap. 16).

(3) In the analytic method, both non-deductive and deductive rules are not a closed set, given once and for all, but an open set that can always be extended by adding new inference rules, when the existing ones are inadequate to obtain a hypothesis or to deduce a consequence from a hypothesis.

(4) In the analytic method, the extensions of non-deductive rules and deductive rules can even be made in the course of solving a problem.

(5) In the analytic method, each extension of non-deductive rules and deductive rules is a development of the method, which grows with the addition of new inference rules.

15.4 Analytic Method and Diagrams

From its Greek origins, mathematics has made essential use of diagrams. This can also be seen as representative of Greek civilization.

Thus, "the Socratic philosopher Aristippus, when cast ashore on the shores of Rhodes by shipwreck noticed some geometrical figures drawn thereon, is said to have exclaimed to his companions: Let us be of good cheer! For, I see traces of men" (Vitruvius, *De architectura*, VI, pref. 1, ed. Krohn).

On the contrary, in the twentieth century, the use of diagrams has been viewed as non-rigorous and alien to the concept of proof.

Thus, Hilbert says that figures "can easily be misleading" (Hilbert 2004a, 75). Therefore, "we will never rely on them" (Hilbert 2004b, 541). Indeed, "a theorem is only proved when the proof is completely independent of the figure. The proof must call step by step on the preceding axioms" (Hilbert 2004a, 75).

However, this negative attitude is contradicted by its very proponents. Thus, even the very first proof in Hilbert (1987) makes essential use of properties obtained from the figure (see Cellucci 2022, Chap. 12).

As Hersh says, "a proof can be words only, of course. It can be, as in Euclid, words and diagrams. Or it can" be "diagram only" (Hersh 1997, 186).

In particular, the use of diagrams is perfectly justified from the point of view of the analytic method. According to it, a hypothesis for solving a problem is obtained non-deductively from the problem and possibly other data already available. The latter include data acquired from a diagram.

15.5 Analytic Method and Demonstration

The analytic method gives rise to a notion of demonstration, the notion of analytic demonstration, which is quite different from the notion of axiomatic demonstration.

An analytic demonstration consists, first, in a non-deductive derivation of a hypothesis from a problem and possibly other data already available, where the hypothesis is a sufficient condition for the solution of the problem, that is, the hypothesis is such that a solution to the problem can be deduced from it, and the hypothesis is plausible. Then, an analytic demonstration consists in a non-deductive derivation of a new hypothesis from the previous hypothesis, in turn considered as a problem, and possibly from other data already available, where the new hypothesis is a sufficient condition for the solution of the problem posed by the previous hypothesis and is plausible. And so on.

Any new hypothesis in this process produces a solution to the problem that deepens our understanding of the problem.

As Friend says, the "analytic conception of proof" leads us "to much more fundamental types of exploration than we would have engaged in had we viewed proofs as axiomatic" (Friend 2014, 205). For, "under this view we look for much more in a proof than truth and rigour" (ibid., 214). Therefore, "it is preferable to think of all proofs as analytic" (ibid., 213).

While axiomatic demonstration does not serve to find solutions to problems and thus to advance knowledge, but only to present, justify, and teach solutions already found, analytic demonstration serves to find solutions to problems and thus to advance knowledge.

Analytic demonstration is the basic constituent of analytic theories. An analytic theory consists of problems and solutions to them provided by analytic demonstrations.

This means that, as Goodman says, "mathematical theories resemble other scientific theories. They are constructed to solve particular problems," and "are not originally deductive structure based on axioms but, rather, informal bodies of reasoning based on conjectures and bold extrapolation" (Goodman 1991, 124).

15.6 Original Formulation of the Analytic Method

The analytic method was used by Hippocrates of Chios to solve the problems of the doubling the cube and the quadrature of the lunule, and also by Hippocrates of Cos to solve the problem of the impact of food on health. (For details, see Cellucci 2022, Chapter 5). But the first formulation of the analytic method was given by Plato.

Plato says that, to solve a problem, you will proceed by "positing the hypothesis" that you "judge to be the strongest" (Plato, *Phaedo*, 100 a 3–4).

Once you have posited the hypothesis, you will not go on until you will have "investigated its consequences, to see whether they are in accord, or are not in accord, with one another" (ibid., 101 d 4–5). This guarantees that the hypothesis is plausible.

But the hypothesis is in turn a problem to be solved. To solve it, you will proceed "positing another hypothesis, whichever should seem best of the higher ones, and so on" (ibid., 101 d 6–7). And so on.

Therefore, solving a problem is "an infinite task" (Plato, *Parmenides*, 136 c 7).

15.7 Analytic Method and Doubling the Square

Plato uses the analytic method to solve both mathematical and philosophical problems. Thus, in *Meno* he considers the following mathematical problem.

Doubling the square: For any given square, "what will be the side of the square double in size?" (Plato, *Meno*, 82 e 1–2).

To solve the problem, Plato posits the following hypothesis:

(I) A diagonal "cuts each square in half" (ibid., 85 a 1).

Hypothesis (I) is a sufficient condition for solving the problem. For, let *ABCD* be the given square.

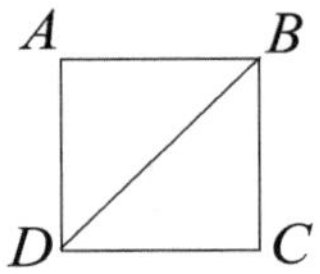

Then, by hypothesis (I), the diagonal *DB* cuts *ABCD* in half, so *ABCD* consists of two halves. Four such halves, arranged into a square, will make a square double in size, because "the relation of four to two" is "double" (ibid., 85 a 8–9). Thus, the question is only how to arrange the four halves into a square. Now, clearly, the only possible way of doing so is to arrange them into the square *DBFH*.

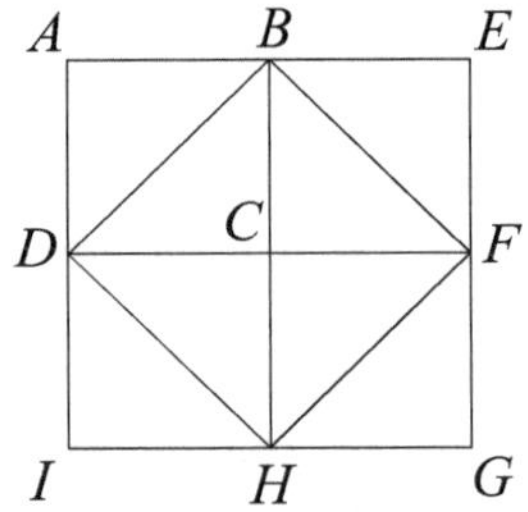

The square *DBFH* is surrounded by four lines, each of which "cuts each of these four" equal squares *ABCD*, *BEFC*, *CFGH*, and *DCHI*, "in half" (ibid., 85 a 5–6). So, each of the four lines is a diagonal of one of the squares in question. Thus, "it is the diagonal that will produce the figure double in size" (ibid., 85 b 5–6). Therefore, for any given square, the side of the square double in size is a diagonal of the given square. This solves the problem.

Hypothesis (I) is plausible. But hypothesis (I) is itself a problem that must be solved. To solve it, Plato posits the following hypothesis:

(II) If two triangles have their corresponding sides equal, then they are equal, because "a thing to which nothing is added and from which nothing is taken away is neither increased nor diminished, but always remains the same in amount" (Plato, *Theaetetus*, 155 a 7–9).

Hypothesis (II) is a sufficient condition for solving the problem. For, the two right-angled isosceles triangles into which a diagonal cuts a square have their corresponding sides equal, so by hypothesis (II) they are equal. Therefore, a diagonal cuts each square in half. Namely, hypothesis (I). This solves the problem. The solution increases the plausibility of hypothesis (I).

Hypothesis (II) is plausible. But hypothesis (II) is itself a problem that must be solved. To solve it, it is necessary to posit another hypothesis. And so on.

The process through which the problem of doubling the square is solved can be schematically represented as follows.

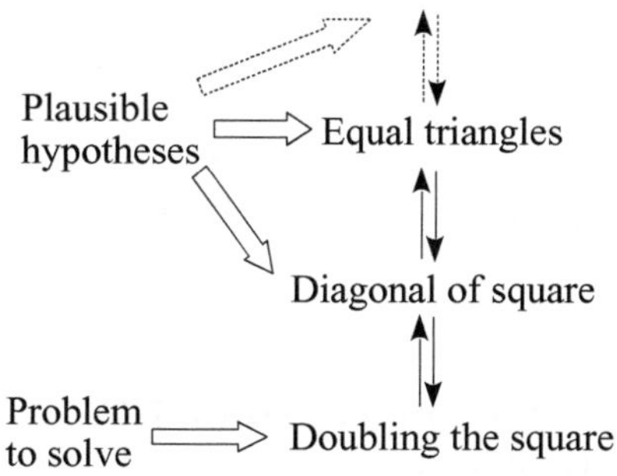

15.8 Analytic Method and Teachability of Virtue

Plato also uses the analytic method to solve philosophical problems. Thus, again in *Meno* he considers the following philosophical problem.

Teachability of virtue: "Is virtue teachable?" (Plato, *Meno*, 70 a 1).

To solve the problem, Plato posits the following hypothesis:

(I) "Virtue is science" (ibid., 87 c 12).

Hypothesis (I) is a sufficient condition for solving the problem. For, every science is teachable, so "if virtue is science, clearly it is teachable" (ibid., 87 c 5–6). From this, by (I), it follows that virtue is teachable. This solves the problem.

Hypothesis (I) is plausible. But hypothesis (I) is itself a problem that must be solved. To solve it, Plato posits the following hypothesis:

(II) "Virtue is something good" (ibid., 87 d 2–3).

Hypothesis (II) is a sufficient condition for solving the problem. For, "science embraces everything that is good" (ibid., 87 d 6–7). From this, by (II), it follows that virtue is science. Namely, hypothesis (I). This solves the problem. So, by "stating the hypothesis that virtue is science, we hypothesize correctly" (ibid., 87 d 7–8). The solution increases the plausibility of hypothesis (I).

Hypothesis (II) is plausible. But hypothesis (II) is itself a problem that must be solved. To solve it, it is necessary to posit another hypothesis. And so on.

The process through which the problem of the teachability of virtue is solved can be schematically represented as follows.

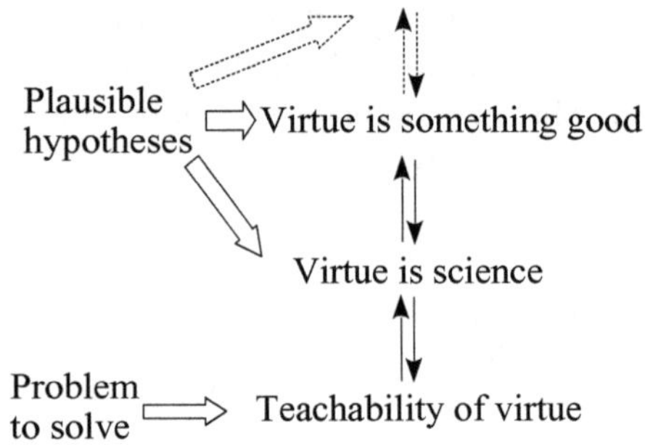

15.9 Limitations of Original Formulation of the Analytic Method

Plato's original formulation of the analytic method, however, has some basic limitations.

(1) Plato does not say how a hypothesis to solve a problem is obtained.

(2) Plato says that the consequences of a hypothesis must be in accord with each other, but this does not guarantee that they are in accord with experience.

(3) Plato says that knowledge is true and infallible because "knowledge is true judgment; to judge truly is infallible" (Plato, *Theaetetus*, 200 e 4–5). But this conflicts with the fact that, as argued in Chap. 11, the aim of science cannot be truth but only plausibility.

The analytic method, as formulated at the beginning of this chapter, does not have these limitations.

It does not have the limitation (1) because, according to it, a hypothesis is obtained from the problem by some non-deductive rule.

It does not have the limitation (2) because, according to it, a hypothesis must be such that the arguments for it are stronger than the arguments against it, on the basis of experience. This guarantees that the consequences of the hypothesis are in accord not only with each other but also with experience.

It does not have the limitation (3) because, according to it, knowledge is not true and infallible but only plausible.

15.10 Knowledge and the Impediment of the Body

It has been said above that a basic limitation of Plato's original formulation of the analytic method is that he says that knowledge is true and infallible. However, a distinction must be made between the knowledge that can be attained by human beings in this life and the knowledge that can be attained by the soul free from the body.

According to Plato, human beings cannot attain true and infallible knowledge in this life, because of the impediment of the body. True and infallible knowledge can be attained only by the soul free from the body. In this life, human beings can only attain plausible knowledge.

Indeed, Plato says that, "as long as we have the body and our soul is contaminated by such an evil, we will never properly acquire what we desire," namely "truth" (Plato, *Phaedo*, 66 b 5–7). So, "if we are ever to have full knowledge of something, we must be separated from the body and contemplate the things by themselves with the soul by itself," because "only then, namely when we have died, in other words when we are outside life, we will get what we desire, wisdom" (ibid., 66 d 8–e 4).

Therefore, somewhat paradoxically, Plato says that knowledge "is for the dead; for then, and not sooner, will the soul be alone by itself apart from the body" (ibid., 66 e 6–67 a 2). Thus the soul will see things "in pure light, being ourselves pure and untainted by this thing which now we carry around with us and call the body, in which we are imprisoned like an oyster in its shell" (Plato, *Phaedrus*, 250 c 3–6). Therefore, "those who pursue philosophy aright, practice nothing but dying and being dead" (Plato, *Phaedo*, 64 a 5–6).

15.11 The Argument of Intuition

Several arguments have been advanced against the analytic method. As we will see, however, they are not valid.

An argument against the analytic method is that, since in the analytic method analysis is not a process of deduction, the analytic method is based on intuition.

Thus, Cornford says that, since "analysis is not a process of deduction," in analysis the activity of the "mind is not demonstration but intuition" (Robinson 1936, 467). Then "Proclus was right when he described analysis in terms reminiscent of the upward path of dialectic in Plato's Divided Line, for that upward path is a series of intuitions" (ibid.).

This argument, however, is not valid. It depends on the assumption that the activity of the mind is either deduction or intuition. But this assumption is unfounded, because the activity of the mind is not restricted to deduction and intuition, there are non-deductive inferences. In particular, in terms of the analytic method as formulated at the beginning of this chapter, the upward path of dialectic in Plato's Divided Line is not a series of intuitions but a series of non-deductive inferences.

15.12 The Argument of Establishment of Hypotheses

Another argument against the analytic method is that, since in the analytic method a hypothesis is not true but only plausible, the analytic method cannot establish a hypothesis. A hypothesis can be established only by exhibiting it as a consequence of what we already know.

Thus, Dummett says that, in mathematics, "if we suppose the hypothesis true, we seek for a proof of it, and it remains a mere hypothesis, whose assertion would therefore be unwarranted, until we find one" (Dummett 1993, 431). And, "if the hypothesis is to be established, this must be done" by "exhibiting it as a consequence of what we already know" (ibid.).

This argument, however, is not valid. For, if a hypothesis can be established only by exhibiting it as a consequence of what we already know, this implies that all new knowledge will be contained in the existing knowledge. This is absurd.

15.13 The Argument of Infinite Regress

Another argument against the analytic method is that, since in the analytic method in order to solve the problem constituted by a hypothesis it is necessary to refer back to another hypothesis, the analytic method involves an infinite regress.

Thus, Tarski says that, in the analytic method, "in order to establish the validity of a statement, it is necessary to refer back to other statements, and (if no vicious circle is to occur) this leads" to "an infinite regress" (Tarski 1994, 110). This makes knowledge impossible, because an infinite regress is "a process which can never be brought to an end" (ibid.).

This argument, however, is not valid. It assumes that, since the capacities of our intellect are finite and therefore do not allow us to travel an infinite path, the analytic method is not humanly usable and therefore knowledge becomes impossible with it.

But this is not the case. In the analytic method, going back from one hypothesis to another leads to a deepening of our understanding of the problem. Each new hypothesis, being plausible, is justified but is not absolutely justified, and therefore cannot be infallible. This does not make knowledge impossible, it only makes infallible knowledge impossible. Therefore, what is at stake in the infinite regress argument is not the possibility of knowledge, but only the fallibility of knowledge.

15.14 The Argument of Non-Finality of Solutions

Another argument against the analytic method is that, since in the analytic method no solution to a problem is final, the analytic method conflicts with the fact that solutions to mathematical problems are final, because mathematical results last forever.

Thus, Rota says that, “once solved, a mathematical problem is forever finished: no later event will disprove a correct solution” (Rota 1997, 93).

This argument, however, is not valid. As Davis says, “there is no finality in the creation, formulation and solutions of problems” (Davis 2006, 164). Even in mathematics, “a solved problem is still not completely solved but leads to new and profound challenges” (ibid., 177). In fact, “discovering a sense in which” this is the case “is one important direction that mathematical research takes” (ibid.).

The analytic method indicates a sense in which a solved problem is still not completely solved but leads to new and profound challenges. According to it, a hypothesis for solving a problem is in turn a problem that must be solved.

15.15 The Argument of the Only True Hypothesis

Another argument against the analytic method is that it does not provide a solid solution to a problem because the same conclusion can follow from different hypotheses and only one of them is the true one, but the analytic method does not tell us which of the hypotheses is the true one. Only the axiomatic method can tell us this, because a proposition is made certain only by the axiomatic method, by showing that the proposition follows from assumptions that are certain.

Thus, Leibniz argues that a hypothesis “is like the key to a cryptograph” (Leibniz 1969, 283). Just as “it is possible to write a letter intentionally so that it can be understood by means of several different keys, of which only one is the true one, so the same effect can have several causes” (ibid.). And the same conclusion can follow from different hypotheses, of which only one is the true one. In order to determine which of the hypotheses is the true one, it is not enough to assume that “the greater the number of events that can be explained by” a hypothesis, “the more probable it is” (ibid.). For, a false hypothesis explains any fact, so “no firm demonstration can be made from the success of hypotheses” (ibid.). To determine which of the hypotheses is the true one, another criterion is needed, but the analytic method does not provide it.

It can be even be “useful to apply less perfect hypotheses as substitutes for truth until a better one occurs, that is, one which explains the same phenomena more happily or more phenomena with equal felicity” (ibid.). But these hypotheses are only substitutes for truth. Truth can only be reached through the axiomatic method, because "demonstration is reasoning by which some proposition is made certain. This is achieved whenever it is shown that the proposition necessarily follows from certain suppositions (which are assumed to be certain)" (Leibniz 1969, 187). And a proposition can be made certain only through the axiomatic method, by showing that it follows from assumptions that are certain.

This argument, however, is not valid. No hypothesis can be said to be true because, as argued in Chap. 11, a criterion of truth is impossible. A hypothesis can only be plausible, and there is a criterion of plausibility. To determine whether a hypothesis is plausible, we see if the arguments for the hypothesis are stronger than those against it, on the basis of experience.

Also, truth cannot be reached though the axiomatic method. For, by Gödel's second incompleteness theorem, a proposition cannot be made certain by showing that the proposition follows from assumptions that are certain.

15.16 Characters of the Analytic Method

The analytic method has the following characters.

(1) In the analytic method there are no principles, everything is a hypothesis. The problem and the other data already available are the only basis for solving a problem.

(2) The solution to a problem by the analytic method involves two ways, an upward way from the problem to hypotheses by non-deductive rules, and a downward way from hypotheses to the problem by deductive rules. But the two ways are one and the same. Thus, in the analytic method, as Heraclitus says, "the way up and the way down are one and the same" (Heraclitus 22 B 60, ed. Diels-Kranz).

(3) The hypotheses for solving a problem by the analytic method need not belong to the same field as the problem, they can belong to other fields. So, the search for a solution to a problem is not carried out in a closed, predetermined space.

(4) The hypotheses for solving a problem by the analytic method are not global but local, since they are aimed at a specific problem. Being local, the hypotheses for solving a problem by the analytic method can be efficient because they can be closely linked to the peculiarity of the problem.

(5) The hypotheses for solving a problem by the analytic method are neither true nor certain, because they can only be plausible. But this does not diminish the value of the analytic method, because plausible knowledge is the best we can achieve, and, on the other hand, without plausible knowledge there is no knowledge at all.

(6) A problem can have different solutions by the analytic method, using different hypotheses. For, a problem can be looked at from different perspectives, which can lead to different solutions to the problem.

(7) Solving a problem by the analytic method is both a process of discovery and a process of justification. It is a process of discovery, because it involves finding hypotheses by non-deductive rules. It is a process of justification, because it involves showing that the hypotheses are plausible.

(8) Solving a problem by the analytic method involves a comparison with experience. For, hypotheses must be plausible, so the arguments for them must be stronger than the arguments against them, on the basis of experience.

(9) Intuition has no role in solving a problem by the analytic method. It has no role in the formulation of hypotheses, because hypotheses are obtained from the problem, and possibly other data already available, by some non-deductive rule, so not by intuition but by inference. It has no role in the justification of hypotheses, because their plausibility is established by showing that the arguments for them are stronger than arguments against them, on the basis of experience, so not by intuition but by inference.

(10) No solution to a problem by the analytic method is final, every solution is always provisional. For, at any moment new data can emerge with which a hypothesis is incompatible, so a plausible hypothesis can become implausible. If this happens, one analyzes the reasons why the hypothesis is implausible, and on that basis modifies the hypothesis or replaces it with a new hypothesis.

15.17 The Axiomatic Method

Although the arguments against the analytic method are not valid, in the past century the method of mathematics has not been identified with the analytic method but with the axiomatic method.

More precisely, the method of mathematics has been identified, not with the traditional axiomatic method, but with the modern axiomatic method.

The traditional axiomatic method is the method according to which, to provide the final presentation and the complete logical grounding of knowledge in a given field, propositions are deduced from the axioms for that field, where the axioms are true, in the sense that there are things, specified in advance, of which the axioms are true.

Instead, the modern axiomatic method is the method according to which, to provide the final presentation and the complete logical grounding of knowledge in a given field, propositions are deduced from the axioms for that field, where the axioms are arbitrarily chosen propositions, subject only to the condition that they must be consistent.

Thus, the modern axiomatic method differs from the traditional axiomatic method because the axioms are not required to be true but can be arbitrarily chosen propositions, subject only to the condition that they must be consistent.

The axiomatic method can be schematically represented as follows.

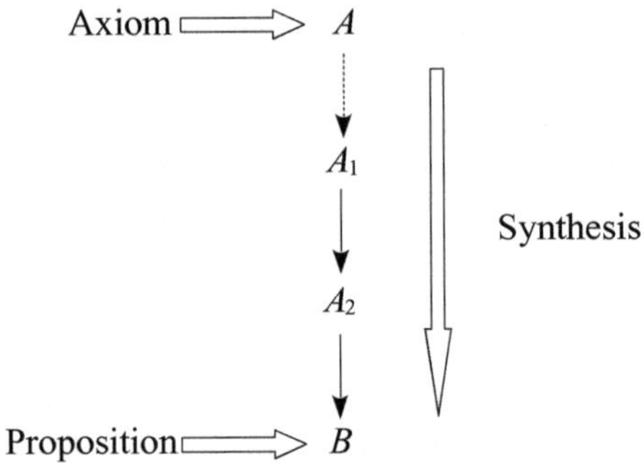

15.18 Origin of Modern Axiomatic Method

The modern axiomatic method developed gradually from the second half of the nineteenth century to the early decades of the twentieth century, and is best known from Hilbert.

Hilbert says that mathematicians must "be free to do as" they please "in giving characteristic marks" (Hilbert 1980, 39). That is, mathematicians must be completely free in choosing the axioms. The choice of the axioms is only subject to the condition that the axioms must not contradict one another. For, "if the arbitrarily given axioms do not contradict one another with all their consequences, then they are true and the things defined by the axioms exist. This" is "the criterion of truth and existence" (ibid., 39–40).

The condition that the axioms are subject to the condition that they must be consistent implies that modern axiomatic method essentially depends on intuition, because the consistency of the axioms must be established by a proof based on intuition.

Indeed, Hilbert says that the consistency of the axioms must be established by a proof based on the "intuitive mode of thought" (Hilbert 1996e, 1150). For, only the intuitive mode of thought can avoid "any dubious or problematic mode of inference" (Hilbert 1996d, 1139). Only a consistency proof based on the intuitive mode of thought can guarantee that "mathematical statements are in fact incontestable and ultimate truths" (Hilbert 1996c, 1121).

The purpose of modern axiomatic method is not to discover new knowledge, but only to provide the final presentation and the complete logical grounding of knowledge in a given field.

Indeed, Hilbert says that the purpose of "the axiomatic exploration of a mathematical truth" is not to find "new or more general propositions connected to that truth" (Hilbert 1902–1903, 50). It is, instead, "to determine the position of that" truth "within the system of known truths," in "such a way that it can be clearly said which conditions are necessary and sufficient for giving a foundation of that truth" (ibid.). Therefore, the purpose of the axiomatic method is "the final presentation and the complete logical grounding of our knowledge," an aim for which "the axiomatic method deserves the first rank" (Hilbert 1996a, 1093).

Hilbert's formulation of modern method has been dominant in the past century.

An example of this is Bourbaki, who says that "'mathematical truth' resides" uniquely "in logical deduction starting from premises arbitrarily set by axioms" (Bourbaki 1994, 17). Their choice is only subject to the condition that the axioms must not contradict one another, otherwise "every theorem is both true and false in the theory, and the theory is consequently of no interest" (Bourbaki 2004, 12). The purpose of the axiomatic method is not to find new theorems but only "to provide" a final presentation and "a solid foundation for the whole body of modern mathematics" (ibid., v).

In fact, as Thom says, "it is characteristic that no new theorem of any importance came out of the immense effort at systematization of Nicolas Bourbaki" (Thom 1971, 697–698)

15.19 All-Pervasiveness of Modern Axiomatic Method

Hilbert argues that the method of mathematics is the modern axiomatic method because the latter is the method of development of mathematical science.

Indeed, Hilbert says that "the development of mathematical science" takes place "in two ways that constantly alternate," that is, "the adjunction of new axioms," and "the derivation of new 'provable'" propositions "from the axioms by means of" deductive "inference" (Hilbert 1996c, 1132).

But Hilbert does not limit himself to arguing that the modern axiomatic method is the method of mathematics, he claims that the modern axiomatic method is the method of all scientific thought and even the foundation of all modern culture, therefore the modern axiomatic method is all-pervasive.

Indeed, Hilbert says that the modern axiomatic method is "the indispensable tool, appropriate to our minds, for all exact research in any field whatsoever" (Hilbert 1996c, 1120). In fact, "anything at all that can be the object of scientific thought," as soon as "it is ripe for the formation of a theory," becomes "dependent on the axiomatic method" (Hilbert 1996b, 1115). And anything at all that can be the object of scientific thought, insofar as it is dependent on the axiomatic method, is "thereby indirectly" dependent "on mathematics, as soon as it is ripe for the formation of a theory" (ibid.). So, "in the sign of the axiomatic method, mathematics is summoned to a leading role in science" (ibid.). More generally, through the axiomatic method, "our entire modern culture, in so far as it rests on the penetration and utilization of nature, has its foundation in mathematics" (Hilbert 1996f, 1163).

15.20 Original Motivation of Modern Axiomatic Method

It is widely believed that the change from the traditional axiomatic method to the modern axiomatic method was due to the birth of non-Euclidean geometries. and abstract algebras.

But this belief is unfounded. For example, as we have seen in Chap. 3, the creators of non-Euclidean geometries, from Lobatchevsky to Riemann, did not think that the axioms of their geometries were arbitrarily chosen propositions, on the contrary, they thought that they were true of the physical space.

Instead, the change from the traditional axiomatic method to the modern axiomatic method was due to a somewhat late impact of Romanticism on mathematics, because a basic characteristic of Romanticism was the claim of an absolutely free creativity of the mind. In fact, several mathematician in the second half of the nineteenth century thought that axioms were absolutely free creations of the human mind, subject only to the requirement of consistency.

For example, Cantor said that "the essence of mathematics lies precisely in its freedom" (Cantor 1996, 896). Mathematics is "in its development entirely free and is only bound in the self-evident respect that its concepts must" be "consistent with each other" (ibid.).

15.21 Modern Axiomatic Method and Reality

An important implication of the modern axiomatic method is a separation between theory and reality.

From the seventeenth century until the nineteenth century, there was a close relation between theory and reality. In contrast, with the development of the modern axiomatic method, there started a separation between theory and reality.

Indeed, Hilbert says that, with the modern axiomatic method, a theory "becomes completely detached from concrete reality," it "has no longer anything to do with real objects," it is "a pure thought construction, of which one cannot say that it is true or false" (Hilbert 2013, 435).

The separation between theory and reality also affects mathematics. For, it makes mathematics an end in itself, with no thought of what objective it might serve.

As Kline says, with the separation between theory and reality, mathematics is "turned inward; it feeds on itself," it becomes "an end in itself with no thought of what objective it might serve" (Kline 1980, 304). Thus "research begets research, which begets research," one "dare not ask for meaning and purpose. Mathematics must not be tainted by reality. The ivy has grown so thick that the researchers within can no longer see the world outside" (ibid., 302).

This leads mathematicians to create mathematics with no application to reality.

Even Bourbaki admits that the separation between theory and reality has led mathematicians to create "a whole crop of monster-structures, entirely without applications" (Bourbaki 1950, 230, footnote). So, there has been the "temptation to conclude that these were the only results that could be expected from the axiomatic method" (ibid.).

This raises doubts about Hilbert's affirmation that the method of mathematics is the modern axiomatic method. But even more is the case. As will be argued in Chap. 17, the limitative results of mathematical logic refute the affirmation that the method of mathematics is the modern axiomatic method. A fortiori, Hilbert's claim that, through the modern axiomatic method, our entire modern culture has its foundation in mathematics, is unfounded.

In what follows, 'axiomatic method' will stand for 'modern axiomatic method', unless otherwise specified.

15.22 Implicit Uses of the Analytic Method

As said above, although the arguments against the analytic method are not valid, in the past century the method of mathematics has not been identified with the analytic method, instead, it has been identified with the axiomatic method.

In fact, mathematicians make no reference to the analytic method in their published works. This does not mean, however, that they do not use it. In fact, mathematicians often proceed in a way that is a, perhaps unconscious, application of the analytic method.

An example of this is provided by the solution to Fermat's conjecture: There are no positive integers x, y, z such that $x^n + y^n = z^n$, for $n > 2$.

According to the current view, Fermat's conjecture was solved by Wiles and Taylor (1995). But this is not correct. Fermat's conjecture was solved by Ribet (1990) using as hypothesis the Taniyama-Shimura conjecture: Every elliptic curve over the rational numbers is modular. For, Ribet demonstrated that the Taniyama-Shimura conjecture is a sufficient condition for solving Fermat's conjecture.

But the Taniyama-Shimura conjecture was in its turn a problem that had to be solved. It was solved by Wiles and Taylor using hypotheses from various parts of mathematics, from differential geometry to complex analysis. Thus, what Wiles and Taylor solved was not Fermat's conjecture, but Taniyama-Shimura conjecture.

The process through which Fermat's conjecture was solved can be schematically represented as follows.

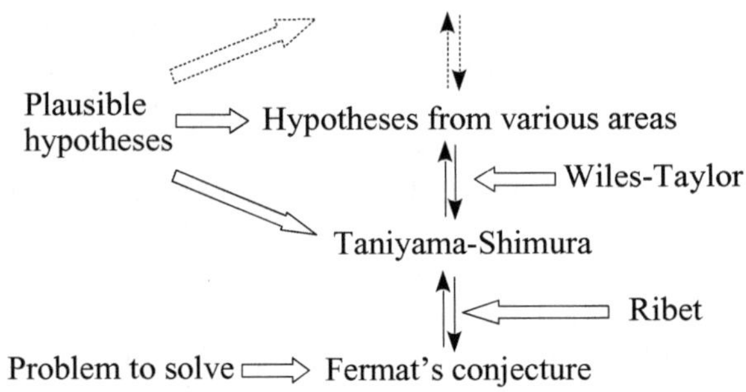

Most mathematicians would argue that Fermat's conjecture was not solved by Ribet, because Ribet used a hypothesis, the Taniyama-Shimura conjecture, which at the time had not been demonstrated yet. Since Wiles and Taylor solved the Taniyama-Shimura conjecture, the solution of Fermat's conjecture must be attributed to them.

But this argument is not valid. For, Wiles and Taylor solved the Taniyama-Shimura conjecture using hypotheses that ultimately depended on the axioms of set theory. Now, the axioms of set theory are a hypothesis that to this very day has not been demonstrated yet. So, if Ribet did not solve Fermat's conjecture because he used a hypothesis, the Taniyama-Shimura conjecture, that at the time had not been demonstrated yet, then Wiles and Taylor did not solve Fermat's conjecture because they used hypotheses that ultimately depended on a hypothesis, the axioms of set theory, that to this very day has not been demonstrated yet.

Saying that what Wiles and Taylor solved was not Fermat's conjecture but the Taniyama-Shimura conjecture, in no way diminishes the value of their achievement. Indeed, Lang says that the Taniyama-Shimura "conjecture is one of the most important of the century" (Lang 1995, 1301). On the other hand, Rav says that Fermat's conjecture "is insignificant (historical interest apart)" (Rav 1999, 25).

15.23 Analytic Method and Abduction

Some people claim that there is a connection between the analytic method, or method of analysis, and abduction.

Thus, Niiniluoto says that "abduction can be viewed as a special case of the method of analysis" (Niiniluoto 1999, 240).

This depends on the fact that, according to a widespread opinion, abduction is a method of discovery.

Thus, Blachowicz says that "abduction" is "a form of inference that is the basis for scientific discovery" (Blachowicz 1996, 156).

But this is not valid. For, while the analytic method is a method of discovery, abduction is not a method of discovery.

Indeed, abduction is the inference rule:

$$(\text{Abd})\frac{B \quad A \rightarrow B}{A}$$

The rule Abd is non-ampliative, because the conclusion A is contained in the premiss $A \rightarrow B$.

Since Abd is non-ampliative, Abd does not produce new knowledge. Therefore, abduction cannot be a method of discovery.

As Peirce says, the conclusion A "cannot be abductively inferred, or if you prefer the expression, cannot be abductively conjectured until its entire content is already present in the" major "premiss" (Peirce 1931–1958, 5.189). That is, until its entire content is already present in $A \rightarrow B$. Therefore, "quite new conceptions cannot be obtained from abduction" (ibid., 5.190).

Since abduction is not a method of discovery, the analytic method must not be confused with abduction.

15.24 Abduction and Detective Stories

An example of confusion of the analytic method with abduction is the belief that abduction is the method used by Sherlock Holmes.

Thus, Carson says that "Sherlock Holmes is renowned as a great detective" and "his particular skills are said to have been in deduction. But that is false" because "Holmes' distinctive investigative skills involved abduction" (Carson 2009, 193).

Hanzal says that Sherlock "Holmes' claims about deduction have been disputed for decades. It has been repeatedly pointed out that the most common type of his inferences is not deduction (nor induction), but abduction" (Hanzal 2024, 166).

But this is not valid. The method used by Sherlock Holmes is not abduction but the analytic method.

As Franchella says, the "long tradition of analysis is the method which Sherlock Holmes referred to" (Franchella 2011, 233).

In fact, Conan Doyle says that "in solving a problem," the "grand thing is to be able to reason backward," namely "analytically" (Conan Doyle 2015, 268). Most people can only "reason forward," namely "synthetically" (ibid.). But "there are few people" who, "if you told them a result, would be able to evolve from their own inner consciousness what the steps were which led up to that result. This power is what I mean when I talk of reasoning backwards, or analytically" (ibid., 268–269).

15.25 Non-Validity of Arguments for Ampliativity of Abduction

Some arguments have been proposed for the ampliativity of abduction, but they are not valid.

Thus, Josephson and Tanner argue that abduction is ampliative because it "can 'interpret' the given data in a new vocabulary" (Josephson and Tanner 1994, 13). Indeed, suppose that, from the premisses, 'If the patient has hepatitis, then the patient has jaundice' and 'The patient has jaundice', we infer the conclusion, 'The patient has hepatitis'. Thus "we have introduced into the conclusion a new term, 'hepatitis', which is from the vocabulary of diseases and not part of the vocabulary of symptoms. By introducing this term, we make conceptual connections with the typical progress of the disease, and ways to treat it, that were unavailable before" (ibid.).

This argument, however, is not valid. The conceptual connection is not made by the introduction into the conclusion of a new term, 'hepatitis', which is from the vocabulary of diseases and not part of the vocabulary of symptoms. The conceptual connection is made by the process that produces the premiss 'If the patient has hepatitis, then the patient has jaundice', which connects hepatitis with jaundice. Therefore, abduction does not introduce a new term in the conclusion.

15.26 Analytic Method and Reductio ad Absurdum

Some people claim that there is a connection between the analytic method and reductio ad absurdum, that is, the rule: From hypothesis A both B and $\neg B$ can be deduced, therefore $\neg A$.

Thus, Knorr says that reductio ad absurdum "invites comparison to the method of 'reduction' used, for instance, by Hippocrates" of Chios in dealing with the problem "of the cube duplication" (Knorr 1993, 359). That is, reductio ad absurdum invites comparison to the analytic method.

In a sense, the analytic method is related to reductio ad absurdum. For, in the latter, to solve a problem of the form $\neg A$, we assume the hypothesis A, so we do not start from already given principles or results, but from something assumed as a hypothesis. Therefore, like the analytic method, reductio ad absurdum is a procedure by hypothesis.

On the other hand, however, between the analytic method and reductio ad absurdum there are substantial differences.

(1) The analytic method involves two ways, an upward way from problems to hypotheses through non-deductive rules, and a downward way from hypotheses to problems through deductive rules. Conversely, reductio ad absurdum only involves a downward way because, to demonstrate $\neg A$, we assume A and deduce both B and $\neg B$ from it.

(2) In the analytic method, hypotheses are not known from the beginning, finding them is the goal of the research. Conversely, in reductio ad absurdum, the hypothesis A is known from the beginning, because it is the opposite of the desired conclusion $\neg A$, which is known from the beginning.

(3) In the analytic method, hypotheses must be plausible. On the contrary, in reductio ad absurdum, the hypothesis A cannot be plausible, because it is the opposite of the conclusion $\neg A$, which is meant to be plausible.

15.27 Original Motivation of Reductio ad Absurdum

Reductio ad absurdum serves to establish something which is contrary to immediate experience. It is not surprising, then, that it had a central role in the Eleatic philosophy, whose doctrines were contrary to immediate experience.

Thus, Parmenides claims: "Being is ungenerated" (Parmenides 28 B 8.3, ed. Diels-Kranz). To establish this claim, which is contrary to immediate experience, Parmenides uses reductio ad absurdum. Indeed, he assumes A: Being is generated. Now, if being is generated, then "it is now, all together" (ibid. 28 B 8.5). Then B: Being is. On the other hand, if being is generated, it came into being, and, "if it came into being, then it is not" (ibid. 28 B 8.20). Then $\neg B$: Being is not. So, from A, both B and $\neg B$ can be deduced. Therefore, by reductio ad absurdum, $\neg A$: Being is ungenerated.

Zeno claims: "Beings are not many" (Zeno 29 B 3, ed. Diels-Kranz). To establish this claim, which is contrary to immediate experience, Zeno uses reductio ad absurdum. Indeed, he assumes A: "Beings are many" (ibid.). Then "necessarily they are as many as they are, and not more and not less. But, if they are as many as they are," then B: "Beings are limited" (ibid.). On the other hand, if beings are many, then "between one being and another there will always be other beings, and between one of these and another there will be still others," hence $\neg B$: "Beings are unlimited" (ibid.). So, from A, both B and $\neg B$ can be deduced. Therefore, by reductio ad absurdum, $\neg A$: Beings are not many.

References

Blachowicz, James. 1996. Ampliative abduction. *International Studies in the Philosophy of Science* 10:141–157.

Bourbaki, Nicholas. 1950. The architecture of mathematics. *The American Mathematical Monthly* 57 (4): 221–232.

Bourbaki, Nicholas. 1994. *Elements of the history of mathematics*. New York: Springer.

Bourbaki, Nicholas. 2004. *Elements of mathematics: Theory of sets*. New York: Springer.

Cantor, Georg. 1996. Foundations of a general theory of manifolds: A mathematico-philosophical investigation into the theory of the infinite. In *From Kant to Hilbert: A source book in the foundations of mathematics*, ed. William Ewald, vol. 2, 881–920. Oxford: Oxford University Press.

Carson, David. 2009. The abduction of Sherlock Holmes. *International Journal of Police Science and Management* 11:193–202.

Cellucci, Carlo. 2022. *The making of mathematics: Heuristic philosophy of mathematics*. Cham: Springer.

Conan Doyle, Arthur. 2015. *A study in scarlet*. New York: HarperCollins.

Davis, Philip. 2006. *Mathematics and common sense: A case of creative tension*. Natick: A K Peters.

Dummett, Michael. 1993. *The seas of language*. Oxford: Oxford University Press.

Franchella, Miriam. 2011. Where the most logical mind may be at fault. In *Sherlock Holmes and philosophy: The footprints of a gigantic mind*, ed. Josef Steiff, 229–236. Chicago: Open Court.

Friend, Michèle. 2014. *Pluralism in mathematics: A new position in philosophy of mathematics*. Dordrecht: Springer.

Goodman, Nicolas. 1991. Modernizing the philosophy of mathematics. *Synthese* 88:119–126.

Hanzal, Tomáš. 2024. Deduction, abduction, and creativity. *Acta Analytica* 39:163–182.

Hersh, Reuben. 1997. *What is mathematics, really?* Oxford: Oxford University Press.

Hilbert, David. 1902–1903. Über den Satz von der Gleichheit der Basiswinkel im gleichschenkligen Dreieck. *Proceedings of the London Mathematical Society* 35:50–68.

Hilbert, David. 1980. Letter to Frege, 29 December 1899. In Gottlob Frege, *Philosophical and mathematical correspondence*, 38–41. Oxford: Blackwell.

Hilbert, David. 1987. *Grundlagen der Geometrie*. Stuttgart: Teubner.

Hilbert, David. 1996a. On the concept of number. In *From Kant to Hilbert: A source book in the foundations of mathematics*, ed. William Ewald, vol. 2, 1092–1095. Oxford: Oxford University Press.

Hilbert, David. 1996b. Axiomatic thought. In *From Kant to Hilbert: A source book in the foundations of mathematics*, ed. William Ewald, vol. 2, 1107–1115. Oxford: Oxford University Press.

Hilbert, David. 1996c. The new grounding of mathematics: First report. In *From Brouwer to Hilbert*, ed. William Ewald, vol. 2, 1117–1134. Oxford: Oxford University Press.

Hilbert, David. 1996d. The logical foundations of mathematics. In *From Brouwer to Hilbert*, ed. William Ewald, vol. 2, 1134–1148. Oxford: Oxford University Press.

Hilbert, David. 1996e. The grounding of elementary number theory. In *From Kant to Hilbert: A source book in the foundations of mathematics*, ed. William Ewald, vol. 2, 1149–1157. Oxford: Oxford University Press.

Hilbert, David. 1996f. Logic and the knowledge of nature. In *From Kant to Hilbert: A source book in the foundations of mathematics*, ed. William Ewald, vol. 2, 1157–1165. Oxford: Oxford University Press.

Hilbert, David. 2004a. Die Grundlagen der Geometrie. In *David Hilbert's lectures on the foundations of geometry 1891–1902*, ed. Michael Hallett and Ulrich Majer, 72–123. Berlin: Springer.

Hilbert, David. 2004b. Grundlagen der Geometrie. In *David Hilbert's lectures on the foundations of geometry 1891–1902*, ed. Michael Hallett and Ulrich Majer, 540–602. Berlin: Springer.

Hilbert, David. 2013. *Lectures on the foundations of arithmetic and logic 1917–1933*. Heidelberg: Springer.

Josephson, John, and Michael Tanner. 1994. Conceptual analysis and abduction. In *Abductive inference: Computation, philosophy, technology*, ed. John Josephson and Susan Josephson, 5–30. Cambridge: Cambridge University Press.

Kline, Morris. 1980. *Mathematics: The loss of certainty*. Oxford: Oxford University Press.

Knorr, Wilbur. 1993. *The ancient tradition of geometric problems*. Mineola: Dover.

Lakatos, Imre. 1978. *Philosophical Papers*. Cambridge: Cambridge University Press.

Lang, Serge. 1995. Some history of the Shimura-Taniyama conjecture. *Notices of the American Mathematical Society* 42:1301–1309.

Leibniz, Gottfried. 1969. *Philosophical papers and letters*. Dordrecht: Kluwer.

Mach, Ernst. 1976. *Knowledge and error: Sketches on the psychology of enquiry*. Dordrecht: Springer.

Niiniluoto, Ilkka. 1999. Abduction and geometrical analysis: Notes on Charles S. Peirce and Edgar Allan Poe. In *Model-based reasoning in scientific discovery*, ed. Lorenzo Magnani, Nancy Nersessian, and Paul Thagard, 239–254. New York: Kluwer.

Peirce, Charles. 1931–1958. *Collected papers*. Cambridge: Harvard University Press.

Rav, Yehuda. 1999. Why do we prove theorems? *Philosophia Mathematica* 7:5–41.

Ribet, Kenneth. 1990. From the Taniyama-Shimura conjecture to Fermat's last theorem. *Annales de la Faculté Mathematiques et des Sciences de Toulouse* 11:116–139.

Robinson, Richard. 1936. Analysis in Greek geometry. *Mind* 45:464–473.

Rota, Gian-Carlo. 1997. *Indiscrete thoughts*. Boston: Birkhäuser.

Tarski, Alfred. 1994. *Introduction to logic and to the methodology of deductive sciences*. Oxford: Oxford University Press.

Thom, René. 1971. 'Modern' mathematics: An educational and philosophic error? *American Scientist* 59 (6): 695–699.

Wiles, Andrew, and Richard Taylor. 1995. Ring-theoretic properties of certain Hecke algebras. *Annals of Mathematics* 141:553–572.

Chapter 16
Knowledge and Inference

Abstract Developing the view that philosophy is acquisition of knowledge also requires answering the question: What is the role of non-deductive rules and deductive rules in knowledge? The role of non-deductive rules is to obtain hypotheses to solve problems, the role of deductive rules is to deduce solutions to problems from hypotheses. This depends on the fact that non-deductive rules are ampliative, deductive rules are non-ampliative. Several arguments have been advanced against the claim that deductive rules are non-ampliative, but this chapter argues that they are not valid.

Keywords Role of non-deductive and deductive rules in knowledge · Non-ampliativity of deductive rules · Arguments against non-ampliativity of deductive rules · Utility of deductive rules · Paradox of inference · Characterization of non-deductive and deductive rules

16.1 Non-Deductive and Deductive Rules

Developing the view that philosophy is acquisition of knowledge also requires answering the question: What is the role of non-deductive rules and deductive rules in knowledge?

The role of non-deductive rules is to find hypotheses to solve problems, the role of deductive rules is to deduce consequences from hypotheses. Thus non-deductive rules and deductive rules have essentially different roles in knowledge.

They have essentially different roles in knowledge because non-deductive rules and deductive rules are essentially different.

Indeed, as already said in Chap. 15, non-deductive rules are ampliative, namely the conclusion is not contained in the premisses. On the contrary, deductive rules are non-ampliative, namely the conclusion is contained in the premisses. Just as a juicer cannot extract more from an orange than it contains, so a deductive rule cannot extract more from its premisses than they contain.

C. Cellucci, *A Philosophy for the Future*, Synthese Library 508,
https://doi.org/10.1007/978-3-031-98443-3_16

16.2 Non-Ampliativity of Deductive Rules

The claim that deductive rules are non-ampliative has already been made in antiquity.

Thus, Epicurus says: "We will not say that syllogism can make us know anything new" (Epicurus, *De natura*, XXVIII, 17, I, ed. Sedley).

The claim has been widely reaffirmed in the modern and contemporary periods.

Thus, Descartes says that logicians cannot "form any syllogism leading to a true conclusion unless" they "already knew the very same truth which is deduced in the syllogism," so they "can learn nothing new from such form of reasoning" (Descartes 1996, X, 406).

Locke says that syllogism "discovers no new proofs," it is only "the art of marshalling, and ranging the old ones we have already" (Locke 1975, 679). It "fails our reason in that part" which "we most need its help in; and that is the finding out of proofs, and making new discoveries" (ibid.). Syllogism "at best, is but the art of fencing with the little knowledge we have, without making any addition to it" (ibid.). In fact, "a man knows first, and then he is able to prove syllogistically. So that syllogism comes after knowledge, and then a man has little or no need of it" (ibid.).

Kant says that deduction "teaches us nothing at all about the content of cognition," so the idea "of using it as a tool (organon) for an expansion and extension of its information, or at least the pretension of so doing, comes down to nothing but idle chatter" (Kant 1998, A61/B86).

Mill says that, in a deduction, the conclusion is "merely a repetition of the same, or part of the same, assertion, which was contained in the first" (Mill 1963–1986, VII, 158). Thus "there is in the conclusion no new truth, nothing but what was already asserted in the premisses" (ibid., VII, 160).

De Morgan says that deduction is subject "to the great rule of all search after truth, that nothing is to be asserted as a conclusion, more than is actually contained in the premises" (De Morgan 1835, 99).

Peirce says that, in deduction, "certain facts are first laid down in the premisses," and "part or all of them" will be thrown "into a new statement" that "will be the conclusion of" the deductive "inference" (Peirce 1931–1958, 2.680). Therefore, deduction "is evidently entirely inadequate to the representation" of inference "which goes out beyond the facts given in the premisses" (ibid., 2.681).

Poincaré says that deduction "remains incapable of adding anything to the data that are given it," the data are contained in the premisses, "and that is all we should find in the conclusions" (Poincaré 2015, 31).

Russell says that deduction "consists merely in saying in other words part or the whole of what is said in the premisses" (Russell 1997, 360).

Wittgenstein says that, "if one proposition follows from another, then the latter says more than the former, and the former less than the latter" (Wittgenstein 2001, 5.14). Thus "there can never be surprises in logic" (ibid., 6.1251). In fact, "in logic process and result are equivalent. (Hence the absence of surprise)" (ibid., 6.1261).

16.3 Arguments Against Non-Ampliativity of Deductive Rules

Although the claim that deductive rules are non-ampliative has been widely reaffirmed even in the contemporary period, several arguments have been advanced against it.

These arguments are intended to show that, like non-deductive rules, deductive rules are ampliative, so non-deductive rules and deductive rules are not essentially different.

However, as will argued below, the arguments against the claim that deductive rules are non-ampliative are not valid.

16.4 The Argument of Labour

An argument against the claim that deductive rules are non-ampliative is that, even if the conclusion of a deductive argument is contained in the premisses, we need labour to extract it from the premisses. The conclusion is contained in the premisses as plants are contained in their seeds, not as beams are contained in a house.

Thus, Frege argues that, even if the conclusions of deductive arguments "are in a way contained covertly in the whole set" of premisses "taken together," this "does not absolve us from the labour of actually extracting them and setting them out in their own right" (Frege 1960, 23). Indeed, "what we shall be able to infer from" the premisses "cannot be inspected in advance; here, we are not simply taking out of the box again what we have just put into it. The conclusions we draw" from the premisses may "extend our knowledge, and ought therefore, on Kant's view, to be regarded as synthetic" (ibid., 100–101). The conclusions are contained in the premisses "as plants are contained in their seeds, not as beams are contained in a house" (ibid., 101). Therefore, the conclusions can "extend our knowledge" (ibid.).

This argument, however, is not valid. For, as Turing argues, "we are always able to obtain from the rules of formal logic" a mechanical "method of enumerating" the proofs from the given axioms and hence "the propositions proved by its means" (Turing 1939, 215). An "enumeration of proofs is for instance obtained from an enumeration of all possible sequences of symbols by striking out those which do not pass the test" (Turing 2004, 212). We can "then imagine that all proofs take the form of a search through this enumeration for the theorem for which a proof is desired" (Turing 1939, 215).

Of course, this is a time consuming method. Therefore, in practice "we do not really want to make proofs by hunting through enumerations for them, but by hitting on one and then checking up to see that it is right. However, this method is always" theoretically "replaceable by the longer" mechanical "method" (Turing 2004, 212). Therefore, we do not need labour to obtain the conclusion from the premisses, it is a mechanical task.

Moreover, it is incorrect to say that the conclusion is contained in the premisses as plants are contained in their seeds, not as beams are contained in a house. For, plants can develop from seeds only by absorbing water from the soil and harvesting energy from the sun, therefore using something which is not contained in the seeds. On the contrary, the conclusion is deduced from the premisses without using anything not contained in the premisses.

16.5 The Argument of Novelty

Another argument against the claim that deductive rules are non-ampliative is that the conclusion of a deductive argument, even when it only involves individuals mentioned in the premisses, tells us something about them that the premisses did not tell us.

Thus, Russell argues that "it is an old debate among philosophers whether deduction ever gives new knowledge. We can now see that in certain cases, at least, it does do so" (Russell 1998, 44). Thus, "if we already know that two and two always make four, and we know that Brown and Jones are two, and so are Robinson and Smith, we can deduce that Brown and Jones and Robinson and Smith are four" (ibid.). This "is new knowledge, not contained in our premisses, because the general proposition, 'two and two are four', never told us there were such people as Brown and Jones and Robinson and Smith, and the particular premises do not tell us that there were four of them" (ibid.).

This argument, however, is not valid. For, the premiss, 'Brown and Jones are two', told us that there were such people as Brown and Jones and that there were two of them. And the premiss, 'Robinson and Smith are two', told us that there were such people as Robinson and Smith and that there were two of them. Then this, together with the premiss that two and two always make four, told us that there were such people as Brown and Jones and Robinson and Smith, and that there were four of them.

In fact, Russell eventually recognizes that his argument is not valid. Indeed, he says that "deduction has turned out to be much less powerful than was formerly supposed; it does not give new knowledge, except as to new forms of words for stating truths in some sense already known" (Russell 2009, 140). Thus "if, from a theorem *A*, you deduce a theorem *B*, it must be the case that *B* repeats *A* (or part of it) in other words" (Russell 1997, 360).

16.6 The Argument of Trivialization

Another argument against the claim that deductive rules are non-ampliative is that, if they were non-ampliative, mathematics would reduce to triviality. For then, as soon as we know the axioms of a theory, we should thereby know all its theorems, but this is nonsense.

Thus, Dummett argues that, if deductive rules were non-ampliative, then as soon as we know "the axioms of a mathematical theory, we should thereby know all the theorems. Obviously, this is nonsense: deductive reasoning has here been justified at the expense of its power to extend our knowledge and hence of any genuine utility" (Dummett 1991, 195). In particular, "Kant made the mistake of supposing that what is arrived at by analysis cannot be new; and so he relegated analytic truths to the rank of trivialities" (ibid., 199).

This argument, however, is not valid. For, our mind is incapable of making even relatively short deductions on its own.

Thus, Hadamard says: "I have several times happened to overlook results which" were "immediate consequences of other ones which I had obtained" (Hadamard 1954, 50). In particular, "two theorems, important to the subject" of my doctoral thesis, "were such obvious and immediate consequences of the ideas contained therein that, years later, other authors imputed them to me, and I was obliged to confess that, evident as they were, I had not perceived them" (ibid., 51).

Since our mind is incapable of making even relatively short deductions on its own, we do not know all the theorems of a mathematical theory a soon as we know the axioms of the theory. This has nothing to do with ampliativity.

16.7 The Argument of Formation of New Concepts

Another argument against the claim that deductive rules are non-ampliative is that, to grasp that the premisses of a deductive argument entail the conclusion, in general it is necessary the formation of new concepts.

Thus, Dummett argues that, to grasp that the premisses of a deductive argument entail the conclusion, "it is not enough merely to grasp the thoughts expressed by" the premisses and the conclusion, it is necessary "the formation of new concepts" (Dummett 1991, 198). Thus, "to grasp that 'Every university that appoints as president a professor of that university solves its financial problems' and 'Harvard has appointed a Harvard professor president' together entail 'Harvard will solve its financial problems', it is necessary to perceive the minor premiss as containing the predicate 'x (university) has appointed a professor of x university president' (ibid., 197). This requires the formation of "a new concept, that of an internally appointed university president" (ibid., 196).

This argument, however, is not valid. For, the concept of an internally appointed university president is already contained in the premiss: Every university that appoints as president a professor of that university solves its financial problems.

16.8 The Argument of Indirect Means

Another argument against the claim that deductive rules are non-ampliative is that a deductive argument provides an indirect means for seeing that the conclusion is true.

Thus, Dummett argues that a "deductive argument" provides "an effective procedure for rearranging the operations accompanying any given observations which yielded a pattern amounting to recognition of the truth of the initial premisses, so as to yield one constituting recognition of the truth of the conclusion" (Dummett 1991, 176). For example, consider "Euler's proof that anyone who, in the course of one journey, crossed every bridge at Konigsberg crossed at least one of them at least twice. The proof shows us an effective means, given observations establishing that someone had crossed every bridge, to make observations establishing that he crossed some bridge twice" (ibid.). The proof "shows how we could arrive" at "the position of recognising directly the truth of the conclusion" if "we began by being in a position to recognise directly the truth of the premisses. By doing this," the proof "provides us with an indirect means of seeing that the conclusion is true" (ibid., 177).

This argument, however, is not valid. For, the proof that provides us with an indirect means of seeing that the conclusion is true is not a deductive argument. Indeed, it requires introducing new concepts, not contained in the direct recognition of the truth of the premisses. That is, it requires introducing the concept of a graph to represent the bridges of Königsberg without knowing the details of the bridges' physical construction, the concepts of vertex of a graph, and the concept of degree of a vertex.

16.9 The Argument of New Individuals

Another argument against the claim that deductive rules are non-ampliative is that, in a deductive inference, the passage from premisses to conclusion may involve introducing new individuals, so intuitions in Kant's sense, therefore a deductive inference is synthetic in the sense Kant had in mind, namely it is ampliative.

Thus, Hintikka argues that often, in a deductive inference, "the transition from the premisses to the conclusion cannot be accomplished without the help of individuals which are not considered in the premisses" (Hintikka 1973, 192). For example, "the inference from 'everyone is mortal' to 'Socrates is mortal' might very well be conceived of as being synthetic in the sense Kant had in mind," namely ampliative, because it introduces in the conclusion the term 'Socrates' which is "a representative of an individual, and hence an intuition in Kant's sense" (ibid., 194).

This argument, however, is not valid. The argument assumes that, according to Kant, in syllogisms a singular judgment, such as ‘Socrates is mortal’, must be treated differently from a universal judgment, such as ‘everyone is mortal’. But this assumption is not valid, because Kant says that, “in the employment of judgments in syllogisms, singular judgments can be treated like those that are universal” (Kant 1998, A71/B96). For, the predicate holds of the concept of the subject “without exception, just as if the latter were a generally valid concept with the predicate applying to the whole of what is signified” (ibid.). Thus, the singular judgment “God is without errors” can be treated as the universal judgment “All that is God is without errors” (Kant 1900-, Reflexion 3080, XVI, 648). Similarly, the singular judgment ‘Socrates is mortal’ can be treated as the universal judgment ‘All that is Socrates is mortal’. Therefore, the inference from ‘everyone is mortal’ to ‘Socrates is mortal’ does not introduce an intuition. (On this question, see Capozzi 2020).

16.10 The Argument of Validity

Another argument against the claim that deductive rules are non-ampliative is that, in a valid deductive argument, if the premisses are true, then the conclusion is true, so if we know that the premisses are true, we can infer that the conclusion is true. Therefore, deductive rules extend knowledge. Conversely, with a non-deductive argument, if we know that the premisses are true, all that we can infer is only that the conclusion has some probability. Therefore, non-deductive rules do not extend knowledge, they only yield reasonable belief.

Thus, Swinburne argues that, “if an argument is a valid deductive one, then, if the premisses are true, the conclusion must be. If we know the premisses to be true, and see that the argument is a valid deductive one, then the argument gives us knowledge of the conclusion” (Swinburne 1974, 6). Conversely, with a non-deductive argument, “if we know the premisses to be true and see that the argument is a correct” non-deductive “one, all that we can conclude is something about the probability of the conclusion,” thus a correct non-deductive argument “yields only probable knowledge or reasonable belief” (ibid.). Therefore, non-deductive arguments do not extend knowledge, only valid “deductive” arguments “extend knowledge” (ibid., 7).

This argument, however, is not valid. The argument assumes that knowledge, in order to be such, must be true, but this conflicts with the fact that, as argued in Chap. 11, the aim of science is not truth but plausibility. Therefore, the premisses of a valid deductive argument can only be plausible, like the premisses of a non-deductive argument. Moreover, as will be argued below, both valid deductive arguments and non-deductive arguments are not plausibility-preserving, that is, they are not such that, if the premisses are plausible, the conclusion is necessarily plausible.

16.11 The Argument of Unpredictability of Deducibility

Another argument against the claim that deductive rules are non-ampliative is that, by the undecidability of first-order logic, we are unable to predict whether a conclusion P is deducible from given premisses Γ.

Thus, Mondadori argues that, by the "undecidability of first-order logic," even if "any consequence P of a given set Γ of propositions can be extracted 'effortlessly' by applying the semi-decision procedure," nevertheless "the fact that P follows from Γ remains 'surprising', because we are unable to predict even whether there will be an answer, until the very moment the answer arrives (if it arrives)" (Mondadori 2000, 131), This is "in contrast with the thesis of the non-ampliative character of the inference" (ibid.).

This argument, however, is not valid. For, it does not concern the ampliativity of deduction, but the existence of a deduction. What provides new information is that there exists a deduction of P from Γ. But the question of the ampliativity of deduction is not whether there exists a deduction of P from Γ, it is instead whether, given a deduction of P from Γ, P contains new information not already implicitly contained in Γ.

16.12 Utility of Deductive Rules

From what has been said above it is clear that the arguments against the claim that deductive rules are non-ampliative are not valid. Therefore, it seems fair to conclude that deductive rules are non-ampliative. This raises the question: What is their utility?

An answer can be given by observing that a deduction makes explicit all or part of what is contained in the premisses of the deduction. It thus facilitates the comparison of the premisses with experience, and so helps to determine whether the premisses are plausible. This is the utility of deductive rules.

That a deduction facilitates the comparison of the premisses with experience depends on the fact that, as said above, our mind is incapable of making even relatively short deductions on its own. To make them, it must use some deductive rules.

This does not mean, however, that the conclusion of a deduction can go beyond what is contained in the premisses of the deduction.

In this connection, a distinction must be made between objective novelty and psychological novelty. The conclusion of a deduction can be psychologically unexpected, and therefore can have psychological novelty, because our mind is incapable of making even relatively short deductions on its own. But it cannot go beyond what is contained in the premisses of the deduction, and therefore cannot have objective novelty.

As Hempel says, "a mathematical theorem, such as the Pythagorean theorem in geometry, asserts nothing that is objectively or theoretically new as compared with the postulates from which it is derived, although its content may well be psychologically new in the sense that we were not aware of its being implicitly contained in the postulates" (Hempel 2001, 21).

16.13 The Paradox of Inference

As said above, the role of non-deductive rules in knowledge is to find hypotheses to solve problems, and they can have this role because they are ampliative, namely such that the conclusion is not contained in the premisses.

On the other hand, non-deductive rules are not truth-preserving, namely they are not such that, if the premisses are true, then the conclusion is true. This is because non-deductive rules are ampliative, and ampliative inference rules cannot be truth-preserving.

That ampliative inference rules cannot be truth-preserving follows from the so-called paradox of inference. The paradox is often credited to Cohen and Nagel (see Cohen and Nagel 1934, 173). But in fact, it is due to Bosanquet.

Indeed, Bosanquet says that "the problem of inference is something of a paradox," the "paradox of inference" (Bosanquet 1895, 137). For, on the one hand, "inference consists in asserting as fact or truth, on the ground of certain given facts or truths, something which is not included in those data" (ibid.). On the other hand, "we have not got inference unless the conclusion, (i) is necessary from the premisses, and (ii) goes beyond the premisses" (Bosanquet 1895, 137).

That is, we have not got an inference unless (i) the conclusion is true if the premisses are true, and (ii) the conclusion is not contained in the premisses.

Then, the paradox of inference expresses that an inference rule cannot be both (i) truth-preserving, and (ii) ampliative.

From the paradox of inference it follows that non-deductive rules, being ampliative, cannot be truth-preserving.

16.14 The Case of Abduction

However, from the paradox of inference it does not follow that deductive rules, being non-ampliative, are necessarily truth-preserving. An example of this is abduction.

As argued in Chap. 15, the abduction rule Abd is non-ampliative, because the conclusion A is contained in the premiss $A \rightarrow B$.

Since Abd is non-ampliative, it cannot be a non-deductive rule, so Abd must be a deductive rule.

On the other hand, however, Abd is not truth-preserving. For, if B is true and A is false, then both premisses B and $A \rightarrow B$ are true but the conclusion A is false.

Thus, Abd is a deductive rule that is not truth-preserving. Therefore, the fact that an inference rule is non-ampliative does not imply that the inference rule is truth-preserving.

16.15 Classification of Inference Rules

Since Abd is a deductive rule and is not truth-preserving, this means that not all deductive rules are truth-preserving.

We say that an inference rule is valid if it is truth-preserving, non-valid if it is not truth-preserving.

Then, all non-deductive rules are not truth-preserving and hence are non-valid. On the other hand, Abd is a deductive rule but, like non-deductive rules, is non-valid. On the contrary, many deductive rules, for example modus ponens, are truth-preserving and hence are valid (see below).

Together with what has been said earlier in this chapter, this leads to the following classification of inference rules.

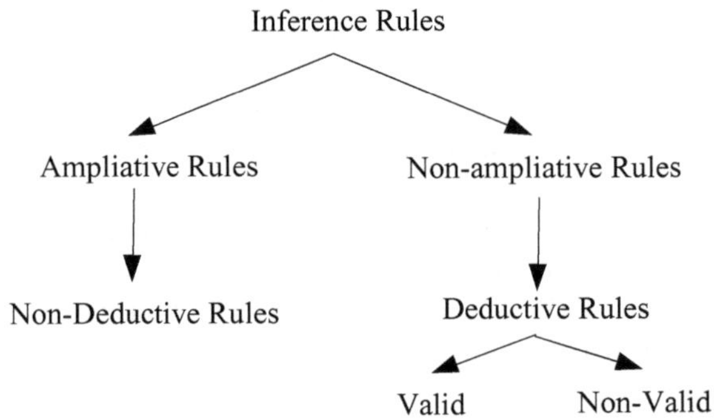

16.16 The Tortoise Argument

It is widely believed that, since valid deductive rules are truth-preserving and non-deductive rules are not truth-preserving, valid deductive rules are justified and non-deductive rules are not justified.

However, the limitations of this belief were already highlighted by Carroll, using a variant of the Achilles and the tortoise argument applied to "a little bit of the argument in" the first proposition of Euclid's *Elements*, "just two steps, and the conclusion drawn from them" (Carroll 1995, 691). Let us call the two steps and the conclusion *A*, *B*, and *Z*:

(*A*) Things that are equal to the same are equal to each other.

(*B*) The two sides of this triangle are things that are equal to the same.

(*Z*) The two sides of this triangle are equal to each other.

Achilles tells the tortoise that she must grant that "Z follows logically from *A* and *B*, so that any one who accepts *A* and *B* as true, must accept *Z* as true" (ibid.). But the tortoise replies that she accepts *A* and *B* but does not accept *Z* because she does not accept the hypothetical:

(*C*) If *A* and *B* are true, *Z* must be true.

Then Achilles tells the tortoise that she must grant that *Z* follows logically from *A* and *B* and *C*. But the tortoise replies that she accepts *A* and *B* and *C* but does not accept *Z* because she does not accept the hypothetical:

(*D*) If *A* and *B* and *C* are true, *Z* must be true.

Then Achilles tells the tortoise that she must grant that Z follows logically from *A* and *B* and *C* and *D*. But the tortoise replies that she accepts *A* and *B* and *C* and D but does not accept *Z* because she does not accept the hypothetical:

(E) If *A* and *B* and *C* and *D* are true, *Z* must be true.

And so on, ad infinitum.

The tortoise comments on this by saying to Achilles: "Would you mind, as a personal favour, considering what a lot of instruction this colloquy of ours will provide for the logicians of the nineteenth century?" (ibid., 693).

16.17 The Asymmetry View

Contrary to the tortoise's expectation, however, the colloquy between Achilles and the tortoise has failed to provide instruction not only for the logicians of the nineteenth century but also for those of the twentieth and the twenty-first centuries.

Although the tortoise argument challenges the view that valid deductive rules are justified, even today it is widely believed that, while non-deductive rules are not justified, valid deductive rules are justified.

Thus, Salmon says that valid deductive rules are justified because, if we reflect upon a valid deductive rule, we "cannot conceive the possibility of any situation in which its use would lead from true premises to a false conclusion" (Salmon 1965, 268). So "we can find no grounds whatever for withholding the judgment" that the deductive rule in question "is truth preserving" (ibid.). Conversely, non-deductive rules are not justified because "we can, without difficulty, imagine all sorts of states of affairs in which practically all—if not absolutely all—of our future" non-deductive "inferences with true premises turn out to have false conclusions" (ibid.).

The view that, while non-deductive rules are not justified, valid deductive rules are justified, may be called the asymmetry view.

16.18 Origin of the Asymmetry View

The origin of the asymmetry view can be attributed to Hume.

According to him, there are two types of justification: intuitional and inferential. And there are two types of inferential justification: demonstrative and inductive or, as Hume says, by experience.

Now, the ground for an inductive inference "is not intuitive" (Hume 2007b, 27). Induction cannot be justified demonstratively, "since it implies no contradiction that the course of nature may change" (ibid., 25). Induction cannot be justified inductively or, as Hume says, 'by experience', since to say that its justification is by experience "is begging the question. For all inferences from experience suppose, as their foundation, that the future will resemble the past, and that similar powers will be conjoined with similar sensible qualities" (ibid., 27). It "is impossible, therefore, that any arguments from experience can prove this resemblance of the past to the future; since all these arguments are founded on the supposition of that resemblance" (ibid.). To claim that an argument from experience can prove it "must be evidently going in a circle" (ibid., 26).

Since induction can have neither an intuitional, nor a demonstrative, nor an inductive justification, induction cannot be justified at all.

On the other hand, deduction cannot have a demonstrative justification, because it would have to use those very same deductive rules that are being justified, and "the same principle cannot be both the cause and effect of another" (Hume 2007a, I, 63). Deduction cannot have an inductive justification, because induction itself cannot be justified. However, deduction has an intuitional justification, because its correctness is "discoverable at first sight," so the justification of deduction falls "under the province of intuition" (ibid., I, 50).

16.19 Untenability of the Asymmetry View

The asymmetry view, however, is untenable because the fact that valid deductive rules are truth-preserving does not provide a justification for them. For, the argument by which a valid deductive rule is shown to be truth-preserving is circular.

For example, let us consider modus ponens, that is, the inference rule:

$$(\mathrm{MP})\frac{A \quad A \to B}{B}.$$

The rule MP is non-ampliative, because the conclusion B is contained in the premiss $A \to B$, therefore MP is a deductive rule.

But MP is truth-preserving. This is shown by the following argument. Suppose that A and $A \to B$ are true. By the truth table of $\to$, if A is true, then, if $A \to B$ is true, then B is true. From this, by two applications of MP, we conclude that B is true.

Thus MP, being a deductive rule that is truth-preserving, is valid. However, the argument by which MP is shown to be truth-preserving involves two applications of MP, therefore it is circular.

This is even clearer if the argument is rewritten as follows, where T(A) stands for 'A is true' and TT($\rightarrow$) stands for 'By the truth table of $\rightarrow$':

$$\text{(MP)}\ \dfrac{\mathrm{T}(A \rightarrow B) \qquad \text{(MP)}\ \dfrac{\mathrm{T}(A) \quad \overbrace{\mathrm{T}(A) \rightarrow (\mathrm{T}(A \rightarrow B) \rightarrow \mathrm{T}(B))}^{\mathrm{TT}(\rightarrow)}}{\mathrm{T}(A \rightarrow B) \rightarrow \mathrm{T}(B)}}{\mathrm{T}(B)}.$$

Since the argument involves two applications of MP, the argument does not provide a justification for MP.

Against this it could be argued that the two applications of MP are in the metalanguage, while the MP that is shown to be truth-preserving applies to expressions in the language. Therefore, the argument is not circular.

Thus, Dummett says: "The circularity complained of is not the ordinary gross circularity that consists of including the conclusion to be reached among the initial premisses of the argument," but "only that at least one of the inferential steps in the argument must be taken in accordance with that law" (Dummett 1991, 202). In fact, "the philosopher does not seriously doubt the validity" of MP, so he "does not need to be persuaded of the truth of the conclusion; what he is seeking is an explanation of its being true" (ibid.). While "a gross circularity is as damaging to an explanatory argument as to a suasive one," this kind of circularity "need do it no harm at all" (ibid.).

But this argument is not valid. For, by an application of Abd in the metalanguage, we could show that Abd applied to expressions in the language is truth-preserving.

Indeed, suppose that B and $A \rightarrow B$ are true. Then, by the truth table for $\rightarrow$, $(A \rightarrow B) \rightarrow B$ is true. Also, by the truth table for $\rightarrow$, if A is true, then $(A \rightarrow B) \rightarrow B$ is true. From the fact that $(A \rightarrow B) \rightarrow B$ is true and that, if A is true, then $(A \rightarrow B) \rightarrow B$ is true, by an application of Abd we may conclude that A is true. Therefore, Abd is truth-preserving.

This is even clearer if the argument is rewritten as follows:

$$\text{(Abd)}\ \dfrac{\mathrm{TT}(\rightarrow)\ \dfrac{\mathrm{T}(B) \quad \mathrm{T}(A \rightarrow B)}{\mathrm{T}((A \rightarrow B) \rightarrow B)} \qquad \overbrace{\mathrm{T}(A) \rightarrow \mathrm{T}((A \rightarrow B) \rightarrow B)}^{\mathrm{TT}(\rightarrow)}}{\mathrm{T}(A)}.$$

This conflicts with the fact that, as already said, if B is true and A is false, then both premisses B and $A \rightarrow B$ are true but the conclusion A is false, therefore Abd is not truth-preserving.

Since the argument by which a valid deductive rule is shown to be truth-preserving is circular, the fact that valid deductive rules are truth-preserving does not provide a justification for them. Therefore, the asymmetry view is untenable.

16.20 Characterization of Non-Deductive and Deductive Rules

In Chap. 15 and above, non-deductive rules and deductive rules have been characterized in terms of ampliativity. Non-deductive rules are inference rules that are ampliative, deductive rules are inference rules that are non-ampliative.

However, non-deductive rules and valid deductive rules are commonly characterized in terms of truth-preservation. Non-deductive rules are inference rules that are not truth-preserving, valid deductive rules are inference rules that are truth-preserving.

Thus, Salmon says that a non-deductive rule is an inference rule whose "conclusion is not necessitated by its premises," namely, "the conclusion could be false even if the premises are true" (Salmon 1967, 8). Therefore, a non-deductive rule is an inference rule that is not truth-preserving. On the other hand, a valid deductive rule is an inference rule "whose premises necessitate its conclusion," namely, "the conclusion cannot be false if the premises are true" (ibid.). Therefore, a valid deductive rule it is an inference rule that is "truth-preserving" (ibid.).

But the characterization of non-deductive rules and valid deductive rules in terms of truth-preservation depends on the characterization in terms of ampliativity.

Even Salmon admits this. For, he says that, once valid deductive rules "have been characterized in terms of their basic property of necessary truth preservation, it is natural to ask how they achieve this very desirable trait" (Salmon 1967, 8). The answer is that valid deductive rules "purchase necessary truth preservation by sacrificing any extension of content. The conclusion" of a valid deductive rule "says no more than do the premises—often less. The conclusion cannot be false if the premises are true because the conclusion says nothing that was not already stated in the premises. The conclusion is a mere reformulation of all or part of the content of the premises" (ibid.). Admittedly, "in some case the reformulation is unanticipated and therefore psychologically surprising, but the conclusion cannot augment the content of the premises," the rule is "nonampliative; an ampliative" rule "has a conclusion with content not present either explicitly or implicitly in the premises" (ibid.).

Since the characterization of non-deductive rules and valid deductive rules in terms of truth-preservation depends on that in terms of ampliativity, the latter is primary. (For more on this, see Cellucci 2013, Chapter 18).

16.21 Failure of Plausibility-Preservation

The characterization of non-deductive rules and valid deductive rules in terms of truth-preservation is motivated by the assumption that the aim of science is truth. If the aim of science is truth, then valid deductive rules must be truth-preserving.

But, as argued in Chap. 11, the view that the aim of science is truth is inadequate, it must be replaced with the view that the aim of science is plausibility. If the aim of science of plausibility, then deductive rules should be plausibility-preserving, that is, they should be such that, if the premisses are plausible, the conclusion is plausible.

This, however, runs into the difficulty that, while non-deductive rules are not truth-preserving and valid deductive rules are truth-preserving, both non-deductive rules and valid deductive rules are not plausibility-preserving. This can be seen as follows.

(1) Non-deductive rules are not plausibility-preserving.

Thus, let us consider the rule of induction: Several things of type *A* are *B*, therefore all things of type *A* are *B*.

Suppose we have observed several ravens and they are black. Then the statement 'Several ravens are black' is plausible. From it, by the rule of induction we can infer 'All ravens are black'. But this statement is not plausible, because there are white ravens.

Therefore, the rule of induction is not plausibility-preserving.

(2) Valid deductive rules are not plausibility-preserving.

Thus, let us consider the conjunction introduction rule: *A* and *B*, therefore $A \wedge B$.

Suppose that *A* is quantum mechanics and *B* Einstein's general relativity. Then, as already said in Chap. 11, *A* and *B* are both confirmable but mutually incompatible. Since *A* and *B* are both confirmable, they are both plausible, but since they are mutually incompatible, their conjunction $A \wedge B$ cannot be plausible.

Therefore, the conjunction introduction rule is not plausibility-preserving.

16.22 Inference Rules and Knowledge

The fact that both non-deductive rules and valid deductive rules are not plausibility-preserving has an important consequence.

It implies that hypotheses obtained by non-deductive rules and conclusions deduced from them by valid deductive rules can be not plausible, and therefore not productive of knowledge.

The analytic method accounts for this fact. For, according to it, hypotheses obtained from plausible premisses by non-deductive rules can be not plausible. That they are plausible can be established only through a comparison between the arguments for them and the arguments against them on the basis of experience.

Also, conclusions deduced from plausible hypotheses can be not plausible. That they are plausible can be established only through a comparison between the arguments for them and the arguments against them on the basis of experience.

Therefore, to produce knowledge, both non-deductive rules and valid deductive rules must be complemented by a procedure for establishing that the conclusions of these rules are plausible.

References

Bosanquet, Bernard. 1895. *The essentials of logic: Being ten lectures on judgment and inference*. London: Macmillan.

Capozzi, Mirella. 2020. Singular terms and intuitions in Kant: A reappraisal. In *Kant's philosophy of mathematics*, ed. Carl Posy and Ofra Rechter, vol. 1, 103–125. Cambridge: Cambridge University Press.

Carroll, Lewis. 1995. What the Tortoise said to Achilles. *Mind* 104:691–693.

Cellucci, Carlo. 2013. *Rethinking logic: Logic in relation to mathematics, evolution, and method*. Berlin: Springer.

Cohen, Morris, and Ernest Nagel. 1934. *An introduction to logic and scientific method*. London: Routledge.

De Morgan, Augustus. 1835. Review of George Peacock, *A treatise on algebra*. *Quarterly Journal of Education* 9 (91–110): 293–311.

Descartes, René. 1996. *Oeuvres*. Paris: Vrin.

Dummett, Michael. 1991. *The logical basis of metaphysics*. Cambridge: Harvard University Press.

Frege, Gottlob. 1960. *The foundations of arithmetic: A logico-mathematical enquiry into the concept of number*. New York: Harper.

Hadamard, Jacques. 1954. *The psychology of invention in the mathematical field*. Mineola: Dover.

Hempel, Carl. 2001. *The philosophy of Carl G. Hempel: Studies in science, explanation, and rationality*. Oxford: Oxford University Press.

Hintikka, Jaakko. 1973. *Logic, language-games and information: Kantian themes in the philosophy of logic*. Oxford: Oxford University Press.

Hume, David. 2007a. *A treatise of human nature*. Oxford: Oxford University Press.

Hume, David. 2007b. *An enquiry concerning human understanding*. Oxford: Oxford University Press.

Kant, Immanuel. 1900. *Gesammelte Schriften (Akademie Aausgabe)*. Berlin: Königlich Preußischen Akademie der Wissenschaften.

Kant, Immanuel. 1998. *Critique of pure reason*. Cambridge: Cambridge University Press.

Locke, John. 1975. *An essay concerning human understanding*. Oxford: Oxford University Press.

Mill, John Stuart. 1963–1986. *Collected works*. Toronto: University of Toronto Press.

Mondadori, Marco. 2000. Sul carattere ampliativo dell'inferenza. *Rivista di filosofia* 91:127–145.

Peirce, Charles Sanders. 1931–1958. *Collected papers*. Cambridge: Harvard University Press.

Poincaré, Henri. 2015. *The foundations of science: Science and hypothesis—The value of science—Science and method*. Cambridge: Cambridge University Press.

Russell, Bertrand. 1997. Is mathematics purely linguistic? In Bertrand Russell, *Collected papers*, vol. 11, 352–362. London: Routledge.

Russell, Bertrand. 1998. *The problems of philosophy*. Oxford: Oxford University Press.

Russell, Bertrand. 2009. *Human knowledge: Its scope and limits*. London: Routledge.

Salmon, Wesley. 1965. The concept of inductive evidence. *American Philosophical Quarterly* 2:265–270.

Salmon, Wesley. 1967. *The foundations of scientific inference*. Pittsburgh: University of Pittsburgh Press.

Swinburne, Richard. 1974. Introduction. In *The justification of induction*, ed. Richard Swinburne, 1–17. Oxford: Oxford University Press.

Turing, Alan. 1939. Systems of logic based on ordinals. *Proceedings of the London Mathematical Society* 45:161–228.

Turing, Alan. 2004. Letters on logic to Max Newman (c. 1940). In *The essential Turing*, ed. Jack Copeland, 205–216. Oxford: Oxford University Press.

Wittgenstein, Ludwig. 2001. *Tractatus logico-philosophicus*. London: Routledge.

Chapter 17
Knowledge and Logic

Abstract Developing the view that philosophy is acquisition of knowledge also requires answering the question: What is the role of logic in knowledge? Answering this question depends on what logic is. The current dominant view is that there is a single paradigm of logic, logic as justification, consisting in mathematical logic. This chapter, however, argues that this view is not valid, because mathematical logic is inadequate as a means of justification. An alternative to the paradigm of logic as justification is the paradigm of logic as discovery. According to it, logic is a method of discovery and is the analytic method. The latter is adequate as a method of discovery, and so also as a means of justification because in the analytic method justification is part of discovery.

Keywords Aristotle's logic · Mathematical logic · Inadequacy of mathematical logic · The analytic method as logic · Adequacy of the analytic method as logic

17.1 The Single Paradigm View

Developing the view that philosophy is acquisition of knowledge also requires answering the question: What is the role of logic in knowledge?

Answering this question depends on what logic is. The current dominant view is that there is a single paradigm of logic, logic as justification. According to it, logic is a means of justification and is a theory of deduction. The paradigm originated with Aristotle's logic and its current form is mathematical logic. The latter contains Aristotle's logic as a small fragment, it was created to justify mathematics by putting it on a firm foundation, and is adequate as a logic of justification.

Thus, Tarski says that logic is a means of justification and is a theory of deduction, because deduction "serves as an indispensable tool for deriving conclusions from accepted assumptions" (Tarski 1994, xi). Logic "was created by Aristotle" and its current form is "mathematical logic" (ibid., 18, footnote 1). Aristotle's logic "forms only a fragment" of mathematical logic, "moreover a fragment which is entirely insignificant from the standpoint of requirements of other sciences, and of

C. Cellucci, *A Philosophy for the Future*, Synthese Library 508,
https://doi.org/10.1007/978-3-031-98443-3_17

mathematics in particular" (ibid., 17–18). Instead, mathematical logic "is highly relevant from the standpoints of the requirements of mathematics," because it "was created for the purpose of making the foundations of mathematics firmer and more profound" (ibid., ix).

This chapter, however, argues that this view is not valid. Aristotle's logic does not belong to the paradigm of logic as justification, and mathematical logic is inadequate as a means of justification. An alternative to the paradigm of logic as justification is the paradigm of logic as discovery. According to it, logic is a method of discovery and is the analytic method. The analytic method is adequate as a method of discovery, and also as a means of justification, because in the analytic method justification is part of discovery.

17.2 The Paradigm of Logic as Justification

Let us first consider the claim that the paradigm of logic as justification originated with Aristotle's logic.

The current dominant view about Aristotle's logic is that Aristotle's logic is a theory of deduction.

Thus, Boger says that Aristotle's "principal concern" was "with deduction" (Boger 2004, 106).

Stricker says that "Aristotle's logic" is "a general theory of deductive argument" (Striker 2009, xviii).

Smith says that "all Aristotle's logic revolves around one notion: the deduction" (Smith 2017).

In particular, according to the current dominant view, Aristotle's logic is not a logic of discovery.

Thus, Woods says that "Aristotle certainly did not think of the formal theory of the syllogism developed in the *Prior Analytics* as a logic of discovery" (Woods 2012, 575).

But this is unfounded, Aristotle's logic is a method of discovery. As Aristotle makes quite clear, logic must show how to find premisses for solving any problem proposed.

Indeed, Aristotle says that logic must indicate "how to reach for premisses concerning any problem proposed, in the case of any discipline whatever," and, generally, how to find "the way through which we may obtain the principles concerning each subject" (Aristotle, *Analytica Priora*, B 1, 53 a 1–3). Therefore, logic must indicate "by what method we will find premisses about each thing" (ibid., A 27, 43 a 21–22).

In fact, according to Aristotle, logic indicates that premisses will be obtained from the conclusion "either by syllogism or by induction" (Aristotle, *Topica* Θ 1, 155 b 35–36).

With the expression 'by syllogism' Aristotle means: by the procedure for "seeking what the middle term is" (Aristotle, *Analytica Posteriora* B 2, 90 a 1).

This is the procedure stated in *Analytica Priora* A 27–31, which the medievals called 'inventio medii' because by means of it, starting from a given conclusion, one can find the middle term, and hence the premisses, of a syllogism, from which the conclusion can be deduced (for details, see Cellucci 2013, Chapter 7). From the problem, by repeated applications of inventio medii or induction, one arrives at principles. Once arrived at principles, by retracing the steps followed from the problem to principles, one obtains a deduction of a solution to the problem from the principles.

Since Aristotle's logic is a method of discovery, it cannot be the origin of the paradigm of logic as justification.

In this regard, it is significant that historians of logic inspired by mathematical logic consider Aristotle's view of logic irrelevant to logic. According to them, discovery is not a matter of logic.

Thus, Bochenski says: "One can ask two different basic questions: (1) What follows given premisses? (2) From what premisses can a given sentence (conclusion) be deduced? Aristotle primarily considered the first question," namely justification, but "poses also the second," namely discovery, "and tries to show" how "the premisses of a syllogism must be constructed in order to yield a given conclusion," but "we do not need to pursue the details of this theory" (Bochenski 1961, 81). For, discovery "is not essentially a matter of formal logic" (ibid., 80).

17.3 Origin of the Paradigm of Logic as Justification

The origin of the paradigm of logic as justification is not Aristotle's logic but Stoic logic.

Indeed, the Stoics say that the "purpose" of logic "is to show that, when premisses are compounded with one another in certain ways, something may be deduced by necessity from what is posited or conceded" (Alexander of Aphrodisias, *In Aristotelis Analyticorum Priorum Librum Primum Commentarium*, 1.20–2.2, ed. Wallies). Therefore, logic is the science of deduction, "it shows us what is capable of yielding demonstration," and leads "to the formation of correct judgements" (Diogenes Laertius, *Lives of Eminent Philosophers*, 7.45: 76–77, ed. Dorandi).

Since, according to the Stoics, logic is the science of deduction and leads to the formation of correct judgements, Stoic logic belongs to the paradigm of logic as justification, and indeed is the origin of the paradigm, because there is no evidence of any previous occurrence of the paradigm.

From this, it is clear that Aristotle's logic and Stoic logic belong to two different paradigms of logic.

17.4 Two Paradigms of Logic

That Aristotle's logic and Stoic logic belong to two different paradigms of logic is already made clear by Cicero.

Indeed, Cicero says that "all methodical treatment of rational discourse involves two arts, one of discovering and the other of judging" (Cicero, *Topica*, 6, ed. Reinhardt). While "Aristotle came first in both," the "Stoics laboured in the second: in fact they pursued the art of judging diligently," but "completely neglected" the "art of discovering" (ibid.).

Therefore, Aristotle's logic and Stoic logic belong to two different paradigms of logic.

However, while Stoic logic is the origin of the paradigm of logic as justification, Aristotle's logic is not the origin of the paradigm of logic as discovery. The origin of the latter is instead the analytic method, and Aristotle's logic is different from the analytic method.

17.5 The Analytic-Synthetic Method

Not only Aristotle's logic is different from the analytic method, but is the origin of an alternative to the analytic method, the analytic-synthetic method.

The analytic-synthetic method, or method of analysis and synthesis, is the method according to which, to solve a problem, one looks for some hypothesis that is a sufficient condition for solving the problem, namely such that a solution to the problem can be deduced from the hypothesis.

The hypothesis is obtained from the problem, and possibly other data already available, by some non-deductive rule, and must be plausible, namely such that the arguments for it are stronger than the arguments against it, on the basis of experience.

If the hypothesis so obtained is not a principle, the hypothesis is in turn a problem that must be solved, and is solved in the same way. Namely, one looks for another hypothesis that is a sufficient condition for solving the problem posed by the previous hypothesis, it is obtained from the latter, and possibly other data already available, by some non-deductive rule, and must be plausible. And so on, until one arrives at a principle.

When one arrives at a principle, the process terminates. This is analysis.

Then, one retraces the steps followed in analysis, obtaining a deduction of a solution of the problem from the principles arrived at in analysis. This is synthesis.

The analytic-synthetic method can be schematically represented as follows.

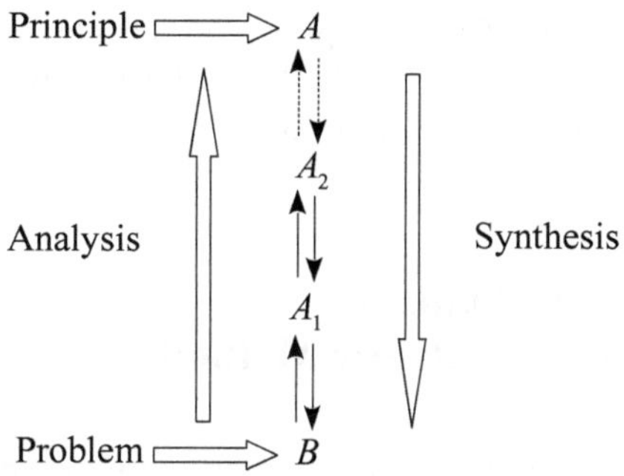

17.6 Original Formulation of the Analytic-Synthetic Method

The analytic-synthetic method was used by some mathematicians, but its first formulation was given by Aristotle.

Aristotle formulates the analytic-synthetic method by saying that "the process of knowledge proceeds from what is more knowable and clearer to us to what is clearer and more knowable by nature" (Aristotle, *Physica*, A 1, 184 a 16–18). And "we must necessarily proceed in this way, namely, starting from the things that are less known by nature but clearer to us, towards those which are clearer and more knowable by nature" (ibid., A 1, 184 a 18–21).

Now, what is more knowable and clearer to us is the problem to solve, while what is clearer and more knowable by nature are the principles. They are "most knowable" by nature because, "by reason of these, and from these, all other things are known, but these are not known by means of the things subordinate to them" (Aristotle, *Metaphysica*, A 2, 982 b 2–4). So, by saying that we must proceed from what is more knowable and clearer to us to what is clearer and more knowable by nature, Aristotle means to say that we must proceed from the problem to principles.

In fact, Aristotle says that, given the problem to solve, we must find "the necessary hypotheses through which the syllogisms come about" (Aristotle, *Topica*, Θ 1, 155 b 29). As we have seen above, according to Aristotle, we will find them either by syllogism or by induction. When the hypotheses are found, we "should not put forward these" hypotheses "right away, but rather should stand off as far above them as possible" (ibid., Θ 1, 155 b 29–30). Namely, we should find other hypotheses from which the previous hypotheses can be deduced. And so on, until we arrive at a principle.

When we arrive at a principle, the process terminates. This is analysis.

Then we retrace the steps followed in the analysis, obtaining a solution to the problem, namely a demonstration from the principles which were the last step in analysis. The demonstration produces knowledge, because "to know is to know through demonstration" (Aristotle, *Analytica Posteriora*, A 2, 71 b 18). This is synthesis.

Since the synthesis starts from the principles which were the last step in the analysis, in the synthesis "the last step in the analysis is the first that comes to be" (Aristotle, *Ethica Nicomachea*, Γ 3, 1112 b 23–24).

17.7 The Two Basic Changes of the Analytic-Synthetic Method

The analytic-synthetic method essentially differs from the analytic method because it admits principles.

Aristotle criticizes the analytic method because it "does not allow one to know anything in an absolute way, but only on the basis of a hypothesis" (Aristotle, *Analytica Posteriora*, A 22, 84 a 5–6). Hypotheses are not true but only plausible, while "scientific knowledge must proceed from premisses that are true" (ibid., A 2, 71 b 20–21). Therefore, it must proceed from principles.

By admitting principles, Aristotle's logic involves two important changes with respect to the analytic method.

(i) The sequence of the premisses starting from the problem terminates, and terminates with principles. According to Aristotle, this is necessary because, if it "did not terminate and there was always something above whatever premiss has been taken, then there would be demonstrations of all things" (ibid., A 22, 84 a 1–2). Including falsehoods.

(ii) Analysis serves to find deductions of given conclusions from principles, namely demonstrations. Therefore, Geminus says that, for Aristotle, "analysis is the discovery of demonstration" (Ammonius, *In Aristotelis Analyticorum Priorum Librum I Commentarium*, 5.28, ed. Wallies).

17.8 The Rationale Behind the Analytic-Synthetic Method

While the analytic method is a method for solving problems, the analytic-synthetic method is a method for finding demonstrations from given principles.

At the origin of Aristotle's replacement of the analytic method with the analytic-synthetic method there was the rise, in Plato's, age, of the view of mathematics as an axiomatic system, namely a deductive system based on indemonstrable principles. Thus, analysis became only a means for finding demonstrations from given principles.

As Lakatos says, with the rise of the view of mathematics as an axiomatic system, "analysis was not any more a venture into the unknown; it was an exercise in mobilizing and ingeniously connecting the relevant parts of the known. The lemmas which were once daring and often falsified conjectures harden into auxiliary theorems" (Lakatos 1978, II, 100).

17.9 Essential Role of Non-Deductive Rules in Discovery

Non-deductive rules play an essential role in discovery, both in the analytic method and in the analytic-synthetic method. For, they are the rules by which the hypotheses are obtained.

As we have seen, the role of non-deductive rules in discovery is acknowledged by Aristotle. For, he says that the premisses will be obtained from the conclusion either by syllogism or by induction.

The role of non-deductive rules in discovery is also acknowledged by Russell. For, he says that "the method in investigating the principles of mathematics is really an inductive method, and is substantially the same as the method of discovering general laws in any other science" (Russell 1973, 274).

Non-deductive rules play an essential role in discovery because, as argued in Chap. 16, they are ampliative.

17.10 Inadequacy of Aristotle's Logic as Method of Discovery

Although Aristotle's logic is a method of discovery, it is inadequate as method of discovery.

A limitation of Aristotle's logic is that Aristotle's procedure of 'inventio medii' requires knowing the essence of things (see Cellucci 2013, Chapter 7). Therefore, Aristotle's logic depends on Aristotle's view of science, that to know a thing is to know its essence.

This makes Aristotle's logic inadequate to the needs of modern science because, as established by Galileo's revolution, science cannot know the essence of things but only some phenomenal properties of things, mathematical in kind. This is the foundation of modern science, so, since Aristotle's logic depends on Aristotle's view of science, Aristotle's logic conflicts with the foundation of modern science.

Another limitation of Aristotle's logic is that the truth of principles can be known only by intuition. For, principles cannot be demonstrated, so "it is intuition, and not discursive thinking, that apprehends the primitive things," namely "it is intuition that apprehends the unchanging and first terms in the order of demonstrations" (Aristotle, *Ethica Nicomachea*, Z 11, 1143 a 36–1143 b 3).

But this is problematic, because there are no rules for intuition. Intuition is merely subjective and unreliable as a basis for acquisition of knowledge.

Thus, intuition led Aristotle to believe that it is part of the essence of bodies that heavier bodies fall faster than light ones, in particular that, “if one weight is twice another, it will take half as long over a given movement” (Aristotle, *De Coelo*, A 6, 274 a 1–2). But this was wrong. As we have seen in Chap. 6, through a simple thought experiment Galileo showed that Aristotle’s belief leads to a contradiction.

In general, the analytic-synthetic method is inadequate as method of discovery (see Cellucci 2013, Chapter 5).

17.11 The Flourishing of the Paradigm of Logic as Discovery

After antiquity, the paradigm of logic as discovery had some uncertain developments, but it flourished in the seventeenth and eighteenth centuries. (On the development of logic from Humanism to Kant, see Capozzi and Roncaglia 2009).

Thus, Descartes says that the paradigm of logic as justification, represented at the time by Scholastic logic, is inadequate to the needs of the scientific revolution because it “contributes nothing whatsoever to the knowledge of truth” (Descartes 1996, X, 406). For, it “does not teach the method by which something has been discovered” (ibid., VII, 156). Therefore, it “is entirely useless for those who wish to investigate the truth of things” (ibid., X, 406). It can only “serve to explain to someone else the things one already knows,” not “to discover them” (ibid., VI, 17).

Thus, there is a need for a new logic that will teach “one to direct his reason to discover the truths of which one is ignorant” (ibid., IX–2, 13–14). The new logic will permit to solve “all problems which can be proposed concerning any sort of quantity, whether continuous or discrete” (ibid., X, 156–157). It will “demonstrate which problems can be solved in this or that way and not in another one, so that almost nothing” will “remain to be discovered” (ibid., X, 157).

We “shall comply with” the new logic “exactly if we will gradually reduce convoluted and obscure propositions to simpler ones” (ibid., X, 379). So we will eventually arrive at the simplest of all propositions, “which can be intuited first and per se, independently of any other one” (ibid., X, 383). This is analysis. Then, “from the intuition of the simplest of all” propositions, “we will try to arrive through the same steps at a knowledge of all the others” (ibid., X, 379). For “all the others can be perceived only by deducing them from those” (ibid., X, 383). This is synthesis. Therefore, the method “is twofold,” by “analysis” and “synthesis” (ibid., VII, 155).

Then, according to Descartes, logic belongs to the paradigm of logic as discovery, and is the analytic-synthetic method.

17.12 The Paradigm of Logic as Discovery and Modern Science

That logic belongs to the paradigm of logic as discovery and is the analytic-synthetic method, was also affirmed by the creators of modern science.

It is widely believed that the rise of modern science was due to the birth of a new scientific method, which replaced Aristotle's method.

Thus, Kline says that "modern science owes its origins and present flourishing state to a new scientific method which was fashioned almost entirely by Galileo Galilei" (Kline 1985, 284).

But this is unfounded. The rise of modern science was not due to the birth of a new scientific method, because the creators of modern science used, and declared to use, the analytic-synthetic method as method of the new science.

In fact, Galileo says that, "when the conclusion is true, by making use of the resolutive method," namely analysis, "one may easily encounter some proposition which is already demonstrated, or arrive at some principle which is known by itself" (Galilei 1968, VII, 75). From the certainty of this proposition or principle "the truth of" the conclusion, which is deduced from it by synthesis, "will acquire strength and certainty" (ibid.VII, 435). Conversely, "if the conclusion is false, one can go on endlessly without ever hitting upon any known truth, if indeed one does not encounter some impossibility or manifest absurdity" (ibid., VII, 75).

Newton says that "as in mathematicks, so in natural philosophy, the investigation of difficult things by the method of analysis, ought ever to precede the method" of synthesis, or "composition" (Newton 1952, 404). By analysis we proceed "from compounds to ingredients" and "in general, "from effects to their causes, and from particular causes to more general ones, till the argument end in the most general" (ibid.). On the other hand, "synthesis consists in assuming the causes discovered, and established as principles, and by them explaining the phaenomena proceeding from them, and proving the explanations" (ibid., 405).

17.13 Historians and the Paradigm of Logic as Discovery

The flourishing of the paradigm of logic as discovery in the seventeenth and eighteenth centuries was an important development of logic.

In this regard it is significant that, as in the case of Aristotle, historians of logic inspired by mathematical logic consider this development irrelevant to logic.

Thus, the Kneales say that "the rise of modern physics (i.e. the physics of Galileo) and the recognition that logic was not an instrument of discovery" led to "the marked decline of interest in formal logic" as a theory of deduction "which occurred during the seventeenth and eighteenth centuries" (Kneale and Kneale 1962, 307). In particular, "logic was divorced at this time from mathematics" (ibid., 308). For, "the mathematical legacy of Greece was chiefly geometrical," but now mathematicians focused their attention on "algebra and analysis" (ibid.). And, while "geometry was conceived from the beginning as a science in which theorems were derived from axioms by principles of logical entailment," so by deduction, algebra and analysis "were not elaborated in axiomatic fashion" (ibid., 309). The "new branches of mathematics were supposed to contain a technique of discovery," so mathematicians "came to think of their new methods as independent of traditional logic" (ibid.). In particular, Descartes "tried to lay down rules for discovery," but so he made no real contribution to logic, therefore in a history of logic "we need not consider the merits of his doctrine" (ibid., 310).

Blanché says that, "as a result of the scientific revolution carried out, most of all, by Galileo," a "radical transformation takes place: the sharp rejection of logic" and its replacement with "the making theoretically explicit of the method practiced by science" (Blanché 1970, 175). The "best representative of this new attitude is Descartes," a philosopher who "has made no real contribution to logic" (ibid.). In fact, in his work "there is strictly nothing to keep for the history of logic" (ibid., 174).

17.14 The Decline of the Paradigm of Logic as Discovery

The flourishing of the paradigm of logic as discovery, however, was short-lived. In the nineteenth century the paradigm declined and fell. This had three main causes.

(I) *The Romantic myth of genius*. According to it, discovery depends on some fortunate cast of intellect, rising above all rules.

Thus, Whewell says that a logic "of discovery is not possible" because, "at each step of the progress of science, are needed invention, sagacity, genius—elements which no" logic "can give. We may hope in vain, as Bacon hoped, for an organon which shall enable all men to construct scientific truths" (Whewell 1847, I, viii). Therefore, "scientific discovery must ever depend upon some happy thought, of which we cannot trace the origin—some fortunate cast of intellect, rising above all rules," indeed "no precepts will elevate a man of ordinary endowments to the level a man of genius" (ibid., II, 20–21).

(II) *The assumption that knowledge is based on intuition*. According to it, discovery is an irrational act because for intuition there are no rules.

Thus, Poincaré says that "to make arithmetic, as to make geometry, or to make any science, something else than pure logic is necessary. To designate this something else we have no word other than intuition" (Poincaré 2015, 215). Discovery in arithmetic, geometry, and science requires "a faculty which makes us see the end from afar, and intuition is this faculty," therefore intuition "is necessary to the explorer for choosing his route" (ibid., 218). Indeed, "intuition is the instrument of invention" (ibid., 219).

(III) *The assumption that discovery is a subjective and psychological process.* According to it, discovery can only be a concern of psychology, not of logic.

Thus, Frege says that the question of how one "arrived at a given proposition," that is, how the proposition was discovered, may "have to be answered differently for different persons" (Frege 1967, 5). Therefore, discovery is a merely subjective, psychological question, it concerns "the psychological genesis" (ibid.). Admittedly, to study the psychological genesis is "a feasible undertaking, but it is not a logical one" (Frege 1979, 146). And we have "always to separate the psychological from the logical, the subjective from the objective" (Frege 1960, xxii).

17.15 Foundation and Justification as the Only Concern of Logic

According to Frege, only foundation and justification can be a concern of logic, in particular logic can only deal with deduction.

Indeed, Frege says: "We are concerned" in logic "not with the way in which" propositions "are discovered but the kind of ground on which their" justification "rests" (Frege 1960, 23). This is not a subjective and psychological question, because "the answer to it is connected with the inner nature of the proposition considered" (Frege 1967, 5). So, the answer to it does not hinge on "the conditions, psychological, physiological and physical, which have made it possible to form the content of the proposition in our consciousness," but "is a judgment about the ultimate ground upon which rests the justification for holding it to be true" (Frege 1960, 3). The judgment involves determining "upon which primitive laws" the proposition "is based" (Frege 1984, 235).

The primitive laws must be true, because a justification is given only by "those grounds of judgment which are truths" (Frege 1979, 3). After determining the primitive laws upon which the proposition is based and ascertaining that they are true, we must deduce the proposition from them. This will provide the required justification for holding the proposition to be true, because to deduce is "to make a judgment because we are cognisant of other truths as providing a justification for it" (ibid.). The laws of deduction are the "laws governing this kind of justification" (ibid.).

In the study of the laws of deduction, logic can "forgo expressing anything that is without significance for the inferential sequence" (Frege 1967, 6). That is, logic can abstain from expressing "anything that it is not necessary for setting up the laws of deduction" (Frege 1979, 5). Thus, "everything necessary for a correct inference is expressed in full, but what is not necessary is generally not indicated" (Frege 1967, 12). Then logic will fail to express all aspects of thinking. But this is inessential, because the task of logic is to provide a proposition with the most secure foundation. Logic is suited to this task, and "one must not condemn it because it is not suited to others" (ibid., 6).

17.16 The Task of Studying the Laws of Deduction

According to Frege, the task of studying the laws of deduction is faced with a difficulty.

The difficulty is that, "if we try to list all the laws governing inference which occur when arguments are conducted in the usual way, we find an almost unsurveyable multitude which apparently has no precise limits" (Frege 1984, 235). Therefore, "it is easy for something to intrude which is not logical in nature" (ibid.). Upon closer inspection, however, we realize that "these inferences are composed of simpler ones" (ibid.). Then, in order to avoid that something which is not logical in nature might intrude, the deduction "must be analysed into the simple logical steps of which it consists" (Frege 2013, Volume I, 1). This is "essential if our trains of thought are to be relied on; for only what is finite and determinate can be taken in at once, and the fewer the number of primitive sentences, the more perfect a mastery can we have of them" (Frege 1979, 39).

The purpose of splitting a deduction into simple steps is not to make deduction efficient, obtaining the conclusion in the speediest way, because "such a resolution of composite modes of inference into their simple components has as a necessary consequence the lengthening of proofs" (Frege 1984, 235). Its purpose is, instead, to "attain to the necessary certainty" (Frege 1960, xxi). Splitting a deduction is indispensable to this purpose because, only when deduction is based on logically simple steps, "one thereby knows precisely the grounds on which each individual theorem rests" (Frege 2013, Volume I, vii). In this way "no presupposition can remain unnoticed" (ibid., Volume I, 1).

17.17 Assumptions of Mathematical Logic

Frege's view that not discovery, but only foundation and justification can be a concern of logic, in particular logic can only deal with deduction, was the basis for his creation of mathematical logic.

Mathematical logic is based on the following three assumptions.

(1) *The method of mathematics is the axiomatic method.* In mathematics we start "from propositions that are accepted as true," the principles, and arrive "via chains of" deductive "inferences to the theorem" (Frege 1979, 204).

(2) *The goal of mathematical logic is to give a firm foundation to mathematics.* Its purpose is to "provide" mathematics "with the most secure foundation" (Frege 1967, 5).

(3) *Mathematical logic deals with what is necessary for the axiomatic method, that is, deductive rules.* For, "there is no" mathematical "mode of inference that cannot be reduced to the general" deductive "inference-modes of logic" (Frege 1984, 113).

17.18 The Reception of Mathematical Logic

The creation of mathematical logic was greeted with emphatic applause.

Thus, Russell said that mathematical logic has made Aristotle's logic "a system which is as definitely antiquated as Ptolemaic astronomy" (Russell 2005, 188). Aristotle's "theory of the syllogism" is "unimportant. Any person in the present day who wishes to learn logic will be wasting his time if he reads Aristotle or any of his disciples" (ibid., 194). In fact, "throughout modern times, practically every advance in science, in logic, or in philosophy has had to be made in the teeth of opposition from Aristotle's disciples" (ibid.). While Aristotle's "logic put thought in fetters," mathematical logic "gives it wings," it has "introduced the same kind of advance into philosophy as Galileo introduced into physics" (Russell 2009, 48). Mathematical logic "has the effect of enlarging our abstract imagination, and providing an infinite number of possible hypotheses to be applied in the analysis of any complex fact. In this respect it is the exact opposite of the logic practised by the classical tradition" (ibid.). It "provides a method which enables us to obtain results that do not merely embody personal idiosyncrasies, but must command the assent of all who are competent to form an opinion" (ibid.).

Tarski said that mathematic al logic serves "the purpose of assuring for mathematics (as well as for logic) the highest possible degree of clarity and certainty" (Tarski 1994, 109). It has been "created for the purpose of making the foundations of mathematics firmer and more profound; this discipline, in spite of its brief existence of barely a century, has already attained a high degree of perfection, and today its role in the totality of our knowledge transcends by far the originally intended boundaries" (ibid., ix). To this end, the assumption that the method of mathematics is the axiomatic method is essential. For, the use of the axiomatic method "eliminates to a large extent the possibility of obscurities and errors, without resorting to an infinite regress" (ibid., 122). The axiomatic method allows us to eliminate any cause for doubt because it replaces "subjective scrutiny of definitions and proofs by criteria of an objective nature, in such a way that the decision regarding the correctness of given definitions or proofs would depend exclusively upon their structure, that is, upon their exterior form" (ibid.). Therefore, the axiomatic method "is justifiably considered the ideal among all methods which are employed in the construction of sciences" (ibid.).

17.19 Mathematical Logic and Closed Systems

The assumption (1) of mathematical logic, that the method of mathematics is the axiomatic method, implies that mathematical theories are closed systems.

A closed system is a system that is based on principles, and whose development consists of deducing conclusions from principles. As argued in Chap. 16, deductive rules are non-ampliative, that is, their conclusion is contained in the premisses. So, all conclusions deduced from principles are contained in them. Therefore, the development of a closed system is completely internal to the system.

Indeed, Frege says that "the whole of mathematics is contained in" the principles "as in a seed. Our only concern is to generate the whole of mathematics from this seed" (Frege 1979, 204–205). Namely, the mathematician's only concern is to deduce the whole of mathematics from the principles. For, all mathematics is contained in the principles.

17.20 Mathematical Logic and Limitative Results

Notwithstanding the emphatic applause with which the rise of mathematical logic was greeted, mathematical logic has serious limitations. Indeed, the three assumptions (1)—(3) on which mathematical logic is based are conclusively refuted by three basic limitative results of mathematical logic itself: Gödel's first incompleteness theorem, Gödel's second incompleteness theorem, and the strong incompleteness theorem for second-order logic.

Assumption (1), that the method of mathematics is the axiomatic method, is refuted by Gödel's first incompleteness theorem because, by the latter, for any consistent, sufficiently strong, formal system, there are sentences of the system that are true but cannot be deduced from the axioms of the system.

Assumption (2), that the goal of mathematical logic is to give a firm foundation to mathematics, is refuted by Gödel's second incompleteness theorem because, by the latter, for any consistent, sufficiently strong, formal system, it is impossible to demonstrate, by absolutely reliable means, that the axioms of the system are consistent, a fortiori that they are true.

Assumption (3), that mathematical logic deals with the laws of deduction, is refuted by the strong incompleteness theorem for second-order logic because, by the latter, there is no consistent set of rules capable of deducing all second-order consequences from given premisses.

17.21 Deductivism

Assumption (1) on which mathematical logic is based is an expression of deductivism, the view that all mathematical reasoning is deductive.

This view is widespread among mathematicians but is refuted by the limitative results.

(1) Deductive reasoning cannot be founded in the sense that it does not lead to any falsity, by a corollary of Gödel's first incompleteness theorem. By it, for any

consistent, sufficiently strong, formal system, there is a consistent extension in which some false sentence is provable. This means that not only it cannot be demonstrated that deductive reasoning does not lead to any falsity, but it can be demonstrated that deductive reasoning, even when it is consistent, can lead to some falsity.

(2) Deductive reasoning cannot be founded in the sense that it does not lead to any contradiction, by Gödel's second incompleteness theorem. By it, the consistency of no sufficiently strong, formal system can be demonstrated in that formal system but only in some proper extension of it. The consistency of this proper extension cannot be demonstrated in this proper extension but only in some proper extension of it. And so on. This means that one cannot prove that deductive reasoning does not lead to any contradiction, because proving it would involve an infinite regress.

(3) Deductive reasoning is inadequate by the strong incompleteness theorem for second-order logic. By it, there is no consistent set of rules capable of deducing all second-order consequences from given premisses.

17.22 Mathematical Logic as a Parody of Mathematical Practice

The three assumptions (1)–(3) on which mathematical logic is based provide a distorted view of mathematics.

As Longo says, they lead to "a parody of mathematical practice: first definitions and axioms, then proofs" (Longo 1991, 119). This contrasts with the fact that "mathematicians first hint results, by wild analogies, deep connections between different areas suggested, for example, by sketchy drawings on a blackboard; later they establish incomplete and only partly formal frames to clean up deductions" (ibid.). Even "when the work is completed, nobody really cares to establish a unique formal frame, as the dynamic of the actual mathematical work soon brings up new issues and new methods, possibly incompatible with any static view of the previous achievements" (ibid., 120–121).

Contrary to what mathematical logic implies, mathematics "proceeds by a maze of back-connections; actual deep proofs rarely fit within one formal frame or use just one language. And when a complete, ad hoc, formal frame and language is later established for a relevant theorem, then the result is already old and has little to say about the current research in mathematics" (ibid., 119–120). The latter "largely uses the flexibility of informal concepts and the use of a variety of languages in a very same proof, in order to establish connections, set bridges and propose new ideas" (ibid., 120).

17.23 Mathematicians and Mathematical Logic

Because of its limitations, mathematical logic is unfit for any purpose relevant to mathematics. This has been argued by several mathematicians.

Thus, Dieudonné says: "All the works of logicians" play "no role in mathematics" (Dieudonné 1981, 22). The trouble with their formal systems is that "no mathematician as far as I know has ever found any use for these systems at all" (Bourbaki 1979, 32). Indeed, "suppose that tomorrow all the works of" mathematical logic "disappeared, as if by magic: well, no mathematician, when he proves a theorem, would notice it," because mathematical logic has "no direct influence on the proofs" of "mathematicians" (Dieudonné 1981, 22).

Halmos says: "Every mathematician finds the famous accomplishments of Gödel and Cohen" about "undecidable propositions" and "the continuum hypothesis both fascinating and admirable" (Halmos 1985, 203). But, "however admirable some of the accomplishments of the logicians may be, the mathematician doesn't need them, cannot use them, in his daily work" (ibid., 204). These "spectacular theorems—Gödel and Cohen—we admire them, but they haven't changed our work, our philosophy, our life" (ibid.).

Several mathematicians have even argued that mathematical logic has lost its connection with reality.

Thus, Rota says: "That magnificent clockwork mechanism that is mathematical logic is slowly grinding out the internal weaknesses of the system" (Rota 1992, 180). Today, "in all circumstances imaginable," logic "shines for its absence" (Rota 1999, 94). If "we are to set the new sciences on firm, autonomous, formal foundations, then a drastic overhaul" of "logic is in order" (Rota 1992, 180).

Davis says: "It would be invidious to mention a specific example of exhaustion of a field when there are people working very happily in it," but this is the case of "mathematical logic," it "has lost its connection to reality and has produced mathematical monsters" (Davis 2006, 176).

17.24 Philosophers and Mathematical Logic

The limitations of mathematical logic have also been underlined by several philosophers.

Thus, Lakatos says: "The subject matter of" mathematical logic "is an abstraction of mathematics" (Lakatos 1976, 1). It leads "to identify mathematics with its formal axiomatic abstraction" (ibid.). According to it, mathematics "starts with a painstakingly stated list of axioms" and "definitions. The axioms and definitions frequently look artificial and mystifyingly complicated" (ibid., 142). Axioms and definition are followed by theorems which "are loaded with heavy-going conditions; it seems impossible that anyone should ever have guessed them" (ibid.). So, mathematical logic "denies the status of mathematics to most of what has been commonly understood to be mathematics, and can say nothing about its growth. None of the 'creative' periods and hardly any of the 'critical' periods of mathematical theories would be admitted into the formalist heaven" of mathematical logic, "where mathematical theories dwell like the seraphim, purged of all the impurities of earthly uncertainty" (ibid., 2). Mathematics "is presented as an ever-increasing set of eternal, immutable truths. Counterexamples, refutations, criticism cannot possibly enter. An authoritarian air is secured for the subject" by "suppressing the primitive conjecture, the refutations, and the criticism of the proof" (ibid., 142). Mathematical logic "hides the struggle, hides the adventure. The whole story vanishes, the successive tentative formulations of the theorem in the course of the proof-procedure are doomed to oblivion while the end result is exalted into sacred infallibility" (ibid.)

Wang says: "A striking phenomenon about mathematical logic is its development into a more and more mathematical discipline," so "the logicians find themselves attracted into the whirl of mathematical activity. At the same time, its contribution to the philosophical understanding of the foundations of human knowledge (and mathematics in particular) appears to decrease rather than increase" (Wang 1974, 28). Thus, "as we understand the nature of mathematical logic better, we find that the early belief in its philosophical relevance was largely an illusion" (ibid.). In fact, "the central place of logic is rather detached from actual knowledge. If we distinguish the three main aspects of knowledge of end results, activities, and progress," mathematical logic is "concerned only with end results" (ibid.).

Macbeth says: "It is by now clear that mathematical logic," with its "proofs in which every step conforms to a recognized rule of that logic, is of no mathematical interest. Such proofs do not advance mathematical understanding; they are not more rigorous than the informal proofs that mathematicians actually produce; and very often they are simply unintelligible. Mathematical logic, it has turned out, is irrelevant to the practice of mathematics—and to our philosophical investigations into the nature of that practice" (Macbeth 2015, 3). In fact, "mathematical reasoning, the reasoning that mathematicians actually engage in, and logical reasoning as understood in mathematical logic," are "very different" because "mathematical reasoning is focused on mathematical ideas while logical reasoning takes account only of logical form" (ibid., 4). While "a fully rigorous proof, in the logician's sense of rigor, is one each step of which conforms to some antecedently specified rule of pure logic and is thoroughly machine checkable, a rigorous proof in the mathematician's sense of rigor is instead one that a mathematician can see to be compelling by focusing on the relevant mathematical ideas and their implications" (ibid.).

17.25 An Admission that Discovery is a Concern of Logic

The limitations of mathematical logic indicate that there is a need for an alternative paradigm of logic. An obvious candidate is the paradigm of logic as discovery.

In this connection it should be noted that, although Frege refers to Kant in several respects, he does not do so with regard to the relation between logic and discovery. For, contrary to Frege, Kant admits discovery as a concern of logic.

As Capozzi observes, "the question is: do we find promising provisional judgments, provided by luminous ideas, only by chance or luck?" (Capozzi 2025, 22). According to Kant, the answer to this question is negative, because discovery is not a matter of chance or luck, it requires "an addition to logic" (ibid.).

In fact, Kant says: "Logic has not yet taken sufficiently to heart the challenge issued to it, that it should also provide rules to direct one in searching for things" (Kant 1996, 592). These rules "can be a guide even to the mathematician in his discoveries, and moreover he often makes use of it" (ibid.).

These rules would constitute a new chapter of logic, devoted to heuristics or logic of discovery.

Kant's admission of discovery as a concern of logic seems to contradict other statements by Kant on the relation between logic and discovery.

Thus, Kant says that logic is "not a universal art of discovery, to be sure, and not an organon of truth—not an algebra with whose help hidden truths can be discovered" (Kant 1992, 534).

But the contradiction is only apparent. For, Kant distinguishes between two parts of logic: pure formal logic and applied logic.

As Capozzi observes, the missing chapter on heuristics "would not belong to the doctrine of elements of pure formal logic. The latter remains a formal canon having no inventive capacity because it has nothing to do with contents" (Capozzi 2025, 22). The "chapter on heuristics would belong to method and would be part of applied logic, that is, the logic that takes into account what pure logic has put aside: how thinking takes place and how it is under various subjective obstacles and conditions" (ibid.). The addition of this chapter would be a "hoped for development of logic" (ibid.).

17.26 The Reaffirmed Rejection of Discovery as a Concern of Logic

While Kant admits discovery as a concern of logic, in the last century the dominant philosophy, analytic philosophy, has reaffirmed Frege's rejection of it.

Thus, Wittgenstein claims that discovery cannot be a concern of logic because "discovery is always unmethodical: you have no method for making the discovery" (Wittgenstein 2016, 46).

The reaffirmed rejection of discovery as a concern of logic is a consequence of the assumption of analytic philosophy that philosophy is not acquisition of knowledge, it aims only to understand what we already know.

Then, if we want to reestablish the relevance of philosophy, just as we must restore the view that philosophy is acquisition of knowledge, so in particular we must restore the view that logic is an instrument for acquiring knowledge. That is, we must restore the paradigm of logic as discovery.

17.27 The Doors Closed by Mathematical Logic

The creation of mathematical logic has closed some doors, in particular the following doors.

(i) *Discovery is a concern of logic*. The creation of mathematical logic has closed this door because, according to it, as Frege argues, discovery is a subjective process, therefore logic cannot concern itself with the way in which propositions are discovered but only with the kind of ground on which their justification rests. This contrasts with the fact that already Greek mathematicians invented a method of discovery, the analytic method, in which there is nothing purely subjective. Therefore, discovery can be a concern of logic.

(ii) *Logic arises from method*. The creation of mathematical logic has closed this door because, as Frege argues, the laws of logic "are the most general laws, prescribing how to think wherever there is thinking at all" (Frege 2013, Volume I, XV). Therefore, logic is a precondition of method and hence cannot arise from method. This contrasts with the fact that, as Dewey observes, "as the methods of the sciences improve, corresponding changes take place in logic" (Dewey 1938, 14). Therefore, logic arises from method.

(iii) *The laws of logic are not the whole of reason*. The creation of mathematical logic has closed this door because, as Frege says, "if beings were even found whose laws of thought directly contradicted" the laws of logic "so that their application often led to opposite results," then "I would say: here we have a hitherto unknown kind of madness" (Frege 2013, Volume I, XVI). Thus, human beings are rational if only if they obey the laws of logic, so logic is the whole of reason. Then, emotion and feeling are excluded from the sphere of reason and rationality. This contrast with the fact that, as Damasio says, "certain aspects of the process of emotion and feeling are indispensable for rationality" (Damasio 1995, xiii). Therefore, the laws of logic are not the whole of reason.

17.28 Reopening the Doors Closed by Mathematical Logic

The doors closed by the creation of mathematical logic must not remain closed. On the contrary, due to the limitations of mathematical logic, they must be reopened.

As van Benthem says, "we cannot keep singing hymns to Gödel and Tarski forever—unless we are already in Heaven" (van Benthem 2008, 40). In fact, "I have never been able to understand this defense of the status quo" by which "innovative 'agents provocateurs' like Lakatos in his mind-opener *Proofs and refutations* were subjected to torrents of abuse, rather than the lively interest in new ideas one would expect from a vigorous community" (ibid.). On the contrary, "I have always found many outspoken critics" of mathematical "logic (Blanshard, Perelman, Toulmin, Lakatos) extremely interesting and well-worth reading, and a useful reminder of the many doors our Founding Fathers have closed historically—doors that could be opened again now" (ibid.).

17.29 The Analytic Method as Logic

Reopening the doors closed by the creation of mathematical logic requires a change in the concept of logic.

According to mathematical logic, a logic is a set of valid deductive rules.

This concept of logic is motivated by the assumption that the aim of science is truth.

In fact, Frege says that "all sciences have truth as their goal," therefore "to discover truth is the task of all sciences" (Frege 1984, 351).

If the aim of science is truth, then a logic must consist of a set of truth-preserving inference rules. Since the only truth-preserving inference rules are valid deductive rules, this implies that a logic must consists of a set of valid deductive rules.

In fact, Frege says that "the goal of logic" is "to set up" the "laws of valid inference" (Frege 1979, 3).

However, as argued in Chap. 11, the view that the aim of science is truth is not valid. It must be replaced with the view that the aim of science is plausibility. Accordingly, instead of saying that a logic is a set of valid deductive rules, one might think of saying that a logic is a set of plausibility-preserving inference rules.

But this runs into the difficulty that, as argued in Chap. 16, both non-deductive rules and valid deductive rules are not plausibility-preserving. Therefore, non-deductive rules and valid deductive rules must be complemented by a procedure for establishing that the hypotheses obtained by non-deductive rules and the conclusions deduced from them by valid deductive rules are plausible.

This implies that a logic cannot consist of a set of plausibility-preserving inference rules. It must consist of a set of non-deductive rules and valid deductive rules together with a procedure for establishing that the hypotheses obtained by non-deductive rules and the conclusions deduced from them by valid deductive rules are plausible. In other words, a logic must consist of the analytic method.

17.30 Implications of the Analytic Method as Logic

Taking the analytic method as logic reopens the doors closed by the creation of mathematical logic.

(i) *Discovery is a concern of logic.* Taking the analytic method as logic implies the discovery is a concern of logic, because the analytic method is a method of discovery.

(ii) *Logic arises from method.* Taking the analytic method as logic implies that logic arises from method, because the analytic method is a method.

(iii) *The laws of logic are not the whole of reason.* Taking the analytic method as logic implies that the laws of logic are not the whole of reason, because the analytic method is a rational method and does not reduce to the laws of logic.

17.31 Overcoming the Three Main Causes of the Decline

Taking the analytic method as logic overcomes the three main causes of the decline of the paradigm of logic as discovery in the nineteenth century.

(I) *The Romantic myth of genius.* Taking the analytic method as logic implies that discovery does not depend on some fortunate cast of intellect, rising above all rules. For, the analytic method proceeds on the basis of rules.

(II) *The assumption that knowledge is based on intuition.* Taking the analytic method as logic implies that discovery is not an irrational act since for intuition there are no rules. For, intuition plays no role in the analytic method.

(III) *The assumption that discovery is a subjective and psychological process.* Taking the analytic method as logic implies that discovery cannot be only a concern of psychology, not of logic. For, there is nothing subjective and psychological in the analytic method as logic. Discovery is based on hypotheses obtained by non-deductive rules, so by logic, and the plausibility of hypotheses is established by comparing the arguments for and against them, so by logic.

17.32 Assumptions of the Analytic Method as Logic

The analytic method as logic is based on the following three assumptions, which are stated parallel with the assumptions (1)–(3) of mathematical logic, for comparison.

(1) The method of mathematics and science is the analytic method.

(2) The goal of the analytic method as logic is to be an instrument for acquiring knowledge in mathematics and science.

(3) The analytic method as logic deals with what is necessary for the analytic method, that is, non-deductive rules, deductive rules, and a procedure for establishing that the conclusions of these rules are plausible.

17.33 The Analytic Method as Logic and Open Systems

The assumption (1) of the analytic method as logic, that the method of mathematics and science is the analytic method, implies that mathematical and scientific theories are open systems.

An open system is a system which initially consists only of a problem and possibly other data already available, and whose development consists in obtaining more and more hypotheses from the problem and the other data by non-deductive rules, checking that the hypotheses are plausible, and deducing conclusions from the hypotheses.

Non-deductive rules are ampliative, namely their conclusion is not already contained in the premisses. So, the hypotheses are not contained in the problem and the other data, and need not belong to the same field as the problem and the other data, they may belong to other fields. Therefore, the development of an open system is not wholly internal to the system.

That the development of an open system is not wholly internal to the system implies that mathematics and science involve interactions between open systems.

In fact, an open system does not contain a complete representation of the knowledge of a given field but only a partial one, so, to acquire knowledge not available to it, an open system must interact with other open systems.

Through interactions with other open systems, new relations are established between the problem to be solved and other problems, so the problem acquires new significance and meanings. In general, a problem is all the more significant and important as it can enter into relation with other problems. Moreover, a problem need not have a single solution. it can have multiple solutions, and each solution establishes new relations between the problem and other problems.

17.34 The Analytic Method as Logic and Limitative Results

The three assumptions of the analytic method as logic (1)–(3) are unaffected and even confirmed by the three basic limitative results of mathematical logic.

Assumption (1) is unaffected and even confirmed by Gödel's first incompleteness theorem. For, in the analytic method as logic, the solution to a problem of a given field is obtained by means of hypotheses that may belong to other fields. And, by Gödel's first incompleteness theorem, solving a problem of a given field may require hypotheses that belong to other fields.

Assumption (2) is unaffected and even confirmed by Gödel's second incompleteness theorem. For, the goal of the analytic method as logic is to be an instrument for acquiring knowledge in mathematics and science, and such knowledge can only be plausible. And, by Gödel's second incompleteness theorem, knowledge in mathematics and science can only be plausible.

Assumption (3) is unaffected and even confirmed by the strong incompleteness theorem for second-order logic. For, in the analytic method as logic, deductive rules are not a closed set given once and for all, but an open set that can always be extended. And, by the strong incompleteness theorem for second-order logic, deductive rules are not a closed set given once and for all, but an open set that can always be extended.

17.35 The Incompletability of Mathematics

Taking the analytic method as logic permits us to interpret the so-called incompletability of mathematics positively.

Gödel says that the incompleteness theorems establish what "might be called the incompletability or inexhaustibility of mathematics" (Gödel 1986–2002, III, 305). In particular, it is the second incompleteness theorem "which makes the incompletability of mathematics particularly evident" (ibid., III, 309).

Since Gödel takes mathematical logic as logic, he interprets the incompletability of mathematics negatively, as expressing an impossibility.

Indeed, Gödel says that the second incompleteness theorem "makes it impossible that someone should set up a certain well-defined system of axioms and rules and consistently make the following assertion about it: All of these axioms and rules I perceive (with mathematical certitude) to be correct, and moreover I believe that they contain all of mathematics. If someone makes such a statement he contradicts himself" (ibid., III, 309). This is because, "if he perceives the axioms in consideration to be correct, he also perceives (with the same certainty) that they are consistent. Hence he has a mathematical insight not derivable from his axioms" (ibid.). Then, "no well-defined system of correct axioms can contain all of mathematics proper," if "by mathematics proper is understood the system all true mathematical propositions" (ibid.). For, "the proposition which states the consistency of the system is true, but not demonstrable in the system" (ibid.).

On the contrary, taking the analytic method as logic, we can interpret the incompletability of mathematics positively, as expressing that mathematics is a problem solving process in which, beyond every hypothesis, we seek deeper hypotheses.

Proust says that "the only true voyage" would "not be to go towards new lands, but to have other eyes" (Proust 1947, 69).

Seeking deeper hypotheses allows us to take such a voyage, because it makes us see the problem with new eyes.

17.36 An Argument Against the Analytic Method as Logic

Against the analytic method as logic it could be argued that a logic cannot consist of a method because, to make a method, some logical tools are needed. But to have these logical tools they must be made, and to make them some method is needed. In turn, to make this method, some logical tools are needed. And so on. Therefore, the claim that logic consists of a method leads to an infinite regress.

But this argument is not valid. A response to it is implicit in Spinoza's iron-and-hammer argument.

Spinoza says that, against the claim that men have the power to work iron, it could be argued that, "to work iron, a hammer is needed, and to have a hammer, it must be made. For this purpose, there is need of another hammer and other tools, and again to get these there is need of other tools, and so on" (Spinoza 1925, II, 13). Therefore, the claim that men have the power to work iron leads to an infinite regress.

But, Spinoza argues, this is not valid. For "at first, with the tools they were born with, men succeeded" in "making some very simple things; and when these were made they made other more complex things" (ibid., II, 13–14). Thus, "advancing gradually from the simplest works to the making of tools, and from tools to other works and other tools, they have reached a point where they can make very many complex things" (ibid., II, 14). Therefore, the claim that men have the power to work iron does not lead to an infinite regress.

Similarly, it can be argued that, at first, with the inborn intellectual powers resulting from natural selection, men succeeded in making some very simple logical tools; and, when these were made, they had a method by which they made other more complex logical tools. Thus, advancing gradually from the inborn intellectual powers to the making of logical tools, and from logical tools to other logical tools, they have reached a point where they can make complex logical tools, such as those provided by the analytic method as logic. Therefore, the claim that logic consists of a method does not lead to an infinite regress.

17.37 Adequacy of the Analytic Method as Logic

As we have seen above, taking the analytic method as logic can overcome the three main causes of the decline of the paradigm of logic as discovery in the nineteenth century. Moreover, the three assumptions of the analytic method as logic (1)–(3) are unaffected and even confirmed by the three basic limitative results of mathematical logic.

From this it seems fair to conclude that the analytic method as logic is adequate as a method of discovery.

We have said above that, to reestablish the relevance of philosophy, just as we must restore the view that philosophy is acquisition of knowledge, so in particular we must restore the view that logic is an instrument for acquiring knowledge. Therefore, we must restore the paradigm of logic as discovery.

Since the analytic method as logic is adequate as a method of discovery, to restore the paradigm of logic as discovery we can take the analytic method as logic.

17.38 Logic as Productive of Knowledge

Taking the analytic method as logic leads to a view of logic that agrees with Aristotle's view that logic must be productive of knowledge, and disagrees with the view of mathematical logic that logic cannot be productive of knowledge, it can only organize existing knowledge.

As already Aristotle made quite clear, a logic that purports to be productive of knowledge cannot consist only of valid deductive rules because valid deductive rules are non-ampliative, it must also consist of non-deductive rules because only non-deductive rules are ampliative.

Moreover, as argued above, a logic that purports to be productive of knowledge cannot consist only of a set of non-deductive rules and valid deductive rules, it must include a procedure for establishing that the hypotheses obtained by non-deductive rules and the conclusions deduced from them by valid deductive rules are plausible.

Taking the analytic method as logic satisfies these conditions because the analytic method consists of a set of non-deductive rules and valid deductive rules, together with a procedure for establishing that the hypotheses obtained by non-deductive rules and the conclusions deduced from them by valid deductive rules are plausible.

References

Blanché, Robert. 1970. *La logique et son histoire, d'Aristote à Russell*. Paris: Armand Colin.

Bochenski, Joseph. 1961. *A history of formal logic*. Notre Dame: University of Notre Dame Press.

Boger, George. 2004. Aristotle's underlying logic. In *Handbook of the history of logic*, ed. Dov Gabbay and John Woods, vol. 1, 101–246. Amsterdam: Elsevier.

Capozzi, Mirella. 2025. Kant on language: semiotics and heuristics. In *Kant on language*, ed. Luigi Filieri and Konstantin Pollok, 9–26. Cambridge: Cambridge University Press.

Capozzi, Mirella, and Gino Roncaglia. 2009. Logic and philosophy of logic from Humanism to Kant. In *The development of modern logic*, ed. Leila Haaparanta, 78–158. Oxford: Oxford University Press.

Cellucci, Carlo. 2013. *Rethinking logic: Logic in relation to mathematics, evolution, and method*. Berlin: Springer.

Damasio, Antonio. 1995. *Descartes' error: Emotion, reason, and the human brain*. New York: Avon Books.

Davis, Philip. 2006. *Mathematics and common sense: A case of creative tension*. Natick: A K Peters.

Descartes, René. 1996. *Œuvres*. Paris: Vrin.
Dewey, John. 1938. *Logic: The theory of inquiry*. New York: Holt, Rinehart and Winston.
Dieudonné, Jean. 1979. The Bourbaki choice. *Mathematical Medley* 7 (2–3): 28–36.
Dieudonné, Jean. 1981. Logica e matematica nel 1980. In *La nuova ragione: Scienza e cultura nella società contemporanea*, ed. Paolo Rossi, 15–25. Bologna: il Mulino.
Frege, Gottlob. 1960. *The foundations of arithmetic: A logico-mathematical enquiry into the concept of number*. New York: Harper.
Frege, Gottlob. 1967. *Begriffsschrift*, a formula language, modeled upon that of arithmetic, for pure thought. In *From Frege to Gödel: A source book in mathematical logic, 1879–1931*, ed. Jean van Heijenoort, 5–82. Cambridge: Harvard University Press.
Frege, Gottlob. 1979. *Posthumous writings*. Oxford: Blackwell.
Frege, Gottlob. 1984. *Collected papers on mathematics, logic, and philosophy*. Oxford: Blackwell.
Frege, Gottlob. 2013. *Basic laws of arithmetic*. Oxford: Oxford University Press.
Galilei, Galileo. 1968. *Opere*. Florence: Barbera.
Gödel, Kurt. 1986–2002. *Collected works*. Oxford: Oxford University Press.
Halmos, Paul. 1985. *I want to be a mathematician: An automathography*. New York: Springer.
Kant, Immanuel. 1992. *Lectures on logic*. Cambridge: Cambridge University Press.
Kant, Immanuel. 1996. *Practical philosophy*. Cambridge: Cambridge University Press.
Kline, Morris. 1985. *Mathematics for the nonmathematician*. Mineola: Dover.
Kneale, William, and Martha Kneale. 1962. *The development of logic*. Oxford: Oxford University Press.
Lakatos, Imre. 1976. *Proofs and refutations: The logic of mathematical discovery*. Cambridge: Cambridge University Press.
Lakatos, Imre. 1978. *Philosophical papers*. Cambridge: Cambridge University Press.
Longo, Giuseppe. 1991. Notes on the foundations of mathematics and of computer science. In *Nuovi problemi della logica e della filosofia della scienza*, ed. Giovanna Corsi and Giovanni Sambin, vol. 2, 117–127. Bologna: CLUEB.
Macbeth, Danielle. 2015. Reasoning in mathematics and machines: The place of mathematical logic in mathematical understanding. In *Proceedings of the AISB 2015 Symposium on social aspects of cognition and computing*, ed. Gordana Dodig-Crnkovic, Yasemin Erden, and Raffaela Giovagnoli, 3–12. Canterbury: University of Kent.
Newton, Isaac. 1952. *Opticks, or a treatise of the reflections, refractions, inflections and colours of light*. Mineola: Dover.
Poincaré, Henri. 2015. *The foundations of science: Science and hypothesis—The value of science—Science and method*. Cambridge: Cambridge University Press.
Proust, Marcel. 1947. *À la recherche du temps perdu XII. La prisonnière*, Part 2. Paris: Gallimard.
Rota, Gian-Carlo. 1992. Husserl. In *Discrete thoughts: Essays on mathematics, science, and philosophy*, ed. Mark Kac, Gian-Carlo Rota, and Jacob Schwartz, 175–181. Boston: Birkhäuser.
Rota, Gian-Carlo. 1999. *Lezioni napoletane*. Naples: La Città del Sole.
Russell, Bertrand. 1973. *Essays in analysis*. London: Allen and Unwin.
Russell, Bertrand. 2005. *History of Western philosophy*. London: Routledge.
Russell, Bertrand. 2009. *Our knowledge of the external world*. London: Routledge.
Smith, Robin. 2017. Aristotle's logic. In *Stanford Encyclopedia of Philosophy*. Stanford: Center for the Study of Language and Information (CSLI).
Spinoza, Baruch. 1925. *Opera*. Heidelberg: Winters.
Striker, Gisela. 2009. Introduction—Commentary. In Aristotle, *Prior analytics*, Book I, xi–xiii, 67–246. Oxford: Oxford University Press.
Tarski, Alfred. 1994. *Introduction to logic and to the methodology of deductive sciences*. Oxford: Oxford University Press.
van Benthem, Johan. 2008. Interview. In *Philosophy of Mathematics: 5 Questions*, ed. Vincent Hendricks and Hannes Leitgeb, 29–43. New York: Automatic Press/VIP.
Wang, Hao. 1974. *From mathematics to philosophy*. London: Routledge.

Whewell, William. 1847. *The philosophy of the inductive sciences, founded upon their history*. London: Parker.

Wittgenstein, Ludwig. 2016. *Lectures, Cambridge 1930–1933: From the notes of G. E. Moore*. Cambridge: Cambridge University Press.

Woods, John. 2012. A history of the fallacies in Western logic. In *Handbook of the history of logic*, ed. Dov Gabbay, Francis Pelletier, and John Woods, vol. 11, 513–610. Amsterdam: Elsevier.

Chapter 18
Knowledge and Problem Solving

Abstract Developing the view that philosophy is acquisition of knowledge also requires answering the question: Is knowledge problem solving by the analytic method? A widely held view is that the answer is negative. This depends on the fact that it is assumed that the method of mathematics is the axiomatic method and the method of science is the hypothetico-deductive method. But this chapter argues that these assumptions are not valid. The method of mathematics, science, and knowledge in general is the analytic method, therefore knowledge is problem solving by the analytic method. Then the view that philosophy is acquisition of knowledge implies that the method of philosophy is the analytic method, therefore philosophy is problem solving by the analytic method.

Keywords Mathematics as problem solving by the analytic method · Science as problem solving by the analytic method · Knowledge as problem solving by the analytic method · Philosophy as problem solving by the analytic method

18.1 The Question of Method

Developing the view that philosophy is acquisition of knowledge also requires answering the question: Is knowledge problem solving by the analytic method?

A widely held view is that the answer is negative. This depends on the fact that it is currently assumed that the method of mathematics is the axiomatic method and the method of science is the hypothetico-deductive method.

But this chapter argues that these assumptions are not valid. The method of mathematics, science, and knowledge in general, is the analytic method, therefore knowledge is problem solving by the analytic method.

Then the view that philosophy is acquisition of knowledge implies that the method of philosophy is the analytic method, therefore philosophy is problem solving by the analytic method.

C. Cellucci, *A Philosophy for the Future*, Synthese Library 508,
https://doi.org/10.1007/978-3-031-98443-3_18

18.2 Mathematics as Problem Solving by the Analytic Method

A widely held view among mathematicians is that the method of mathematics is the axiomatic method.

Thus, Mac Lane says that "the traditional nature of a proof remains. It is a deduction from suitable axioms," the nature of "proof is eternal" (Mac Lane 1984, 44).

Naylor and Sell say that "the axiomatic method" not only "is the method of mathematics," but in fact "it is mathematics" (Naylor and Sell 2000, 6).

As Nathanson argues, this depends on the fact that "we mathematicians have a naive belief in truth. We prove theorems. Theorems are deductions from axioms. Each line in a proof is a simple consequence of the previous lines of the proof, or of previously proved theorems. Our conclusions are true, unconditionally and eternally" (Nathanson 2008, 773).

But the view that the method of mathematics is the axiomatic method is not valid. By what has been said in Chap. 17, the view is refuted by the three basic limitative results of mathematical logic. The view also has several other shortcomings (see Cellucci 2022, Chapter 10).

In fact, as Rota says, the axiomatic method is not the method of mathematics but only a "method of presentation of mathematics" (Rota 1997, 112). So, it is only "a style of exposition" (ibid., 142). We must guard against "confusing mathematics with the axiomatic method for its presentation," in particular, we must guard against thinking that "mathematicians use the axiomatic method in solving problems and proving theorems," and hence that "the axiomatic method is a basic instrument of discovery" (ibid., 96).

Instead, the method of mathematics is the analytic method. Again by what has been said in Chap. 17, the view that the method of mathematics is the analytic method is not refuted and is even confirmed by the three basic limitative results of mathematical logic, and does not have the other shortcomings of the view that the method of mathematics is the axiomatic method.

Since the method of mathematics is the analytic method, mathematics is problem solving by the analytic method.

18.3 The Nature of Mathematical Objects

That mathematics is problem solving by the analytic method has an implication for the nature of mathematical objects.

It implies that mathematical objects are hypotheses human beings make to solve mathematical problems by the analytic method.

The hypotheses characterize the identity of mathematical objects but do not imply that they exist. To say that mathematical objects exist it is necessary to show that the hypotheses are plausible.

However, that mathematical objects exist does not mean that they exist in the physical or in a superphysical world. They exist in the same sense as all mental-cultural human products, that is, they exist as part of a culture. A mathematician introduces some new mathematical objects as hypotheses to solve a specific mathematical problem by the analytic method, other mathematicians find them useful for solving other mathematical problems by the analytic method, and so on. Thus, the new mathematical objects introduced by a mathematician to solve a specific mathematical problem become part of a culture.

Moreover, hypotheses do not characterize the identity of mathematical objects completely and conclusively. Mathematical objects are not closed in themselves, they are open to interactions with other mathematical objects from which they can gain new determinations, so mathematical objects have an open-ended character, as all elements of a culture. (For more on mathematical objects, see Cellucci 2022, Chapter 9).

18.4 Science as Problem Solving by the Analytic Method

A widely held view among scientists is that the method of science is the hypothetico-deductive method.

Thus, Einstein says that "the intuitive grasp of the essentials or a large complex of facts leads the scientist to the postulation of a hypothetical basic law, or several such basic laws. From the basic law" he "derives his conclusion as completely as possible in a purely logically deductive manner" (Einstein 2002, 108), The conclusions derived from the basic law are then "compared to experience, and in this manner provide criteria for the justification of the assumed basic law. Basic law" and "conclusions together form what is called a 'theory'" (ibid.). The "truth of a theory can never be proven. For one never knows if future experience will contradict its conclusion" (ibid.). In fact, "the greatest advances in natural science" all "originated in this manner," and their basis has this "hypothetical character" (ibid., 107–108).

But the hypothetico-deductive method cannot be the method of science, because it cannot provide any rational account of how a hypothesis or theory is obtained. Thus, Einstein can only say that the hypothetical basic laws are reached through an intuitive grasp, so through a subjective, arbitrary and irrational source.

Moreover, the hypothetico-deductive method cannot account for theory change, the process through which a theory comes to replace another. For, according to it, a theory has no rational connection with a previous one, except that it withstands more severe tests.

Instead, the method of science is the analytic method. The analytic method does not have the shortcomings of the hypothetico-deductive method. For, according to it, a hypothesis is obtained by some non-deductive rule, so by a rational procedure. And, when a hypothesis becomes implausible, one analyzes the reasons why it has become implausible and, on this basis, one modifies it or replaces it with a new hypothesis.

Since the method of science is the analytic method, science is problem solving by the analytic method.

18.5 The Nature of Scientific Objects

That science is problem solving by the analytic method has implications for the nature of scientific objects.

It implies that, like mathematical objects, scientific objects are hypotheses human beings make to solve scientific problems by the analytic method.

The hypotheses characterize the identity of scientific objects but do not imply that they exist.

To say that scientific objects exist it is necessary to show that the hypotheses are plausible.

18.6 Inadequacy of the Big Data View

Against the view that science is problem solving by the analytic method it could be argued that it is anachronistic because today we are faced with massive data, therefore the traditional idea of a method of science is becoming obsolete. It must be replaced with the Big Data view, according to which one can throw data into a computer and let statistical algorithms find patterns.

Thus, Anderson says that, since today we are "faced with massive data," the old idea of a method of science "is becoming obsolete" (Anderson 2008). We are at "the end of theory," because "the data deluge makes the scientific method obsolete" (ibid.). The "new availability of huge amounts of data, along with the statistical tools to crunch these numbers, offers a whole new way of understanding the world. Correlation supersedes causation, and science can advance even without coherent models, unified theories" (ibid.). In fact, "correlation is enough," so "we can stop looking for models. We can analyze the data without hypotheses about what it might show. We can throw the numbers into the biggest computing clusters the world has ever seen and let statistical algorithms find pattern where science cannot" (ibid.). So, "no semantic or causal analysis is required" (ibid.).

But this argument is not valid. For, data do not speak for themselves, they only gain meaning if they are interpreted. Interpreting them requires a theory or model to extract information from them, and hence requires a method. Even supporters of the Big Data view, like Berman, recognize that, "for Big Data projects, holding a prior theory or model is almost always necessary; otherwise, the scientist is overwhelmed by the options" (Berman 2013, 147).

Moreover, the statistical algorithms which are supposed to find patterns in data are unreliable. Calude and Longo "document the danger of allowing the search of correlations in big data to subsume and replace the scientific approach" (Calude and

Longo 2017, 600). This "is mathematically wrong" (ibid., 609). For, "the overwhelming majority of correlations are spurious. In other words, there will be regularities, but, by construction, most of the time (almost always, in the mathematical sense), these regularities cannot be used to reliably predict and act" (ibid.). Indeed, "the bigger the database which one mines for correlations, the higher is the chance to find recurrent regularities and the higher is the risk of committing such fallacies" (ibid.). Therefore, "the more data, the more arbitrary, meaningless and useless (for future action) correlations will be found in them. Thus, paradoxically, the more information we have, the more difficult is to extract meaning from it" (ibid., 600). In fact, "big data analytics cannot replace science and, symmetrically, no theory can be so good to supplant the need for data and testing" (ibid., 610).

Contrary to the Big Data view, the traditional idea of a method of science is by no means becoming obsolete, it is more important than ever.

Thus, Rovelli says that the question of the method of science "is particularly relevant today in science, and particularly in physics," because "in fundamental theoretical physics, it is thirty years that we fail. There hasn't been a major success in theoretical physics in the last few decades" (Rovelli 2014, 215). This "might be in part because of the wrong ideas we have about science, and because methodologically we are doing something wrong, at least in theoretical physics, and perhaps also in other sciences" (ibid.).

18.7 The Big Data View and Logic

The Big Data view requires a complete break with deduction, it calls for inductive reasoning. According to it, inductive reasoning is the non-scientific part of human thought and is the exact opposite of deductive thinking, so either we reason in the context of a deductive approach, or else we firmly choose an inductive approach. While deductive reasoning always comes to an end, inductive reasoning generally produces no finished status, the results of inferences are likely to alter the inferences already made, and it is possible to continue the reasoning indefinitely. Nevertheless, inductive reasoning produces an imperfect but useful knowledge

Thus, Malle says that Big Data view "requires a complete break from Cartesian logic. It calls for the non-scientific part of human thought: inductive reasoning" (Malle 2013, n.p.). The "analysis of huge amounts of data mainly focuses on finding correlations," such correlations will be found by inductive algorithms, and "induction allows algorithms to reproduce observed phenomena by generalizing beyond their scope," but "without trying to make models out of them" (ibid.). An "algorithm designed within an inductive approach," is not "intended to test an existing hypothesis," and yet it may achieve "a specified aim" (ibid.). This "is far from Cartesian principles," because Cartesian logic is based on deduction, and induction and deduction are opposed to each other, in fact, inductive reasoning is "the exact opposite of deductive thinking" (ibid.). So "either we reason in the context of a deductive approach, or else we firmly choose an inductive approach" (ibid.). While "deductive reasoning always comes to an end, inductive reasoning generally produces no finished status. The results of inferences are likely to alter the inferences already made. It is possible to continue the reasoning indefinitely" (ibid.). Nevertheless, inductive reasoning "produces an imperfect but useful knowledge" (ibid.).

In terms of the analytic method as logic, these claims appear to be completely unjustified.

To begin with, as we have seen in Chap. 17, as a matter of historical fact, Cartesian logic is not based on deduction

Moreover, it is unjustified to say that inductive reasoning is the non-scientific part of human thought. According to the analytic method as logic, inductive reasoning is a main constituent of the method. Indeed, induction is one of the non-deductive rules by which hypotheses are obtained, so inductive reasoning is an important tool for scientific discovery.

Also, it is unjustified to say that inductive reasoning is the exact opposite of deductive thinking, so either we reason in the context of a deductive approach, or else we firmly choose an inductive approach. According to the analytic method as logic, induction and deduction work side by side. For, in the analytic method, every step upward by which a hypothesis is obtained from a problem by a non-deductive rule, is accompanied by a step downward by which a solution to the problem is deduced from the hypothesis.

Furthermore, it is unjustified to say that, while deductive reasoning always comes to an end, inductive reasoning generally produces no finished status, the results of inferences are likely to alter the inferences already made, and it is possible to continue the reasoning indefinitely. According to the analytic method as logic, like deductive reasoning, the non-deductive reasoning by which a hypothesis is obtained from a problem always comes to an end. What does not come to an end is the method as a whole, which is an ongoing process. But it is an ongoing process because, when a hypothesis has been found that is a sufficient condition for solving the problem and is plausible, the hypothesis, in turn, is a problem that must be solved. For, although the hypothesis provides a solution to the problem, the hypothesis may not provide a complete understanding of all aspects of the problem. Moreover, the hypothesis may raise new questions not involved in the problem.

18.8 Knowledge as Problem Solving by the Analytic Method

The method, not only of mathematics and science, but of knowledge in general, is the analytic method.

As we have seen in Chap. 7, Aristotle claims that from knowledge nothing results apart from contemplation. But this is unwarranted. Contrary to Aristotle's claim, human beings desire knowledge because knowledge plays a vital role, as it allows them to solve problems essential for their survival.

More generally, knowledge allows human beings to solve problems arising from many of their needs. Thus, knowledge allows human beings to solve problems arising from needs of social or economic life, such as commercial or financial transactions, the construction of edifices and churches, bridges and dams, ships and aircrafts. Or, problems arising from the need to investigate the environment, and the physical world in general. Or even problems arising from the need for beauty, which often leads scientists and mathematicians to ask and seek answers to questions which give them great aesthetic satisfaction.

The best available method for solving these problems is the analytic method, so the method of knowledge is the analytic method. Therefore, knowledge is problem solving by the analytic method.

18.9 Knowledge as an Infinitely Renewing Task

That knowledge as problem solving by the analytic method means that knowledge is an infinitely renewing task.

Indeed, in the analytic method, to solve a problem we find a hypothesis and check that it is plausible, to solve the problem posed by the hypothesis we find another hypothesis and check that it is plausible. And so on. Thus, solving a problem is a potentially infinite process.

This process is potentially infinite because, as argued in Chap. 15, although a hypothesis provides a solution to the problem, the hypothesis may not provide a complete understanding of all aspects of the problem. Moreover, the hypothesis may raise new questions not involved in the problem. This leads to an ever-renewed transition from one hypothesis to another, more radical one, which results in a deeper understanding of the problem.

Therefore, problem solving is a process which unfolds from the problem, not to a conclusive answer, but to ever more radical hypotheses, which open the way to a boundless deepening of the problem, because every more radical hypothesis leads to broader and deeper developments.

18.10 Inadequacy of Knowledge as Problem Solving by Trial and Error

Against the view that knowledge is problem solving by the analytic method, it could be argued that the analytic method requires non-deductive rules, in particular induction, but there is no such thing as induction, so problem solving cannot proceed by the analytic method. On the contrary, problem solving always proceeds by trial and error. There are no rules for successful trials, the success of trials depends very largely on the number and variety of the trials.

Thus, Popper says that, not only knowledge, but all "life is problem-solving" (Popper 1979, 148). Problem-solving does not proceed by induction, because "there is no such thing as induction" (Popper 1999, 54). On the contrary, "problem-solving always proceeds by the method of trial and error" (Popper 1979, 242). That is, it always proceeds "trying out solutions to our problems and then discarding the false ones as erroneous" (Popper 1999, 3). There are no rules for successful trials, "the demand for a theory of successful thinking cannot be satisfied" because "success depends on many things—for example on luck" (Popper 2002, 50). We can only say that the success of trials "depends very largely on the number and variety of the trials: the more we try, the more likely it is that one of our attempts will be successful" (ibid., 419–420).

But this view is inadequate. For, the claim that problem solving always proceeds by the method of trial and error, conflicts with the fact that the method of trial error is very inefficient. Indeed, because of our physical and time constraints, the number of trials we can make is very small compared to all possible ones, so the probability that a single trial will be successful is very low. This makes it inexplicable why, in fact, mathematicians and scientists often arrive at plausible hypotheses, and therefore why mathematics and science are successful. For example, it has been estimated that "upwards of a quarter million theorems are published a year" (Hoffman 1998, 204). This would be inexplicable if mathematicians proceeded by the method of trial and error.

Moreover, the claim that the success of trials depends very largely on the number and variety of the trials, amounts to admitting that the method of trial and error depends on induction. This contradicts the claim that there is no such thing as induction.

Induction not only exists, but plays an essential role in knowledge and survival.

As Quine ironically says, "creatures inveterately wrong in their inductions have a pathetic but praiseworthy tendency to die before reproducing their kind" (Quine 1969, 126).

The view that knowledge is problem solving by the analytic method does not have the shortcomings of Popper's view. In the analytic method hypotheses are obtained by non-deductive rules, including induction. So, contrary to the method of trial and error, in the analytic method the search for hypotheses does not proceed blindly but is guided by non-deductive rules.

18.11 *Meno*'s Paradox

Against the view that knowledge is problem solving by the analytic method it could also be argued that it conflicts with *Meno*'s paradox, according to which searching for a solution to a problem is either unnecessary or impossible, therefore there cannot be a method for solving problems.

Indeed, according to *Meno*'s paradox, "a man cannot search either for what he knows or for what he does not know. He cannot search for what he knows, because he knows it, and in that case has no need to search for it; for what he does not know, because he does not know what to search for" (Plato, *Meno*, 80 e 2–5).

Plato's solution to *Meno*'s paradox is that "our learning is nothing but recollection," therefore "we must have learned at some previous time what we now recollect" (Plato, *Phaedo*, 72 e 5–7). Indeed, "when people are questioned, if you put the questions well, they themselves answer correctly about everything. And yet they could not do this, if they had not some knowledge and right reason within them" (ibid., 73 a 7–10).

Therefore, according to Plato, a man can search for what he does not know because it is already written in his memory, and it is a matter of extracting it from there.

18.12 Innate Knowledge

Plato's solution to *Meno*'s paradox assumes that most of human knowledge is innate. This assumption is shared by Popper.

Popper even says that "our knowledge is 99 per cent, or let us say 99.9 per cent, biologically innate" (Popper 1999, 54).

That human being have some innate knowledge is confirmed by the fact that, as we have seen in Chap. 10, all human beings have a set of cognitive systems that are phylogenetically ancient, innate, and universal across human beings.

This, however, does not mean, as Plato and Popper assume, that most of our knowledge is innate.

Innate knowledge is a result of natural selection and natural selection is slow, so innate knowledge is very limited. Therefore, although human being have some innate knowledge, Plato's and Popper's assumption that most of human knowledge is innate is untenable.

18.13 Tacit Knowledge

Another solution to *Meno*'s paradox is Polanyi's, which is in terms of tacit knowledge rather than innate knowledge.

Indeed, Polanyi argues that "the *Meno* shows conclusively that if all knowledge is explicit, i.e., capable of being clearly stated, then we cannot know a problem or look for its solution" (Polanyi 2009, 22). However "problems nevertheless exist, and discoveries can be made by solving them, we can know things, and important things" (Ibid.). From this we can conclude that not all knowledge is explicit, there must also be tacit knowledge. In particular, "the kind of tacit knowledge that solves the paradox of the *Meno* consists in the intimation of something hidden, which we may yet discover" (ibid., 22–23).

Polanyi's distinction between explicit and tacit knowledge is foreshadowed by Plato's distinction between having knowledge and possessing knowledge.

Indeed, Plato says: "Having seems to me different from possessing. For example, if someone buys a cloak and becomes its owner but does not wear it, we would not say that he has it, but rather that he possesses it" (Plato, *Theaetetus* 197 b 8–10). In the same way, "it is possible to possess knowledge but not have it" (ibid., 197 c 1–2).

Polanyi's solution, however, is not valid. For, tacit knowledge is either innate or acquired. If it is innate, Polanyi's solution runs into the difficulty of Plato's solution. If it is acquired, it must be searched for, but then it is subject to Meno's paradox.

18.14 *Meno*'s Paradox is Not a Paradox

A more satisfactory solution to *Meno*'s paradox is obtained by recognizing that *Meno*'s paradox is not really a paradox. This can be accounted for if knowledge is problem solving by the analytic method.

Indeed, on the one hand, what a man knows are solution to problems already found. But in the analytic method no solution to a problem is final because, at any moment, new data can emerge with which a hypothesis is incompatible. When this occurs, the man must analyze why the hypothesis is incompatible with the new data, and must modify the hypothesis or replace it with a new hypothesis. Thus, the man must investigate the problem further.

Therefore, it is unjustified to say that a man cannot search for what he knows since he knows it, and in that case he has no need to search for it.

On the other hand, what a man does not know is a solution to a problem. But the fact that he does not know it does not mean that he does not know what he is to search for. He knows that he is to search for a solution to the problem, and he is to search for it by the analytic method.

Therefore, it is unjustified to say that a man cannot search for what he does not know, since he does not know what to search for.

From this we may conclude that both alternatives of *Meno*'s paradox do not arise, and hence *Meno*'s paradox is not really a paradox.

18.15 An Extension of Inference, Hypothesis, and Plausibility

What has been said above about knowledge as problem solving by the analytic method is not limited to human knowledge, but also applies to knowledge of non-human organisms.

Since, however, non-human organisms do not have a language, at least a language consisting of propositions, this requires an extension the concepts of inference, hypothesis, and plausibility. The extension is as follows.

(1) Premisses and conclusions of inferences are not necessarily propositions, they can be data.

(2) Hypotheses are not necessarily propositions, they can be data.

(3) Saying that hypotheses are plausible means that the arguments or data for them are stronger than the arguments or data against them, on the basis of experience.

18.16 The Nature of Problems

That knowledge is problem solving by the analytic method implies that problems are the starting points and paths to knowledge. This raises the questions: What are problems? How are problems posed? How are problems solved?

What are problems? The term 'problem' comes from the Greek 'problema', which has several meanings. One of them is 'question to be solved'.

Thus, Plato says: "We must take over the question and try to solve it as one does with a problem" (Plato, *Theaetetus* 180 c 5–6).

That knowledge is problem solving by the analytic method refers to this meaning of 'problema', because it implies that knowledge comes from solving questions.

18.17 Problem Posing

How are problems posed? This question is important because, with an inappropriate formulation, the search for a solution to a problem may have difficulty in starting, or may take a wrong direction. The formulation of a problem suggests which data to consider and which to disregard, and what type of hypothesis to look for, therefore it is said that a well posed problem is already half solved.

Cantor even says: "Ars proponendi quaestionem pluris facienda est quam solvendi [The art of problem posing must be held more valuable than that of problem solving]" (Cantor 1932, 31).

To formulate a problem, one must first analyze it. The analysis of the problem provides the terms of the problem, that is, the conditions that must be met when one tries to solve it.

An appropriate formulation of a problem does not guarantee that one will be able to solve it, because the means to solve it may have not yet been developed. But even when a problem cannot be solved for this reason, an appropriate formulation of a problem by itself produces a growth of knowledge, since it suggests that there might be certain relations between the data. The solution to the problem, when it eventually becomes available, establishes that such relations actually exist and shows why they exist.

Moreover, even if an appropriate formulation of a problem does not guarantee that one will be able to solve it, the confidence in its solvability produces an intellectual tension that drives one to tackle it intensely. One can hope to solve only those problems whose solution is passionately desired and sought with great intellectual tension.

The intellectual tension plays an essential role in the very choice of problems to investigate. It serves as a guide to discriminate what is interesting from what is not, what is worth investigating from what is not.

18.18 Problem Solving

How are problems solved? An appropriate answer to this question is that they are solved by the analytic method, which from the antiquity is the main method for problem solving.

To solve a problem, first we examine it in order to understand it. We consider the problem from different perspectives, highlighting different aspects of it. We describe what general features a solution to the problem should have. We investigate whether there is any solved problem to which our problem may be related. Then we formulate a hypothesis on the basis of the general features that a solution to the problem should have. The hypothesis is obtained by some non-deductive rule. Which non-deductive rule is to be used depends on the problem, but different non-deductive rules may be used to find the hypothesis. If none of the hypotheses that we have obtained permits us to solve the problem, we examine whether this depends on some common assumption which prevents the solution. If we arrive at a hypothesis that permits us to solve the problem, we examine whether the hypothesis is plausible, namely whether the arguments for it are stronger than the arguments against it, on the basis of experience. To this purpose, we draw consequences from the hypothesis and compare the consequences with experience. If the hypothesis is not plausible, we investigate why it is not plausible. Even when the hypothesis is plausible, this is not decisive for its acceptance. For, the latter also depends on other factors, such as fruitfulness, that is, the ability to open new lines of research. When the hypothesis is seen to be not only plausible but also fruitful, it tends to consolidate and become stable. This, however, is not the end of the inquiry. For, the hypothesis is in turn a problem that requires to be solved. To solve this problem, we repeat this procedure for it.

18.19 Knowledge and Ignorance

Knowledge as problem solving by the analytic method is not the opposite of ignorance. For, in the analytic method, one starts from a problem for which no solution is known, so one starts from ignorance. But the formulation of the problem shows the conditions that must be satisfied to solve it, so the formulation of the problem is part of a process that leads to a solution to the problem. Therefore, ignorance is part of a process that leads to knowledge.

This positive view of ignorance contrasts with the standard view of ignorance, going back to Plato, according to which knowledge is the opposite of ignorance, because "knowledge deals with what is, ignorance necessarily with what is not" (Plato, *Republic*, V 477 a 10–11).

The standard view of ignorance leads to "the ignorance norm," according to which we must "inquire as to the answer of a given question only if we do not know that answer" (Whitcomb 2017, 153). So, when we know the answer to a given question, we must not investigate the question further.

On the contrary, in the analytic method no solution to a problem is final because, at any moment, new data can emerge with which some hypothesis is incompatible. In this case, it is necessary to analyze why the hypothesis is incompatible, and to modify it or replace it with a new hypothesis. Therefore, when we know the answer to a given question, we must remain open to investigating the question further.

18.20 Knowledge and Error

Knowledge as problem solving by the analytic method is not the opposite of error.

One might think that error is intrinsically heterogeneous to knowledge because it is non-knowledge, absence of knowledge. Either there is knowledge or there is error, therefore, if there is error, then there cannot be knowledge.

But this is not the case. For, knowledge and error have the same origin, they both arise from formulating a hypothesis by some non-deductive rule to solve a problem.

That the hypothesis yields knowledge or error does not depend on the non-deductive rule but on the plausibility of the hypothesis. Then the difference between knowledge and error is not intrinsic but extrinsic, because it does not depend on the process of formulating the hypothesis but on the relation of the hypothesis with experience.

As Mach says, "knowledge and error flow from the same mental sources, only success can tell the one from the other" (Mach 1976, 84).

Not only error has the same origin as knowledge, but error can produce knowledge.

Indeed, an analysis of the cause of the error can provide indications useful for modifying the hypothesis or formulating a new hypothesis. Thus, an analysis of the cause of Russell's paradox led Zermelo to formulate his axioms for set theory.

An analysis of the cause of the error can even lead to formulate a new concept. For example, an analysis of the cause of Russell's paradox led Zermelo to formulate the iterative concept of set, according to which "a set is something obtainable form the integers" by "iterated application of the operation 'set of'" (Gödel 1986–2002, II, 259). Specifically, a set is a member of the iterative hierarchy of sets which is formed in successive stages, starting from the integers, by transfinite iteration of the power-set operation.

Of course, an analysis of the cause of the error can achieve this only if the process through which hypotheses are formulated is rational.

Popper denies that this process is rational, he claims that hypotheses are "the result of an almost poetic intuition" (Popper 2002, 260). Therefore, from Popper's viewpoint, an analysis of the causes of error cannot provide useful indications for formulating new hypotheses.

Indeed, according to Popper, what is required to formulate new hypotheses is not an analysis of the causes of error, but "the creative ability to produce new guesses, and more new guesses" (Popper 1979, 260). Producing new guesses and more new guesses is not a rational process, it requires originality, and "originality is a gift of the gods" (Popper 2002, 68). Therefore, according to Popper, formulating new hypotheses is a gift of the gods.

Contrary to Popper's view, formulating new hypotheses is not a gift of the gods. As argued in Chap. 15, in the analytic method formulating new hypotheses and finding solutions to problems is a rational process because hypotheses are obtained by non-deductive rules, so by logic, and their plausibility is established by comparing the arguments for and against them, so by logic.

Only if the process through which hypotheses are formulated is a rational one we can assert that error can lead to new knowledge. For, only so it is justified to say that an analysis of the causes of error can provide useful indications for formulating new hypotheses.

18.21 Knowledge and Certainty

Knowledge as problem solving by the analytic method is not certain because it is only plausible, and plausibility does not ensure certainty. However, plausibility is the best we can achieve.

As Plato says, "certain knowledge is either impossible or extremely difficult to come by in this life," we can only "adopt the best and least refutable of human hypotheses, and embarking upon it as upon a raft, run the risk of sailing the sea of life" (Plato, *Phaedo*, 85 c 3–d 2).

Even mathematical knowledge cannot be certain.

Feferman says that mathematics is "the paradigm of certain and final knowledge" (Feferman 1998, 77).

Gowers says that, by the axiomatic method, "any dispute about the validity of a mathematical proof can always be resolved" (Gowers 2002, 40).

But this is not valid. By Gödel's second incompleteness theorem, mathematics cannot be said to be the paradigm of certain and final knowledge, and it is unwarranted to say that by the axiomatic method any dispute about the validity of a mathematical proof can always be resolved.

As Hersh says, "certainty is what many yearn for in childhood, but learn to live without in adult life, including in mathematics" (Hersh 2014, 81). For, "mathematics is human, and nothing human can be absolutely certain" (ibid., 82).

Admittedly, the search for certainty responds to a profound need of human beings who, in their life subjected to insecurity, feel the need to find some secure point of support. But insecurity is a constitutive element of human life and there is no escape from it.

As Russell says, "the demand for certainty is one which is natural to man, but is nevertheless an intellectual vice" (Russell 2009, 26). Uncertainty "is painful, but must be endured if we wish to live without the support of comforting fairy tales" (Russell 2004, 17).

Ultimately, the search for certainty is a way to exorcise the fear of death.

As Galileo says, "those who so much exalt incorruptibility, inalterability, etc., I believe, are reduced to saying these things because of their great desire to live long and because of their terror of death; and they do not consider that if men were immortal, they themselves would not have come into the world. They would deserve to encounter a Medusa's head which would transmute them into statues of jasper or of diamond, and thus make them more perfect than they are" (Galilei 1968, VII, 84).

18.22 Knowledge and Intuition

Knowledge as problem solving by the analytic method does not depend on intuition because, as argued in Chap. 15, intuition plays no part in the analytic method. Since knowledge is problem solving by the analytic method, from this we may conclude that knowledge does not depend on intuition.

This conclusion conflicts with Turing's and Gödel's view that, because of Gödel's incompleteness theorems, knowledge depends on intuition.

(1) Turing argues that, "in pre-Gödel times it was thought by some" that "all the intuitive judgments of mathematics could be replaced by a finite number of" mechanical "rules. The necessity for intuition would then be entirely eliminated" (Turing 1939, 215). But Gödel's first incompleteness theorem demonstrated "the impossibility of finding a formal logic which wholly eliminates the necessity of using intuition" (ibid., 216).

This argument, however, is not valid. It is based on the assumption that either problems can be solved by algorithmic rules or intuition is necessary. Since, by Gödel's first incompleteness theorem, not all problems can be solved by the algorithmic rules of a given formal system, from the assumption Turing concludes that intuition is necessary. But the assumption is not valid because, between the dullness of algorithmic rules and the inscrutability of intuition, there is an intermediate region inhabited by the heuristic methods, which are neither algorithmic nor involve intuition. The analytic method is an example of this.

(2) Gödel argues that his second incompleteness theorem implies that for the "axioms there exists no other rational (and not merely practical) foundation except that" they "can directly be perceived to be true" by "an intuition of the objects falling under them" (Gödel 1986–2002, III, 346–347). In particular, we have "an intuition which is sufficiently clear to produce the axioms of set theory" (ibid., II, 268). This intuition is obtained by "focusing more sharply on the concepts concerned by

directing our attention in a certain way, namely, onto our own acts in the use of these concepts, onto our powers in carrying out our acts, etc." (ibid., III, 383). So we will get an "intuitive grasping of ever newer axioms" (ibid., III, 385).

This argument, however, is not valid. First, the claim that axioms can directly be perceived to be true by an intuition of the objects falling under them, is based on the assumption that knowledge, to be knowledge, must be absolutely certain. Since, by Gödel's second incompleteness theorem, axioms cannot be proved to be true by absolutely reliable means, from the assumption Gödel concludes that they must be perceived to be true by intuition, therefore knowledge depends on intuition. But the assumption is not valid because, as argued above, knowledge cannot be certain but only plausible.

Second, the claim that an intuition which is sufficiently clear to produce the axioms of set theory is obtained focusing more sharply on the concepts concerned by directing our attention in a certain way, is based on the assumption that in this way axioms can be perceived to be true. But the assumption is not valid. Indeed, suppose that, by focusing more sharply on the concept of set Σ, we directly perceive the axioms of a deductive theory S to be true of Σ. Then Σ is a model of S, so S is consistent, hence, by Gödel's first incompleteness theorem, there is a sentence G of S which is true of Σ but unprovable in S. Let $T=S\cup\{\neg G\}$. Since G is unprovable in S, the theory T is consistent, and hence has a model, say Σ'. Since Σ' is a model of S, Σ' is a concept of set. By focusing more sharply on the concept of set Σ', we directly perceive the axioms of T to be true of Σ'. So $\neg G$ is true of Σ'. On the other hand, G is true of Σ, so $\neg G$ is false of Σ. Then the question arises: Is $\neg G$ true or false? Or, in other words: Which of Σ and Σ' is the genuine concept of set? Intuition gives no answer to this question.

18.23 A Priori Knowledge as Problem Solving by the Analytic Method

In the analytic method hypotheses are obtained by non-deductive rules that are ampliative, therefore they contain something that is not contained in the premisses. So, even when the premisses are derived from experience, the hypotheses obtained from them by non-deductive rules go beyond experience, therefore they are not derived from experience.

This suggests to identify a priori knowledge with knowledge as problem solving by the analytic method. Knowledge as problem solving by the analytic method is a priori not in the sense that it is absolutely independent of experience, but in the sense that it is based on hypotheses that are not derived from experience, so it goes beyond experience.

18.24 Conflict with the Standard Concept of A Priori Knowledge

Against a priori knowledge as problem solving by the analytic method it could be argued that it conflicts with the standard concept of a priori knowledge, namely Kant's concept of a priori knowledge. For, although a priori knowledge as problem solving by the analytic method is based on hypotheses that are not derived from experience, nevertheless the hypotheses are plausible, that is, such that the arguments for them are stronger than the arguments against them, on the basis of experience, so they depend for their plausibility on experience. Therefore, a priori knowledge as problem solving by the analytic method does not occur absolutely independently of all experience.

On the contrary, according to Kant, a priori knowledge is knowledge that occurs "absolutely independently of all experience" (Kant 1998, B3). It has the character of "strict universality," that is, "no exception at all is allowed to be possible" (ibid., B4). It has "the character of inner necessity" (ibid., A2). And it is "certain" (ibid.).

Kant's concept of a priori knowledge, however, is not valid. For, no knowledge occurs absolutely independently of all experience, because all knowledge, including mathematical knowledge, depends on experience (as regards mathematical knowledge, see Cellucci 2025). No knowledge has the character of strict universality, because all knowledge is at most plausible, so exceptions are always possible. No knowledge has the character of inner necessity, because all knowledge is the product of our limited experience and our fallible biology, so it is always contingent. No knowledge is certain, because all knowledge is at most plausible, so it is potentially fallible.

Instead, a priori knowledge as problem solving by the analytic method is not subject to these limitations. Indeed, it does not occur independently of all experience because, as already said, the hypotheses on which it is based must be plausible, so they depend for their plausibility on experience. It has no character of strict universality, because it is only plausible, so exceptions are always possible. It has no character of inner necessity, because it can be incompatible with future data, so it is contingent. It is not certain, because it can be fallible.

18.25 Genetically A Priori Knowledge and Valid A Priory Knowledge

Against a priori knowledge as problem solving by the analytic method it could also be argued that it implies that all knowledge is a priori. For, all knowledge is problem solving by the analytic method. But, if all knowledge is a priori, then a priori knowledge as problem solving by the analytic method does not discriminate between a priori knowledge and non-a priori knowledge.

This argument, however, is not valid because, in a priori knowledge as problem solving by the analytic method, 'a priori knowledge' means 'genetically a priori knowledge', not 'valid a priori knowledge'.

Genetically a priori knowledge is knowledge not derived from experience. Valid a priori knowledge is knowledge whose validity does not depend on experience.

A priori knowledge as problem solving by the analytic method is genetically a priori, but not valid a priori. It is genetically a priori, because it is based on hypotheses that are not derived from experience. It is not valid a priori, because the hypotheses on which it is based are plausible, so they depend for their plausibility on experience.

Therefore, a priori knowledge as problem solving by the analytic method discriminates between genetically a priori knowledge and non-valid a priori knowledge.

18.26 A Disputable Concept of A Priori Knowledge

The concept of a priori knowledge as problem solving by the analytic method is different from Popper's concept of a priori knowledge.

Popper says that a priori knowledge is "that kind of knowledge—of fallible or conjectural knowledge—which an organism has prior to sense experience; roughly speaking, it is inborn knowledge" (Popper 1999, 69). A priori knowledge is extremely extensive, because "99 per cent of the knowledge of all organisms is inborn and incorporated in our biochemical constitution" (ibid., 70). The "rest is a modification, a revolutionary overturning of some previous knowledge," so "in the end all knowledge goes back to innate knowledge and to its modification" (ibid., 54).

We are born with inborn expectations, in particular we "expect laws or regularities in" our "surroundings" (ibid., 3–4). These inborn expectations are a priori knowledge, which "is not, however, valid a priori," because "an inborn expectation, no matter how strong and specific, may be mistaken" (Popper 2002, 62). A priori knowledge is only "genetically a priori, i.e. prior to all observational experience" (ibid.)

If "an expectation proved to have been wrong," then for us "a problem arises" (Popper 1999, 4). When this occurs, we attempt "to solve the problem" (ibid.). We attempt to solve it by formulating a hypothesis or theory. We formulate it by trial and error, where trial and error "is the method of trying out solutions to our problem and then discarding the false ones as erroneous. This method assumes that we work with a large number of experimental solutions. One solution after another is put to the test and eliminated" (ibid., 3). Our trials "are always our hypotheses. They stem from us, not from the external world. All we learn from the external world is that some of our efforts are mistaken" (ibid.).

Popper's concept of a priori knowledge, however, is not valid. Indeed, it is unfounded to say that a priori knowledge is inborn knowledge, that inborn knowledge constitutes 99 per cent of the knowledge of all organisms and is incorporated in our biochemical constitution. For, inborn knowledge has been incorporated in our biochemical constitution as a result of natural selection, and, as argued earlier in this chapter, natural selection is very slow, so inborn knowledge is very limited and hence cannot constitute 99 per cent of the knowledge of all organisms. Therefore, a priori knowledge cannot all be inborn knowledge.

Moreover, it is also unfounded to say that, when for us a problem arises because an expectation proved to have been wrong, we attempt to solve the problem formulating a hypothesis or theory by trial and error. For, as argued earlier in this chapter, the method of trial and error is very inefficient, so trying to solve a problem through it is very unlikely to succeed.

The concept of a priori knowledge as problem solving by the analytic method does not have these shortcomings.

For, a priori knowledge as problem solving by the analytic method, in particular scientific knowledge and mathematical knowledge, is very extensive because it is a result of cultural evolution which, compare to natural selection, is very fast.

Moreover, when for us a problem arises because an expectation proved to have been wrong, we attempt to solve the problem formulating a hypothesis not by trial and error but by some non-deductive rule. So the search for hypotheses does not proceed blindly as in trial and error, it is guided by non-deductive rules.

18.27 Philosophy as Problem Solving by the Analytic Method

The fact that the method of knowledge is the analytic method has an important implication also for philosophy.

If the method of knowledge is the analytic method and philosophy is acquisition of knowledge, then the method of philosophy is the analytic method. Therefore, philosophy is problem solving by the analytic method.

From this it appears that the question 'What is philosophy?', raised in Chap. 4, is tightly connected with the question 'What is the method of philosophy?'. And the question 'What should philosophy be?' is tightly connected with the question 'What should the methods of philosophy be?'

That the two questions are tightly connected means that to answer one them is to answer the other one.

18.28 Origin of Analytic Method as Method of Philosophy

The origin of the view that the method of philosophy is the analytic method is Plato.

According to a widespread opinion, Plato contrasts the method of philosophy with the method of mathematics.

Thus, Bostock argues that "Plato thinks of the method of mathematics as one that starts by assuming some hypotheses and then goes 'downwards' from them (i.e. by deduction), whereas the method of philosophy (i.e. dialectic) is to go 'upwards' from the initial hypotheses, finding reasons for them (when they are true), until eventually they are shown to follow from an 'unhypothetical first principle'" (Bostock 2009, 13–14).

But this argument is not valid. Plato does not contrast the method of philosophy with the method of mathematics, on the contrary, he takes the method of philosophy to be the same as the method of mathematics. Indeed, as we have seen in Chap. 15, he uses the analytic method to solve both the mathematical problem of doubling the square and the philosophical problem of teachability of virtue.

In fact, Plato says: "To inquire" a mathematical or philosophical question I proceed "from a hypothesis," in the way "in which geometricians usually proceed in inquiring" (Plato, *Meno*, 86 e 4–5). So, I proceed by the analytic method, because "if we are to believe Hippocrates" of Cos, the physician, we cannot learn anything "unless we follow this method" (Plato, *Phaedrus*, 270 c 3–5).

Therefore, according to Plato, both the method of mathematics and the method of philosophy go 'upwards' and are the same method, the analytic method.

18.29 Support for Analytic Method as Method of Philosophy

In the modern period some major philosophers have supported the view that the method of philosophy is the analytic method.

Thus, Hobbes says: "The knowledge of" universal things "is to be acquired" by "resolution" (Hobbes 1839–1845, I, 69). For, "by resolving continually, we may come to know what those things are" (ibid.). Now, "the resolutive" method "is commonly called analytical method" (ibid., I, 66). Therefore, "the method of attaining to the universal knowledge of things, is purely analytical" (ibid., I, 69).

Kant says: "In in all rational cognitions" the method "is either synthetic or analytic. In the former I go from ground to consequence, in the latter from consequence to ground" (Kant 1992b, 236). In particular "metaphysics" has, "in spite of the great efforts of scholars, remained imperfect and uncertain because the method peculiar to it has been misunderstood. Its method is not synthetic" but "analytic" (Kant 1992a, 294).

18.30 Difference from the Methods of Analytic Philosophy

The method of philosophy as acquisition of knowledge is different from the methods of analytic philosophy.

Williamson says: "Forget the idea of a single method, employed in all and only philosophical thinking," instead, "philosophers use methods of various kinds: they philosophize in various ways. A philosophical community's methodology is its repertoire of such methods" (Williamson 2022, 5).

Williamson refers to the fact that analytic philosophers use the variety of methods considered in Chap. 6. But, as argued there, those methods are inadequate to acquire knowledge. Analytic philosophers use them because they assume that philosophy is not acquisition of knowledge, so analytic philosophy does not need a method for acquiring knowledge.

On the contrary, philosophy as acquisition of knowledge needs a method for acquiring knowledge. The method for acquiring knowledge is the analytic method, therefore the method of philosophy is the analytic method.

18.31 Inadequacy of the Armchair Argument

That the method of philosophy is the analytic method means that the method of philosophy is the same as the method of science.

In contrast, analytic philosophy assumes that philosophy is an armchair discipline, requiring no input from experience but only thought, so the method of philosophy cannot be the same as the method of science.

Thus, Williamson says that "philosophy is far less different from the rest of inquiry than many philosophers like to think," but "that should not lead us to expect a gradual convergence in methodology between philosophy" and "the natural and social sciences, for philosophy also has much in common with the foundational parts of mathematics, the most armchair science of all" (Williamson 2019, 24). In fact, "the traditional methods of philosophy are armchair ones: they consist of thinking, without any special interaction with the world beyond the chair, such as measurement, observation or experiment would typically involve" (Williamson 2022, 3).

But the assumption of analytic philosophy that philosophy is an armchair discipline, requiring no input from experience but only thought, is not valid. As argued in Chap. 5, philosophy is not an armchair discipline like mathematics. So, one cannot appeal to the fact that philosophy is an armchair discipline to argue that the method of philosophy cannot be the same as the method of science, and hence, if the method of science is the analytic method, then the method of philosophy cannot be the analytic method.

18.32 Essential Dependence of Philosophy on Experience

The assumption of analytic philosophy that philosophy is an armchair discipline, requiring no input from experience but only thought, has implausible implications.

For example, as we have seen in Chap. 3, this assumption led Hacker to claim that, by conceptual analysis, philosophy can study the essence of things, therefore philosophy can perform a task that, as argued in Chap. 4, science is unable to perform. But Hacker's claim is unfounded because there is no evidence that there exists a superior source of knowledge that yields knowledge of the essence of things, let alone that philosophy is such a superior source of knowledge.

On the contrary, according to philosophy as problem solving by the analytic method, philosophy does not proceed by mere conceptual analysis, it essentially depends on experience. For, most philosophical problems originate from the world and hence from experience, and are solved by the analytic method using hypotheses which must be plausible, so they depend for their plausibility on experience. Therefore, philosophy depends on experience both for the origin and the solution of philosophical problems.

Our mind is constantly engaged in trying to make the world intelligible to us. Philosophy as problem solving by the analytic method is a means to this end, so it is anchored to experience.

18.33 Philosophy and the Birth of New Sciences

That philosophy is problem solving by the analytic method is the basis for the claim that, if successful, philosophy can open new ways to knowledge and lead to the birth of new sciences.

But how can philosophy do this? The example of Turing considered in Chap. 7 provides indications in this direction.

We look for some problem that is of interest to several people and is beyond the present sciences.

We analyse the problem as thoroughly as possible. The analysis may suggest under what conditions the problem could be solved. Then, from the problem, we can arrive non-deductively at formulating a hypothesis.

We consider what arguments can be provided for the hypothesis and against it on the basis of the existing knowledge. If the arguments for the hypothesis are stronger than those against it, the hypothesis is plausible, so we accept the hypothesis for the moment and deduce a solution to the problem from it.

References

Anderson, Chris. 2008. The end of theory: The data deluge makes the scientific method obsolete. *Wired Magazine*, June 23.

Berman, Jules. 2013. *Principles of big data: Preparing, sharing, and analyzing complex information*. Amsterdam: Elsevier.

Bostock, David. 2009. *Philosophies of mathematics: An introduction*. Oxford: Wiley-Blackwell.

Calude, Cristian, and Giuseppe Longo. 2017. The deluge of spurious correlations in big data. *Foundations of Science* 22:595–612.

Cantor, Georg. 1932. *Gesammelte Abhandungen mathematischen und philosophischen Inhalts*. Berlin: Springer.

Cellucci, Carlo. 2022. *The making of mathematics: Heuristic philosophy of mathematics*. Cham: Springer.
Cellucci, Carlo. 2025. Mathematics and experience. *Foundations of Science* 30:473–487.
Einstein, Albert. 2002. Mathematics and experience. In Albert Einstein, *Collected Papers*, vol. 7, 108–109. Princeton: Princeton University Press.
Feferman, Solomon. 1998. *In the light of logic*. Oxford: Oxford University Press.
Galilei, Galileo. 1968. *Opere*. Florence: Barbera.
Gödel, Kurt. 1986–2002. *Collected works*. Oxford: Oxford University Press.
Gowers, Timothy. 2002. *Mathematics: A very short introduction*. Oxford: Oxford University Press.
Hersh, Reuben. 2014. *Experiencing mathematics: What do we do, when we do mathematics?* Providence: American Mathematical Society.
Hobbes, Thomas. 1839–1845. *English works*. London: Longman.
Hoffman, Paul. 1998. *The man who loved only numbers: The story of Paul Erdös and the search for mathematical truth*. London: Fourth Estate.
Kant, Immanuel. 1992a. *Theoretical philosophy 1755–1770*. Cambridge: Cambridge University Press.
Kant, Immanuel. 1992b. *Lectures on logic*. Cambridge: Cambridge University Press.
Kant, Immanuel. 1998. *Critique of pure reason*. Cambridge: Cambridge University Press.
Mac Lane, Saunders. 1984. Proof is eternal. *Proceedings of the American Philosophical Society* 128:44–47.
Mach, Ernst. 1976. *Knowledge and error: Sketches on the psychology of enquiry*. Cham: Springer.
Malle, Jean-Pierre. 2013. Big data: Farewell to Cartesian thinking? *Paris Innovation Review*, March 15.
Nathanson, Melvyn. 2008. Desperately seeking mathematical truth. *Notices of the American Mathematical Society* 55:773.
Naylor, Arch, and George Sell. 2000. *Linear operator theory in engineering and science*. Springer.
Polanyi, Michael. 2009. *The tacit dimension*. Chicago: The University of Chicago Press.
Popper, Karl. 1979. *Objective knowledge: An evolutionary approach*. Oxford: Oxford University Press.
Popper, Karl. 1999. *All life is problem solving*. London: Routledge.
Popper, Karl. 2002. *Conjectures and refutations: The growth of scientific knowledge*. London: Routledge.
Quine, Willard. 1969. *Ontological relativity and other essays*. New York: Columbia University Press.
Rota, Gian-Carlo. 1997. *Indiscrete thoughts*. Boston: Birkhäuser.
Rovelli, Carlo. 2014. Science is not about certainty. In *The universe*, ed. John Brockman, 214–228. New York: Harper.
Russell, Bertrand. 2004. *History of Western philosophy*. London: Routledge.
Russell, Bertrand. 2009. *Unpopular essays*. London: Routledge.
Turing, Alan. 1939. Systems of logic based on ordinals. *Proceedings of the London Mathematical Society, Series* 2 (45): 161–228.
Whitcomb, Dennis. 2017. One kind of asking. *The Philosophical Quarterly* 67:148–168.
Williamson, Timothy. 2019. Armchair philosophy. *Epistemology and Philosophy of Science* 56:19–25.
Williamson, Timothy. 2022. *The philosophy of philosophy*. Hoboken: Wiley-Blackwell.

Chapter 19
Knowledge and Perception

Abstract Developing the view that philosophy is acquisition of knowledge also requires answering the question: What is perception? A widely held view is that perception is a passive process, determined entirely by the features of the external world, but this view is not valid. Alternatively, this chapter argues that perception is an active process because, like all knowledge, perception is problem solving by the analytic method.

Keywords Inadequacy of perception as a passive process · Perception as problem solving by the analytic method · Arguments against perception as problem solving by the analytic method · Arguments for perception as problem solving by the analytic method · Limitations of the human perceptual system

19.1 Two Views of Perception

Developing the view that philosophy is acquisition of knowledge also requires answering the question: What is perception?

In Chap. 18 it has been argued that knowledge is problem solving by the analytic method. One might think that perception is a counterexample to this because, according to a widely held view, perception is a passive process, determined entirely by the features of the external world.

This view, however, is not valid. Alternatively, this chapter argues that perception is an active process because, like all knowledge, perception is problem solving by the analytic method.

In contemporary philosophy of perception, the view that perception is problem solving by the analytic method has received little attention. On the contrary, in the psychology of perception, the inferential character of perception has been properly acknowledged. The aim of this chapter is to remedy this deficiency of contemporary philosophy of perception.

C. Cellucci, *A Philosophy for the Future*, Synthese Library 508,
https://doi.org/10.1007/978-3-031-98443-3_19

19.2 Inadequacy of Perception as a Passive Process

According to several analytic philosophers, perception is a passive process because it is immediate consciousness of sense-data, so perception is direct, non-inferential.

Thus, Moore says that "sense-data" are the entities "of which I am directly conscious" in perception, such as "the colours" which "I actually see;" the "sounds which I actually hear; the peculiar sort of entity of which I am directly conscious when I feel the pain of a toothache, and which I call 'the pain'; and many others" (Moore 1909–1910, 57). Thus, perception is immediate consciousness of sense-data. Since I am directly conscious of sense-data, I am absolutely certain of perceptions, for example, "there is nothing more certain to me than that I do constantly see one colour at one time" (ibid., 39).

Russell says that "sense-data" are "the things that are immediately known" to us in perception, such as "colours, sounds, smells, hardness, roughness, and so on" (Russell 1998, 4). Thus, perception is immediate knowledge of sense-data, namely knowledge of sense-data of which "we are directly aware, without the intermediary of any process of inference" (ibid., 25). Since we have immediate knowledge of sense-data, "we are certain of our own sense-data," which are then "a solid basis from which to begin our pursuit of knowledge" (ibid., 8).

Ayer says that "there is general agreement among the philosophers who make use of the term 'sense-datum', or some equivalent term, that what we immediately experience is always a sense-datum and never a material thing" (Ayer 1940, 5). It is "legitimate to use the phrase 'direct observation'" only "in such a way that it is only what is designated by the term 'sense-datum', or some equivalent term, that can be said to be directly observable" (ibid., 234). Therefore "any proposition that refers to a material thing must somehow be expressible in terms of sense-data, if it is to be empirically significant" (ibid., 231).

But the claim that perception is a passive process because it is immediate consciousness of sense-data is not valid. For, it assumes that perception is veridical, namely that things are as they appear to be. This conflicts with science, which shows that things are not as they appear to be.

In particular, perception is underdetermined by the sense-data. For example, in the case of vision, the stimuli on the retina are poor and unstable, but we have a rich and stable visual experience. The image on the retina is two-dimensional, but we have a three-dimensional impression. The image on the retina is upside down and with left and right reversed, but we see things upright and with left and right unreversed. The image on the retina of the left eye is different from the image on the retina of the right eye, but we have a single impression. Therefore, vision is underdetermined by the image on the retina.

19.3 Perception as Problem Solving

Perception is not a passive process, because perception is problem solving by the analytic method.

In perception, we form hypotheses by means of non-deductive inferences from the data received from sensory receptors.

The non-deductive inferences are non-propositional and unconscious. They are non-propositional, because their premisses and conclusions are not propositions but data. They are unconscious, because they take place in us at too low level and too high speed to be accessible to direct introspection.

Therefore, the inferences involved in perception are non-deductive, non-propositional, and unconscious.

Moreover, the hypotheses on which perception is based are plausible.

This requires the extension of the concepts of inference, hypothesis, and plausibility already described in Chap. 18 regarding knowledge of non-human organisms.

19.4 Origin of Perception as Problem Solving

That inferences involved in perception are non-deductive, non-propositional, and unconscious, is asserted by von Helmholtz.

Indeed, von Helmholtz says that, although the "conclusions leading to the formation of our sense-perceptions certainly do lack the purifying and scrutinizing work of conscious thinking," they "may be classed as conclusions, inductive conclusions unconsciously formed" (von Helmholtz 2013, 27). While "the 'conclusions' of logicians" are "capable of expression in words," the conclusions leading to the formation of our sense-perceptions "are not; because, instead of words, they only deal with sensations" (von Helmholtz 1995, 198).

This involves an extension of logic.

Indeed, Dilthey argues that "logic is capable of complying with the demands of critical consciousness only by extending its province beyond the analysis of discursive thought" (Dilthey 1989, 166). The "achievement by which the perceptual process transcends what is given to it is an equivalent of discursive thought. The profound notion of unconscious inferences as developed by Helmholtz implies such an extension of logic" (ibid., 167).

However, von Helmholtz assumes that the inferences involved in perception are inductive. This contrasts with the fact that induction does not account for all types of perception. In fact, the inferences involved in perception are non-deductive but not necessarily inductive. The view that perception is problem solving by the analytic method accounts for this because, according to it, the inferences involved in perception are non-deductive inferences.

19.5 Irresistibility of Inferences Involved in Perception

Being unconscious, the inferences involved in perception are irresistible, in the sense that we cannot resist perceiving what we perceive even if we know that it is an illusion.

Thus, because of the moon illusion, we perceive the moon at horizon as larger than the moon at zenith. But, if we take a photo of the moon at horizon and another of the moon at zenith and compare them, we see that the moon at horizon is the same size as the moon at zenith. Nevertheless, once we have seen the two photos, even though our conscious experience tells us that the moon at horizon is the same size as the moon at zenith, we cannot resist perceiving the moon at horizon as larger than the moon at zenith.

Inferences involved in perception are irresistible because, as a result of natural selection, the human brain embodies neural circuits that make inferences through which the human brain categorizes data without our being aware of them.

19.6 Adequacy of Perception as Problem Solving

Unlike the view that perception is a passive process, the view that perception is problem solving by the analytic method is adequate.

In particular, in the case of vision, the view accounts for the fact that the stimuli on the retina are poor and unstable but we have a rich and stable visual experience. This occurs because what we see is not the image on the retina, but the result of forming hypotheses which go beyond the image on the retina. For example, when we see an object moving fast, the eye moves with the object, thus the image of the object is rather stable on the retina, while the background with respect to which the object moves, which we perceive as steady, moves fast on the retina. Therefore, what we see is based on a hypothesis that distinguishes the stability of the image of the object on the retina from the movement of the object in the world.

Also, the view accounts for the fact that the image on the retina is two-dimensional but we have a three-dimensional impression. This occurs because what we see is not the image on the retina, but the result of forming hypotheses which go beyond the image on the retina.

Moreover, the view accounts for the fact that the image on the retina is upside down and with left and right reversed, but we see things upright and with left and right unreversed. This occurs because what we see is not the image on the retina, but the result of forming hypotheses which go beyond the image on the retina. For example, if we wear special glasses with prism lenses that show an upside down image of things, at first we are unable to do anything, in particular we are unable to walk. For, turning the image of the world upside down and with left and right reversed upsets our hypotheses about the external world.

Furthermore, the view accounts for the fact that the image on the retina of the left eye is different from the image on the retina of the right eye, but we have a single impression. This occurs because what we see are not the images on the two retinas, but rather the result of forming the hypothesis that the stimuli on the two retinas are produced by the same object.

19.7 Arguments Against Perception as Problem Solving

Several arguments have been advanced against the claim that perception is problem solving by the analytic method.

(1) *Perception is not knowledge*. Thus, Dewey argues that a "fact 'presented' in perception" is "one directly 'apprehended'," while an object of knowledge is "determined by inference" and so is "known through a logical way," because knowledge consists of "propositions arrived at through inference" (Dewey 1911, 398).

But this argument is not valid. For, it assumes that things presented in perception are directly apprehended, and hence are as they appear to be. This assumptions is not valid, because the sciences show that things presented in perception are not as they appear to be.

(2) *Perception is not inference*. Thus, Pylyshyn argues that vision is not inference, our visual system embodies "(without explicitly representing and drawing inferences from) certain very general constraints on the interpretations that it is allowed to make" (Pylyshyn 2003, 96). These constraints "produce the correct interpretation under specified conditions that frequently obtain in our kind of physical world," so the visual system yields vision "without 'unconscious inference'" (ibid.). The constraints "are based almost entirely on principles that derive from laws of optics and projective geometry" (ibid., 120).

But this argument is not valid. For, to say that our visual system embodies certain very general constraints that produce the correct interpretation of stimuli, is to say that the human brain embodies specialized circuits that make inferences which yield the correct perception.

(3) *Perception is not non-propositional inference*. Thus, Bennett and Hacker argue that, while inferences are "transformations of propositions in accordance with a rule," instead, "perceiving something involves no transformation of propositions" (Bennett and Hacker 2003, 137).

But this argument is not valid. For, it assumes that inferences must be transformations of propositions. But there is no evidence for this assumption, inferences need not be transformations of propositions. As we have seen above, this has already been argued by Dilthey.

(4) *Perception is not unconscious inference, it depends on our knowledge, experience, and theories.* Thus, Hanson argues that at dawn "Tycho sees the sun beginning its journey from horizon to horizon," that is, he sees it "circling our fixed earth," while Kepler sees "the horizon dipping, or turning away, from our fixed local star" (Hanson 1965, 23). Therefore, at dawn Tycho and Kepler see different things. This "difference is due neither to differing visual pictures, nor to any 'interpretation' superimposed on the sensation" (ibid., 11). It "depends on" Tycho's and Kepler's "knowledge, experience, and theories" (ibid., 18).

But this argument is not valid. At dawn, Tycho and Kepler do not see different thing, they see the same thing because our perception does not depend on our knowledge, experience, and theories. Thus, as it appears from the moon illusion considered above, although our knowledge, experience, and theories tell us that the moon at horizon is the same size as the moon at zenith, we perceive the moon at horizon as larger than the moon at zenith.

(5) *The hypotheses on which perceptions are supposed to be based could be answerable only to perceptions themselves.* Thus, Anscombe argues that "a hypothesis is something answerable to evidence, to data. To what data could the perceptual hypotheses" be "answerable, but to perceptions?" (Anscombe 1981, II, 65).

But this argument is not valid. The hypotheses on which perceptions are based are answerable to data from multiple sensory receptors. For example, we observe a stick in water and it appears crooked, so we form the hypothesis that it is crooked, then we verify the hypothesis by touching the stick, and we realize that it is not crooked. Therefore, the hypothesis that the stick is crooked, formed from the image on the retina, is checked through data from touch receptors.

19.8 Arguments for Perception as Problem Solving

There are several arguments for the claim that perception is problem solving by the analytic method. Here are the main ones.

(1) *Size constancy.* Our perceptions of the size of objects are relatively constant even though the size of images of objects on the retina varies greatly with distance. For example, in the following figure, the golfer in the background, although smaller, appears to be of normal size compared to the golfer in the foreground.

This happens because our perceptual system forms the hypothesis that the golfer in the background is of normal size on the basis the fact that the size of an object remains constant as the object gets closer or moves away from us.

(2) *Shape constancy*. Our perceptions of the shape of objects are relatively constant although their image on the retina varies greatly with their orientation in space. For example, the following two identical plates, oriented differently in space, appear to have the same shape to us.

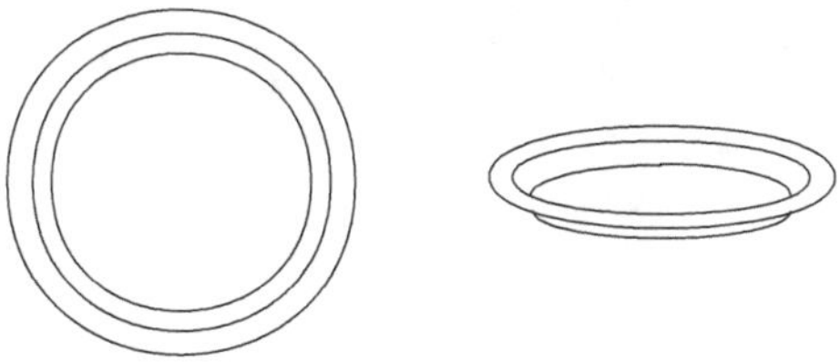

This happens because our perceptual system forms the hypothesis that the two plates have the same shape on the basis of the fact that the shape of an object remains constant as the object is differently oriented in space.

(3) *Colour constancy*. Our perceptions of the colour of objects are relatively constant, even though the image of an object on the retina varies greatly when the object is illuminated by lights of different colours. For example, a white sheet of paper will appear to us white both at midday, when the main illumination is white, and at sunset, when the main illumination is red, although the light signals at the retina change drastically due to the change in the illuminant.

This happens because our perceptual system forms the hypothesis that the colour of the sheet of paper is the same on the basis of the fact that the colour of an object remains constant independently of its illumination.

(4) *Ambiguous figures*. There are figures that can be read in different ways. For example, in the following drawing, the Necker cube, if we focus on the drawing, a face of the cube sometimes appears as the front face, sometimes as the back face.

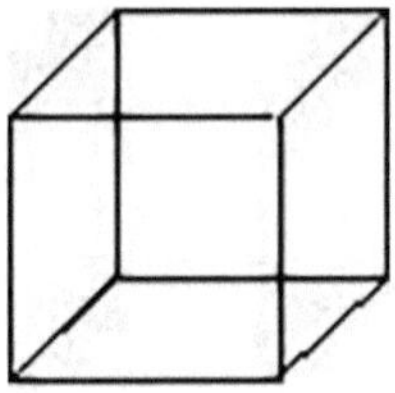

This happens because our perceptual system forms the hypothesis that the drawing represents a cube resting on a flat surface, say a table, on the basis the fact that any real object rests on something. But then, since there is no flat surface in the picture, in particular no table is drawn, our perceptual system forms the alternative hypothesis that the drawing represents a cube mounted on a wall that is coming out towards the viewer. Since the picture provides no indication as to which of the two alternative hypotheses is plausible, the perceptual system entertains one hypothesis first, then the other, and never comes to a conclusion.

(5) *Illusory contours*. If we look at the following picture and are asked what it is, we will answer that it is a tiger, not five anatomical sections of a tiger.

This happens because our perceptual system forms the hypothesis that it is a tiger partially hidden by four vertical objects, say, four tree trunks, on the basis of the fact that this would be our actual visual experience if a tiger were partially hidden by four tree trunks.

(6) *Born blind people who recover sight*. Born blind people who, through surgery, recover sensitivity to visual stimuli, do not acquire the capacity of seeing immediately, but only after long practice.

This happens because they must first learn to associate visual data to objects, making hypotheses on the objects that produce those visual data. Until they learn to do this, they are unable to interpret the data correctly.

(7) *Movement as a trigger for vision*. At least fifty per cent of people can see the movement of their own hand even in total darkness.

This happens because they make hypotheses about the position of their hands on the basis of the sensory signals transmitted by their body movements.

19.9 Perception and Multimodality

Most philosophy of perception has tended to focus on vision, and to assume that the senses are relatively distinct from one another. But perception can be multimodal, namely it can use data from multiple sensory receptors. A clear indication of this comes from speech perception.

As McGurk and McDonald argue, "most verbal communication occurs in contexts where the listener can see the speaker as well as hear him. However, speech perception is normally regarded as a purely auditory process," yet experiments demonstrate that there is an "influence of vision upon speech perception" (McGurk and McDonald 1976, 746).

For example, if watching television news, audio and video get out of synchrony, the newsreader's mouth will appear to be saying something while the audio will be saying something else, and we will experience the visual event and the auditory event as distinct. Our perception of the newsreader's speech is multimodal, and is disrupted when audio and video get out of synchrony.

Many of our perceptions are multimodal, and the data from multiple sensory receptors are integrated to form a coherent and robust perception.

Perception as problem solving by the analytic method can account for how this integration can be accomplished. For, hypotheses can be formed by non-deductive, non-propositional, and unconscious inferences from data originating from multiple sense receptors.

19.10 Limitations of Human Perceptual System

Perception depends on our perceptual system, and our perceptual system has limitations. We have sensory receptors capable of capturing only those aspects of the world that are essential for the survival of human beings. Thus, unlike goldfishes, we are incapable of detecting infrared radiations, unlike bees, we are incapable of detecting ultraviolet radiations, unlike bats we are incapable of ultrasound echolocation.

But, however paradoxical it may sound, that our perceptual system has limitations is essential to our life. For, while a perceptual system like ours, which is subject to these limitations, can deal with the aspects of the world that are essential for the survival of human beings, a perceptual system not subject to these limitations would be unable to do so.

This is pointed out already by Locke, who argues that our senses are adequate "to provide for the conveniences of living" that "are our business in this world" (Locke 1975, 302). Indeed, "were our senses alter'd, and made much quicker and acuter, the appearance and outward scheme of things would have quite another face to us," and "would be inconsistent with our being, or at least well-being, in the part of the universe which we inhabit" (ibid.). For example, if vision "were in any man 1000, or

100,000 times more acute than it is by the best microscope," a man could not "take in but a very small part of any object at once, and that too only at a very near distance" (ibid., 303). He "could not see things he was to avoid, at a convenient distance; nor distinguish things he had to do with, by those sensible qualities others do" (ibid.). Then vision would be inadequate to provide for the conveniences of living.

Since our perceptual system has limitations, we must not assume that properties of objects which depend on the peculiarities of our perceptual system are properties of reality.

19.11 Limitations of Human Cognitive System

Since our perceptual system has limitations, our conceptual system also has limitations, because it depends to a large extent on our perceptual system. Therefore, our entire cognitive system has limitations, meaning by cognitive system the composite of our perceptual system and our conceptual system.

This implies that we cannot have full knowledge of the world. Since our cognitive system has evolved in response to certain challenges of the world, our cognitive apparatus can only deal with those aspects of the world from which such challenges arose. Therefore, the purpose of knowledge cannot be to give a complete picture of the world.

Rather, the purpose of knowledge is to make plausible hypotheses about the world that allow us to solve the problems with which we are faced, starting from that of survival. It is through knowledge that life, from the prokaryotes to human beings, has been able to exist and preserve itself. The prokaryotes that solved the problem of survival using data about the environment obtained through their rudimentary sense organs, were the first to discover that knowledge is essential to life, and indeed a precondition for it.

References

Anscombe, Gertrude. 1981. *The collected philosophical papers*. Oxford: Blackwell.

Ayer, Alfred. 1940. *The foundations of empirical knowledge*. New York: Macmillan.

Bennett, Max, and Peter Hacker. 2003. *Philosophical foundations of neuroscience*. Oxford: Blackwell.

Dewey, John. 1911. Brief studies in realism. I. *The Journal of Philosophy, Psychology and Scientific Method* 8:393–400.

Dilthey, Wilhelm. 1989. *Introduction to the human sciences*. Princeton: Princeton University Press.

Hanson, Norwood. 1965. *Patterns of discovery*. Cambridge: Cambridge University Press.

Locke, John. 1975. *An essay concerning human understanding*. Oxford: Oxford University Press.

McGurk, Harry, and John McDonald. 1976. Hearing lips and seeing voices. *Nature* 264:746–748.

Moore, George. 1909–1910. The subject-matter of psychology. *Proceedings of the Aristotelian Society* 10: 36–62.
Pylyshyn, Zenon. 2003. *Seeing and visualizing: It's not what you think*. Cambridge: The MIT Press.
Russell, Bertrand. 1998. *The problems of philosophy*. Oxford: Oxford University Press.
von Hemholtz, Hermann. 1995. *Science and culture: Popular and philosophical essays*. Chicago: The University of Chicago Press.
von Hemholtz, Hermann. 2013. *Treatise on physiological optics*. Vol. III. Mineola: Dover.

Chapter 20
Knowledge and Explanation

Abstract Developing the view that philosophy is acquisition of knowledge also requires answering the question: What is an explanation? A widely held view is that an explanation of a statement is a deduction of the statement from some other given statements, but this chapter argues that this view is not valid. Alternatively, the chapter argues that an explanation of a problem is a solution to the problem by the analytic method.

Keywords Deductive explanation · Deductive-nomological explanation · Analytic explanation · Inference to the best explanation · Explanation and understanding

20.1 The Relevance of Explanation

Developing the view that philosophy is acquisition of knowledge also requires answering the question: What is an explanation?

Explanation is relevant to knowledge, because providing an explanation of a thing gives full knowledge of the thing.

Wittgenstein says: "What we want is to describe, not to explain" (Wittgenstein 1978, 205). Therefore, "all explanation must disappear, and description alone must take its place" (Wittgenstein 2009, 109).

But this is unfounded. A description alone does not provide full knowledge of a thing, because a description gives information about the 'what' of the thing, while an explanation gives information about the 'how' of the thing. To have information about the 'how' of a thing is important to have full knowledge of the thing, therefore an explanation is relevant to knowledge.

To say that an explanation is relevant to knowledge, however, does not say what an explanation is.

A widely held view is that an explanation of a statement is a deduction of the statement from some other given statements. But this chapter argues that this view is not valid.

C. Cellucci, *A Philosophy for the Future*, Synthese Library 508,
https://doi.org/10.1007/978-3-031-98443-3_20

Alternatively, the chapter argues that an explanation of a problem is a solution to the problem by the analytic method.

20.2 Deductive Explanation

An answer to the question 'What is an explanation?' is deductive explanation, according to which an explanation of a statement is a deduction of the statement from some given principles that are the causes of the statement.

Deductive explanation was proposed by Aristotle. He says that a demonstration of a statement is a deduction of the statement "from premisses that are true" and are "the causes of the conclusion" (Aristotle, *Analytica Posteriora*, A 2, 71 b 20–22). Two types of demonstration can be distinguished, demonstration 'that', or demonstration which "only shows that something is the case," and demonstration 'why', or demonstration which "shows why something is the case" (ibid., A 13, 78 a 37). In a demonstration 'that', "the cause" of the conclusion "is not stated" (ibid., A 13, 78 b 14–15). In a demonstration 'why', the cause of the conclusion is stated, so "a demonstration through the cause is a 'demonstration why'" (ibid., B 16, 98 b 19–20). By 'cause' Aristotle means "the reason why" (ibid., B 3, 194 b 19).

Since in a demonstration 'why' the cause of the conclusion is stated while in a demonstration 'that' it is not stated, a demonstration 'why' is explanatory while a demonstration 'that' is non-explanatory.

20.3 Inadequacy of Deductive Explanation

Deductive explanation, however, is inadequate. For, on the one hand, according to Aristotle, an explanation of a statement is a deduction of the statement from some given principles that are the causes of the statement. On the other hand, according to Aristotle, a demonstration of a statement is a deduction of the statement from premisses that are true and are the causes of the conclusion. But then all demonstrations are explanatory. This conflicts with the assertion that, in a demonstration 'that', the cause of the conclusion is not stated, therefore a demonstration 'that' is non-explanatory.

Against this it could be argued that Aristotle distinguishes between two types of demonstration, direct demonstration and demonstration by reductio ad absurdum. In a direct demonstration the conclusion is reached from premisses that are true and are the causes of the conclusion, so a direct demonstration is explanatory. Conversely, in a demonstration by reductio ad absurdum, "the conclusion is reached from a hypothesis" (Aristotle, *Analytica Priora*, A 44, 50 a 32). And it is reached from a hypothesis whose "falsehood is evident" (ibid., A 44, 50 a 36–37). So, the conclusion is not reached from a premiss that is true and the cause of the conclusion. Therefore, demonstration by reductio ad absurdum is non-explanatory.

This argument, however, contrasts with the fact that there are demonstrations by reductio ad absurdum that are explanatory. An example is provided by the following demonstration that $\sqrt{2}$ is an irrational number.

Suppose that $\sqrt{2}$ is a rational number. Then $\sqrt{2} = m / n$ for some natural numbers m and n, so $m^2 = 2n^2$. Since $m^2 = m \cdot m$, in the prime factorization of m^2 each prime factor occurs an even number of times. Similarly for n^2. Then, in the prime factorization of $2n^2$ the prime factor 2 occurs an odd number of times. So, $m^2 \neq 2n^2$. Contradiction. Therefore $\sqrt{2}$ is an irrational number.

The demonstration shows why $m^2 \neq 2n^2$, so it is explanatory. As Davis and Hersh say, the demonstration "seems to reveal the heart of the matter," it "exposes the 'real' reason" (Davis and Hersh 1981, 299).

20.4 Deductive-Nomological Explanation

A variant of deductive explanation is deductive-nomological explanation, according to which an explanation of a statement is a deduction of the statement from some given universal laws, together with certain given singular statements which express the initial conditions.

Deductive-nomological explanation was proposed by Popper under the name of 'causal explanation'.

Popper says that "to give a causal explanation of an event means to deduce a statement which describes it, using as premises of the deduction one or more universal laws, together with certain singular statements, the initial conditions" (Popper 2005, 38). Then we have "two different kinds of statement, both of which are necessary ingredients" of a complete "explanation," that is, "universal statements, i.e. hypotheses of the character of natural laws," and "singular statements, which apply to the specific event in question," the "'initial conditions'. It is from universal statements in conjunction with initial conditions that we deduce" the event in question, which I call "a specific or singular prediction" (ibid.). The universal laws and the "initial conditions describe what is usually called the 'cause' of the event in question," and "the prediction describes what is usually called the 'effect'" (ibid., 38–39).

Thus, we can say that we have provided an "explanation of the breaking of a certain piece of thread if we have found that the thread has a tensile strength of 1 lb. and that a weight of 2 lbs. was put on it" (ibid., 38). In this explanation we have two different types of statements. On the one hand, we have "the hypothesis: 'Whenever a thread is loaded with a weight exceeding that which characterizes the tensile strength of the thread, then it will break'; a statement which has the character of a universal law of nature" (ibid.). On the other hand, we have "singular statements (in this case two) which apply only to the specific event in question: 'The weight characteristic for this thread is 1 lb.', and 'The weight put on this thread was 2 lbs.'" (ibid.). From the hypothesis together with the two singular statements we deduce "the singular statement 'This thread will break'" (ibid.).

20.5 Inadequacy of Deductive-Nomological Explanation

Deductive-nomological explanation, however, is inadequate because a deduction of a statement from some given universal laws, together with certain given singular statements which express the initial conditions, is neither a necessary nor a sufficient condition for an explanation of the statement.

(1) A deduction of a statement from some given universal laws, together with certain given singular statements which express the initial conditions, is not a necessary condition for an explanation of the statement.

Thus, suppose we want to demonstrate the statement: The sum of consecutive odd numbers starting with 1 is a square. This was first demonstrated by the Pythagoreans with a figure like the following, which shows that if, starting from a point representing the odd number 1, one adds arrays of points representing the odd numbers 3, 5, 7, one obtains squares. The same would hold adding arrays of points representing all other odd numbers.

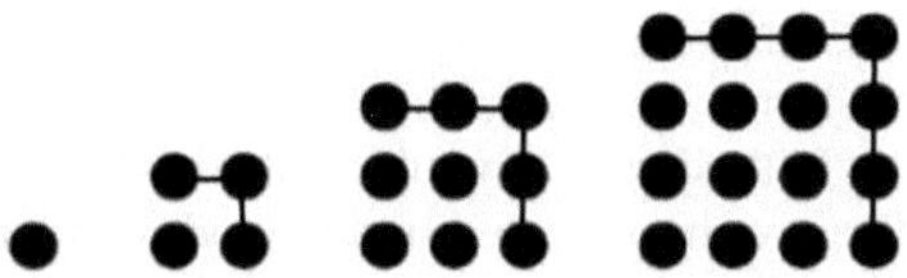

This demonstration explains why the statement holds, but does not consist in a deduction of the statement from some given universal laws, together with certain given singular statements which express the initial conditions,

(2) A deduction of a statement from some given universal laws, together with certain given singular statements which express the initial conditions, is not a sufficient condition for an explanation of the statement.

Thus, suppose we want to find the height of a tree. We walk away from the tree and stop at a point where we see the top of the tree at 45 degree angle. Then we measure our distance from the tree and find that it is 15 metres.

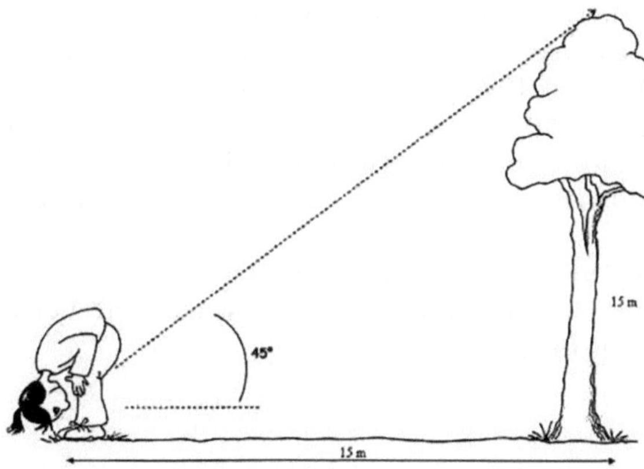

From the universal laws of trigonometry and the initial condition that we saw the top of the tree at 45 degree angle and 15 metres distance, we deduce that the tree is 15 metres high. But this is not an explanation of why the tree is 15 metres high. The real explanation is, rather, that the tree is of a certain species and a certain age.

20.6 Analytic Explanation

Both deductive explanation and deductive-nomological explanation are limited to the context of justification. For, the purpose of a deduction of a statement from some given principles, or from some given universal laws together with certain given singular statements which express the initial conditions, is to justify the statement. But an explanation of a statement may require considering how the statement is discovered, so an explanation of a statement may require taking into account both the context of discovery and the context of justification.

This condition is satisfied by analytic explanation. According to it, an explanation of a problem is a solution to the problem by the analytic method. The hypotheses from which the solution is deduced are the explanations of the problem.

Since the analytic method is both a method of discovery and a method of justification, analytic explanation takes into account both the context of discovery and the context of justification.

20.7 Descartes's Opposition of Analysis to Synthesis

That explanation should take into account both the context of discovery and the context of justification is underlined by Descartes.

Indeed, Descartes says that, although synthesis "demonstrates the conclusion clearly," it is not satisfying "nor appeases the minds of those who are eager to learn, since it does not show how the thing in question was discovered" (Descartes 1996, VII, 156). Only "analysis shows the true way by which a thing was discovered methodically" (Descartes 1996, VII, 155). By analysis, "it is the causes which are demonstrated by the effects," and "the causes from which I deduce" the effects "serve not so much to demonstrate them as to explain them" (ibid., VI, 76). In fact, "there is a great difference between" merely "demonstrating and explaining" (ibid., II, 198).

Thus, according to Descartes, an explanation of a statement must take into account not only the context of justification but also the context of discovery, so not only synthesis but also analysis.

20.8 Kepler's Explanation of Motion of Planets

An example of analytic explanation is Kepler's explanation of the motion of planets. In particular, Kepler wants to explain what is the moving power of the planets, how this moving power acts on them, and why the planets further away from the sun move slower than those closer to the sun.

Kepler's explanation is based on the hypothesis that the moving power of the planets is in the sun, it acts on planets at a distance, and impels each planet more strongly in proportion to how near it is to the sun.

Kepler arrives at this hypothesis by analogy.

Indeed, Kepler starts from the fact that "the source of light is in the sun" (Kepler 1981, 199). Sun's light is poured out throughout the whole world, and yet "does not exist in the intermediate space between the source and the illuminable" (Kepler 1992, 383). And "sun's light also grows thinner with distance from the sun" (Kepler 1981, 201).

From this, by analogy, Kepler arrives at the hypothesis that "the source of the five planets' motion is the sun" (Kepler 1992, 52). Sun's moving power "is poured out throughout the whole world, and yet does not exist" in the intermediate space between the source and the "movable" (ibid., 382). And sun's moving power also "grows weaker with distance" (Kepler 1981, 201).

Then Kepler argues that the hypothesis is plausible. Thus, he says that the hypothesis accounts for the "irregularity" in the planetary paths: a "planet will be slower" when it is "further away from the sun," because it "is moved by a weaker power," and "it will be faster" when "it is closer to the sun," because it is "subject to a stronger power" (ibid., 217).

20.9 Newton's Explanation of Universal Gravitation

Another example of analytic explanation is Newton's explanation of universal gravitation.

Newton's explanation is based on the hypothesis that gravity exists in all bodies universally, and the gravitation toward each of the particles of a body is inversely as the square of the distance of places from those particles.

Newton arrives at this hypothesis by induction.

Indeed, Newton starts from the fact that "all planets gravitate toward one another and also that the gravity toward any one planet, taken by itself, is inversely as the square of the distance of places from the center of the planet" (Newton 1999, 810).

From this, by induction, Newton arrives at the hypothesis that "gravity exists in all bodies universally" (ibid.). And "the gravitation toward each of the individual equal particles of a body is inversely as the square of the distance of places from those particles" (ibid., 811).

Then Newton argues that the hypothesis is plausible. Thus, he says: "That force by which the moon is kept in its orbit, in descending from the moon's orbit to the surface of the earth, comes out equal to the force of gravity here on earth" (ibid., 804). For, "if gravity were different from this force, then bodies making for the earth by both forces acting together would descend twice as fast, and in the space of one second would by falling describe 30⅙ Paris feet, entirely contrary to experience" (ibid.).

20.10 Darwin's Explanation of Living Things

Another example of analytic explanation is Darwin's explanation of living things. In particular, Darwin wants to explain the characters of living things and how these characters came to be.

Darwin's explanation is based on the following two hypotheses: (1) Natural selection produced different species of animals and plants; (2) In any species, as more individuals are produced than can possibly survive, there must in every case be a struggle for existence.

Darwin arrives at hypothesis (1) by analogy.

Indeed, Darwin starts "from what artificial selection had done for domestic animals" (Darwin 2009b, II, 118). Namely, he starts from the fact that breeders used artificial selection to produce different breeds of animals and plants.

From this, by analogy, Darwin arrives at the hypothesis "that selection was the principle of change" (Darwin 1903, I, 118). Namely, natural selection produced different species of animals and plants.

On the other hand, Darwin arrives at hypothesis (2) by induction.

Indeed, Darwin starts from "the doctrine of Malthus" (Darwin 2009a, 50). Namely, he starts from the observation that, as more human beings are produced than can possibly survive, there must be a struggle for existence.

From this, by induction, Darwin arrives at the hypothesis that, in any species, "as more individuals are produced than can possibly survive, there must in every case be a struggle for existence" (ibid.).

Then Darwin argues that the hypotheses (1) and (2) are plausible. Thus, he says that "the close similarity between man and the lower animals in embryonic development, as well as in innumerable points of structure and constitution," the "rudiments which he retains, and the abnormal reversions to which he is occasionally liable, are facts which cannot be disputed" (Darwin 1981, 385–386).

20.11 Bohr's Explanation of the Constitution of Atoms

Another example of analytic explanation is Bohr's explanation of the constitution of atoms.

Bohr's explanation is based on the hypothesis that atoms consist of a nucleus surrounded by a cluster of electrons which rotate around it in fixed quantized orbits.

Bohr arrives at this hypothesis by analogy.

Indeed, Bohr starts from the fact that the solar system consists of the sun surrounded by planets that rotate around it in fixed orbits obeying Keplerian laws.

From this, by analogy, Bohr arrives at the hypothesis that the atom behaves as if it were a minuscule solar system, consisting "of a positively charged nucleus surrounded by a cluster of electrons" (Bohr 1913, 476). The "states of the electron system" can be pictured "as planetary motions obeying Keplerian laws" (Bohr 1963, 85). In particular, the "electrons are arranged at equal angular intervals in coaxial rings rotating round the nucleus" (Bohr 1913, 477).

Then Bohr argues that the hypothesis is plausible. Thus, he says that there is "agreement between the observed and calculated values of the constant entering in the expression" of the frequency of the radiation "for the Balmer series of the hydrogen spectrum" (Bohr 2022, 61).

20.12 Visual Analytic Explanation

A special type of analytic explanation is visual analytic explanation, that is, analytic explanation in which the explanation is given by a figure. In this type of analytic explanation, a hypothesis for solving the problem is obtained non-deductively from the figure, and the figure shows that the hypothesis is plausible.

For example, let us consider the problem of showing that, for all real numbers a and b, $a \cdot b = b \cdot a$. Then, an explanation is given by the following figure.

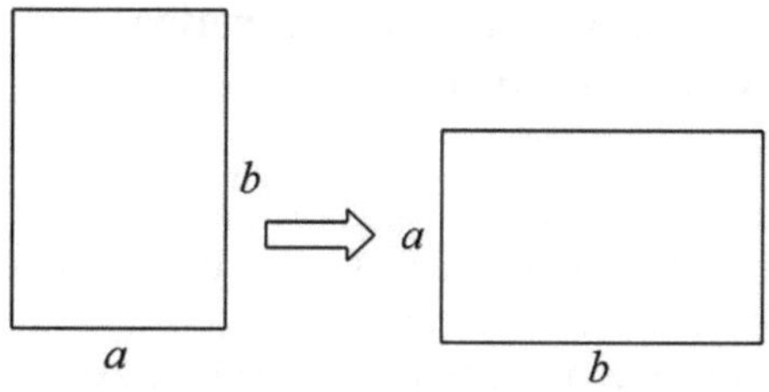

For, from the figure, by analogy, we obtain the hypothesis that, for all real numbers a and b, $a{\cdot}b{=}b{\cdot}a$. The hypothesis is plausible because the figure shows that, rotating a rectangle 90 degrees clockwise, the area of the rectangle is unchanged. Therefore, the figure explains why the hypothesis holds.

20.13 Top-Down and Bottom-Up Explanation

Two types of explanation can be distinguished: top-down explanation and bottom-up explanation.

According to top-down explanation, an explanation of a statement is reached deducing the statement from some other given statements that are the causes of the statement.

According to bottom-up explanation, an explanation of a problem is reached obtaining from the problem, by some non-deductive rule, a hypothesis that provides a solution to the problem.

Deductive explanation and deductive-nomological explanation are top-down explanations, analytic explanation is a bottom-up explanation.

Between top-down explanation and bottom-up explanation there is an important difference.

Since, in a top-down explanation, the statement to be explained is deduced from some other given statements, and deductive rules are non-ampliative, in such explanation the explanans is implicitly given from the beginning. The aim of the explanation is to make it explicit.

On the other hand since, in a bottom-up explanation, the hypotheses that provides a solution to the problem can be obtained from the problem by some non-deductive rules, and non-deductive rules are ampliative, in such explanation the explanans is not implicitly given from the beginning. The aim of the explanation is to discover it.

20.14 Analytic Explanation and Inference to the Best Explanation

In Chap. 15 it has been argued that the analytic method is not related to abduction because, while the analytic method is a method of discovery, abduction, being non-ampliative, cannot be a method of discovery.

Much in the same way it can be argued that analytic explanation is not related to inference to the best explanation, because inference to the best explanation, being non-ampliative, cannot be a method of discovery.

Indeed, Harman says that "'the inference to the best explanation' corresponds approximately to what others have called 'abduction'" (Harman 1965, 88). For, inference to the best explanation is the inference by which "one infers, from the premise that a given hypothesis would provide a 'better' explanation for the evidence than would any other hypothesis, to the conclusion that the given hypothesis is true" (ibid., 89).

That inference to the best explanation corresponds approximately to abduction is clear if abduction and inference to the best explanation are formulated as follows.

Abduction is the rule: *B*, *A* explains *B*, therefore *A*.

Inference to the best explanation is the rule: *B*, *A* explains *B*, no other *C* explains *B* better than *A*, therefore *A*.

Since the conclusion *A* already occurs in the premisses '*A* explains *B*' and 'no other *C* explains *B* better than *A*', inference to the best explanation is non-ampliative, therefore it is not in the context of discovery. Then analytic explanation is not related to inference to the best explanation, because analytic explanation is in the context of discovery.

However, inference to the best explanation is not a genuine explanation. For, as already pointed out in Chap. 6, it does not offer any criterion for deciding when a given hypothesis is a better explanation than another hypothesis.

Harman recognizes that "there is, of course, a problem about how one is to judge that one hypothesis is sufficiently better than another hypothesis" (ibid., 89). But, surprisingly enough, he declares: "I will not, however, say anything more about this problem" (ibid.).

Contrary to Harman, as we have seen in Chap. 6, Williamson proposes a criterion, that is, simplicity. But, as argued there, this criterion is not valid.

Moreover, inference to the best explanation involves that there is a single best explanation of a thing. But, again as argued there, this is not valid because often there are several explanation of a thing which are useful to different purposes.

Analytic explanation does not have these defects. According to it, one hypothesis is a better explanation than another if it is more plausible than the other, that is, the arguments for it are stronger than the arguments for the other on the basis of experience. Therefore, analytic explanation provides a criterion for recognizing whether one hypothesis is a better explanation than another.

Moreover, as argued in Chap. 15, a problem can have different solutions by the analytic method, using different hypotheses. Therefore, there can be several explanations of a thing.

20.15 Explanation and Understanding

From antiquity it has been thought that explanation provides understanding.

Thus, Plato says: "For the person who is unable to give an explanation, will not you say" that "he has no understanding of it?" (Plato, *Respublica*, VII 534 b 3–6).

That explanation provides understanding has been reaffirmed until recently.

Thus, von Wright says: "Practically every explanation, be it causal or teleological or of some other kind, can be said to further our understanding of things" (von Wright 1971, 6).

Kitcher says: "A theory of explanation should show us how scientific explanation advances our understanding" (Kitcher 1981a, 329). In particular, in mathematics, "the goal of explanations is to provide understanding," and we should "approach the topic of mathematical explanation via the concept of understanding—or, more exactly, via the concept of a failure in understanding" (Kitcher 1981b, 473).

Bangu says: "The very point of an explanation is to convey understanding," not subjective understanding but "objective understanding," the "kind of understanding" that "can only be achieved when such a correct, publicly available answer is available" (Bangu 2017, 104–105).

However, while explanation provides understanding, there can be understanding without explanation.

For example, we can understand what type of person someone is without having an explanation of why he is that type of person.

This suggests distinguishing between two types of understanding: understanding that something is the case, or understanding 'that', and understanding why something is the case, or understanding 'why'.

We can have understanding 'that' even if we do not have an explanation. Thus, we can understand that $\sqrt{2}$ is an irrational number even if we do not have an explanation of why $\sqrt{2}$ is an irrational number.

However, we cannot have understanding 'why' if we do not have an explanation. Thus, we cannot understand why $\sqrt{2}$ is an irrational number if we do not have an explanation of why $\sqrt{2}$ is an irrational number.

References

Bangu, Sorin. 2017. Scientific explanation and understanding: Unificationism reconsidered. *European Journal for Philosophy of Science* 7:103–126.

Bohr, Niels. 1913. On the constitution of atoms and molecules. *Philosophical Magazine* 26:1–25, 476–502, 857–875.
Bohr, Niels. 1963. *Essays 1958–1962 on atomic physics and human knowledge*. New York: Interscience Publishers.
Bohr, Niels. 2022. *On the constitution of atoms and molecules*. Cham: Birkhäuser.
Darwin, Charles. 1903. *More letters of Charles Darwin*. London: Murray.
Darwin, Charles. 1981. *The descent of man, and selection in relation to sex*. Princeton: Princeton University Press.
Darwin, Charles. 2009a. *The origin of species by means of natural selection*. Cambridge: Cambridge University Press.
Darwin, Charles. 2009b. *The life and letters of Charles Darwin, including an autobiographical chapter*. Cambridge: Cambridge University Press.
Davis, Philip, and Reuben Hersh. 1981. *The mathematical experience*. Boston: Birkhäuser.
Descartes, René. 1996. *Oeuvres*. Paris: Vrin.
Harman, Gilbert. 1965. The inference to the best explanation. *The Philosophical Review* 74:88–95.
Kepler, Johannes. 1981. *Mysterium Cosmographicum: The secret of the universe*. New York: Abaris Books.
Kepler, Johannes. 1992. *New astronomy*. Cambridge: Cambridge University Press.
Kitcher, Philip. 1981a. Explanatory unification. *Philosophy of Science* 48:507–531.
Kitcher, Philip. 1981b. Mathematical rigor—Who needs it? *Noûs* 15 (4): 469–473.
Newton, Isaac. 1999. *The Principia: Mathematical principles of natural philosophy*. Berkeley: University of California Press.
Popper, Karl. 2005. *The logic of scientific discovery*. London: Routledge.
von Wright, Georg. 1971. *Explanation and understanding*. London: Routledge & Kegan Paul.
Wittgenstein, Ludwig. 1978. *Remarks on the foundations of mathematics*. Oxford: Blackwell.
Wittgenstein, Ludwig. 2009. *Philosophische Untersuchungen—Philosophical investigations*. Oxford: Blackwell.

Chapter 21
Knowledge and Emotions

Abstract Developing the view that philosophy is acquisition of knowledge also requires answering the question: What is the relation of knowledge to emotions? A widely held view is that, while playing a positive part in the biological role of knowledge, emotions play a negative part in the cultural role of knowledge, but this chapter argues that this view is not valid. Alternatively, the chapter argues that emotions play a positive part both in the biological role of knowledge and in the cultural role of knowledge. In particular, emotions play a positive part in science and mathematics.

Keywords Emotions and roles of knowledge · The chariot allegory · Emotions and acquisition of knowledge · Emotions and analytic method · Emotions and science · Emotions and mathematics

21.1 Emotions and Roles of Knowledge

Developing the view that philosophy is acquisition of knowledge also requires answering the question: What is the relation of knowledge to emotions?

A widely held view is that, while playing a positive part in the biological role of knowledge, emotions play a negative part in the cultural role of knowledge, but this chapter argues that this view is not valid.

Alternatively, the chapter argues that emotions play a positive part both in the biological role of knowledge and in the cultural role of knowledge. In particular, emotions play a positive part in science and mathematics.

C. Cellucci, *A Philosophy for the Future*, Synthese Library 508,
https://doi.org/10.1007/978-3-031-98443-3_21

21.2 Emotions and Biological Role of Knowledge

Emotions play a positive part in the biological role of knowledge, because they are means for evaluating the environment and responding to it appropriately, helping to avoid dangers and to exploit favourable opportunities. Only organisms that perceive dangers as unpleasant emotions and favourable opportunities as pleasant emotions can survive. In fact, "there is abundant evidence of 'emotional' reactions" even "in simple organisms" such as "a lone paramecium, a simple unicellular organism" (Damasio 2003, 40).

This does not mean that, every time an organism experiences an emotion, this favours its survival. Not all emotions serve this purpose. Moreover, whether emotions favour survival depends on their intensity and the context in which they arise. Nevertheless certain emotions, if they have an adequate intensity and arise in an appropriate context, play a positive part in regulating and preserving the life of all organisms, from the simplest to human beings.

Natural selection has provided all organisms with means to regulate and preserve life automatically, without having to resort to elaborate forms of thought. Emotions are among these means. They are inborn and inherent in all organisms.

In particular, the human brain has several specialized circuits, each of which serves a specific purpose and is the result of an adaptation. But, if activated all together, they could interfere and hinder each other. To avoid this, some coordination is required.

Emotions have a part in this. They cause the system of specialized circuits to react to certain situations in an orderly manner.

Many of the situations that arise today have occurred many times in the course of the evolutionary history of the species, and repeated encounters with them have led organisms to adopt the most appropriate behaviour towards them. When these situations arise, emotions activate and coordinate the appropriate specialized circuits to solve the problem they pose.

21.3 Emotions and Cultural Role of Knowledge

From antiquity it has been widely believed that, while playing a positive part in the biological role of knowledge, emotions play a negative part in the cultural role of knowledge, because they are obstacles to acquisition of knowledge.

Thus, the Stoics say that an emotion is "a vehement opinion, inherent and deeply implanted," which is an obstacle to acquisition of knowledge because it is the "belief that one knows what one does not know" (Cicero, *Tusculanae disputationes*, IV.26).

Bacon says that "the human understanding" is "subject to influence from the will and the emotions, a fact that creates fanciful knowledge; man prefers to believe what he wants to be true" (Bacon 2004, 44). Thus "emotion marks and stains the understanding in countless ways which are sometimes impossible to perceive" (ibid., 44–45).

Spinoza says that "the mind's essence consists in knowledge" (Spinoza 1985, 613). Therefore, "the power of the mind is defined by knowledge alone" (ibid., 605–606). Emotions "indicate our lack of power and mutilated knowledge" (ibid., 588). Only "so long as we are not torn" by emotions, "the power of the mind by which it strives to understand things is not hindered" (ibid., 601).

Russell says that "emotions are what make life interesting, and make us feel important," and "from this point of view, they are the most valuable element in human existence" (Russell 1995, 175–176). But "when, as in philosophy, we are trying to understand the world, they appear rather as a hindrance. They generate irrational opinions" which "cause us to view the universe in the mirror of our moods, now as bright now as dim, according to the state of the mirror," so "the emotions are on the whole a hindrance to the intellectual life" (ibid.,176).

21.4 The Chariot Allegory

Some attribute to Plato the origin of the view that emotions play a negative part in the cultural role of knowledge because they are an obstacle to acquisition of knowledge.

Thus, Heller says that "feelings, emotions and sentiments" have been "treated as the main obstacles to knowledge," in particular "the model was set by Plato and has been followed ever since" (Heller 2009, 1).

This, however, is not valid, as it can be seen from Plato's chariot allegory.

Plato says that "the soul is like the natural union of a winged two-horse chariot and a charioteer" (Plato, *Phaedrus*, 246 a 6–7). The two horses represent emotions, the charioteer represents reason. Of the two horses, "one is good and of good breed, and the other is bad and of bad breed" (ibid., 246 b 2–3). One of the horses represents good emotions, the other bad emotions. The two horses pull the chariot in opposite directions. Therefore, for the charioteer, "the driving of them is necessarily difficult and troublesome" (ibid., 246 b 4).

But, without the two horses the chariot would not move, and the charioteer could not reach his destination, the heaven's limit, beyond which he can behold the Forms and so reach knowledge. Therefore, according to Plato, without emotions, reason cannot reach knowledge.

21.5 Emotions and Acquisition of Knowledge

The chariot allegory suggests that it is unjustified to believe that emotions are an obstacle to acquisition of knowledge.

Admittedly, emotions can negatively affect acquisition of knowledge, because they can lead us to disregard data that could serve as evidence for a hypothesis we are interested in keeping.

On the other hand, however, emotions can also positively affect acquisition of knowledge because, by leading us to disregard data that could serve as evidence against a hypothesis, they can save us from prematurely abandoning a hypothesis that could be fruitful. This can help advance knowledge. For, abandoning assumptions as soon as an anomaly occurs, can negatively affect acquisition of knowledge.

Emotions can positively affect acquisition of knowledge because they are linked to knowledge. On the one hand, emotions presuppose knowledge. Some facts can trigger emotions in us only if we have knowledge of those facts and their possible effects. On the other hand, emotions affect knowledge. They can provide a first evaluation of the relevance of certain facts to the achievement of a desired goal, and this can lead us to focus on them, pouring available resources into them.

21.6 Emotions and Solving Problems

But emotions play an even more substantial part in knowledge: without them, the drive to seek knowledge is lacking. Knowledge is acquired by solving problems, but we will only solve those problems whose solution we desire, and which therefore arouse an emotion in us. The search for a solution to a problem has always an emotional component that leads us to engage in it. Interest, curiosity, wonder, and surprise are aspects of emotional engagement. We do not get engaged in things that do not intrigue us, or make us wonder, or surprise us.

Of course, emotions are not a substitute for reasoning, but they can play an auxiliary role in it, because they can help increase the efficiency of reasoning by making it faster. Sometimes emotions can also act as a substitute for reasoning. This occurs, for example, when under the impulse of emotions we seize opportunities that might prove advantageous, or avoid making choices that might prove harmful.

Our response to emotions can even give rise to some sort of thought. When we begin to reflect on a problem or hypothesis, some thought can trigger some bodily feelings in us. If the feelings are unpleasant, this can alert us that the problem or hypothesis might have undesirable consequences. Conversely, if the feelings are pleasant, this can encourage us to continue that way.

This response to our bodily feelings is a sort of thought that is often at the very root of an investigation, because something can become a problem for us when we feel puzzled by it. This feeling of puzzlement can lead us to seek a solution to the problem and can act as a springboard for conscious thought.

Even Russell recognizes this function of emotions, because he says: "In all creative work that I have done, what has come first is a problem, a puzzle involving discomfort. Then comes concentrated voluntary thought entailing great effort" (Russell 1949, 112).

21.7 Emotions and Analytic Method

In particular, emotions play a positive part in the analytic method, both in the choice of problems and in the choice of hypotheses.

That emotions play a positive part in the choice of problems, is clear from the fact that we can seek solutions only for a small part of the problems that arise. Emotions can serve us as a guide in choosing which problems to consider and which to ignore. Only if we feel strongly about a problem we can have the drive to get involved in it, and undertake the hard work that may be necessary to find a solution. The drive comes from our needs and desires and from the goals we want to achieve.

That emotions play a positive part in the choice of hypotheses is also clear from the fact that, if the search for a solution to a problem is not driven by emotions, it eventually runs out. The absence of emotions can even have a paralyzing effect because, in solving a problem, we can be faced with so many alternative hypotheses that it would be impracticable to consider them all. Emotions can guide us in choosing which hypothesis to consider and which to discard, so they can act as a shortcut.

When it comes to acquisition of knowledge, what is good to ignore may be as important as what is good to consider. Indecision must come to an end. Through our bodily feelings, emotions can put an end to indecision, guiding us in choosing between alternative hypotheses.

This is confirmed by the findings of the neurosciences, which show that patients who suffer a damage to the ventromedial prefrontal cortex "have an abnormality in their processes of emotion and feeling" which is so severe that they are "unable to decide advantageously on matters pertaining to their own lives" (Bechara et al. 2000, 295).

21.8 Emotions and Science

Since emotions play a positive part in the analytic method, both in the choice of problems and in the choice of hypotheses, they play a positive part in science. Indeed, eliminating emotions from science would mean cancelling one of science's most vital supporting forces. A scientist must have a deep interest in what he does, an uninterested and bored scientist would do little and often not very well.

In fact, several scientists recognize that emotions play a positive part in their work.

Thus, Crick says: "I had always appreciated that the scientific way of life, like the religious one, needed a high degree of dedication and that one could not be dedicated to anything unless one believed in it passionately" (Crick 1988, 17). In fact, if Watson and I were successful in our quest for the structure of DNA, "it was because we passionately wanted to know the details of the structure. This, then, was a powerful force in our favor" (ibid., 70).

Feynman says: "You see? That's why scientists persist in their investigations, why we struggle so desperately for every bit of knowledge, stay up nights seeking the answer to a problem, climb the steepest obstacles to the next fragment of understanding, to finally reach that joyous moment of the kick in the discovery, which is part of the pleasure of finding things out" (Feynman 1999, xiv).

Rubbia says: "We are essentially driven not" by "the success, but by a sort of passion, namely the desire of understanding better, to possess, if you like, a bigger part of the truth" (Rubbia 1997, 197). In fact, "the act of discovery, the act of being confronted with a new phenomenon, is a very passionate and very exciting moment in everyone's life," and "it's very badly transmitted through the books," because "you never see in the scientific books" the "struggle and the passion of the person, who made that discovery" (ibid.).

In particular, among the emotions that play a positive part in the acquisition of scientific knowledge, several scientists assign a prominent role to aesthetic emotions.

Thus, Penrose says that "a beautiful idea has a much greater chance of being a correct idea than an ugly one" (Penrose 1991, 421). The "importance of aesthetic criteria applies not only to the instantaneous judgments of inspiration, but also to the much more frequent judgments that we make all the time in mathematical (or scientific) work" (ibid., 422).

Dirac says that "one has a great confidence in" a "theory arising from its great beauty, quite independently of its detailed successes" (Dirac 1980, 44). Indeed, "one has an overpowering belief that" the theory's "foundations must be correct quite independently of its agreement with observation" (ibid.). In fact, "it is more important to have beauty in one's equations that to have them fit experiment" (Dirac 1963, 47). For, "if one is working from the point of view of getting beauty in one's equations, and if one has really a sound insight, one is on a sure line of progress" (ibid.).

It can be risky to rely on aesthetic emotions in scientific research work, because it can easily lead one astray.

Thus, aesthetic emotions led Dirac to assert that, though renormalization "has been very successful in setting up rules for handling the infinities and subtracting them away," the "resulting theory is an ugly and incomplete one," and hence "cannot be considered as a satisfactory solution of the problem of the electron" (Dirac 1951, 291). Therefore, Dirac stuck to his theory of quantum electrodynamics, which made predictions that were often infinite and hence unacceptable, rejecting renormalization which led to accurate predictions, because the resulting theory seemed ugly to him.

Nevertheless, aesthetic emotions often motivate and effectively guide scientists in their research work.

21.9 Emotions and Mathematics

Emotions also play a positive part in mathematics. Several mathematicians recognize this.

Thus, Hadamard says: "That an affective element is an essential part of every discovery or invention is all too evident, and has been insisted upon by several thinkers; indeed, it is clear that no significant discovery or invention can take place without the will of finding" (Hadamard 1954, 31).

Pólya says: "It would be a mistake to think that solving problems is a purely intellectual affair; determination and emotions play an important role," indeed "to solve a serious scientific problem, will power is needed that can outlast years of toil and bitter disappointments" (Pólya 2004, 93). To solve a problem, one must be ready to face "the varying emotions of the struggle for the solution" (ibid., 94).

Rota says: "It is a frequent experience among mathematicians" that "they cannot solve a problem unless they like it" (Rota 1991, 262). In fact, "motivation and desire are essential components of mathematical reasoning" (Rota 1997, 160).

Even the initial acceptance of a demonstration may not depend on a thorough examination of the details of the demonstration, but rather on a gut feeling which suggests that there is something fundamentally right about the demonstration. Thus, several mathematicians came to believe that Wiles' original demonstration of the Taniyama-Shimura conjecture was basically right, long before the details of the demonstration were checked.

In particular, among the emotions that can influence the acquisition of mathematical knowledge, several mathematicians assign a prominent role to aesthetic emotions.

Thus, Atiyah says: "Mathematicians try to understand the physical world, to unearth the secrets of nature, to search for truth. They do this by creating intellectual edifices of great subtlety and beauty, guided by their aesthetic judgement" (Atiyah 2010, 8).

Gowers says: "It is remarkable how important a well-developed aesthetic sensibility can be, for purely pragmatic reasons, in mathematical research" (Gowers 2007, 37).

Some mathematicians even consider the feeling of beauty as a criterion of mathematical truth.

Thus, Hrbacek and Jech say that "the great amount of research" on questions about large cardinals "performed over the last 40 years has produced the very rich, often very subtle and difficult theory of large cardinals," and although for this theory "no axioms that would be intuitively obvious in the same way as the axioms of" Zermelo–Fraenkel set theory with the axiom of choice "have been proposed," its "aesthetic appeal makes it difficult not to believe that it describes true aspects of the universe of set theory" (Hrbacek and Jech 1999, 270).

As in the case of scientific research work, it can be risky to rely on aesthetic emotions in mathematical research work, because this can easily lead astray.

Thus, Bricard (1896) gave a very elegant demonstration that there are two tetrahedra of equal volume which are neither equidecomposable nor equicomplementable, but the demonstration was wrong.

Nevertheless, aesthetic emotions often motivate and effectively guide mathematicians in research work.

21.10 Emotions and Doubt

Emotions also play a positive part in doubt, which is an engine of knowledge.

Thus, Abelard says: "Dubitando enim ad inquisitione venimus, inquisendo veritatem percipimus [By doubting we come to inquiry, through inquiring we perceive the truth]" (Abelard, *Sic et non*, prologus, col. 1349 B).

Emotions play a positive part in doubt because doubt is a kind of fear, the fear that a hypothesis in which we have invested may turn out to be implausible. When we begin to doubt a hypothesis in which we have invested, we become anxious about its plausibility because of that fear.

Doubt provides a preliminary evaluation of the hypothesis, because it means that we do not feel that the existing arguments are sufficient to establish its plausibility. This plays a positive part in knowledge, because it leads us to reflect on the plausibility of the hypothesis and to look for new evidence for it.

21.11 Emotions and Rationality

As argued in Chap. 14, rationality is the capacity to choose appropriate means to given ends. Then one may expect that logic has a strict connection with rationality, even that logic is an important part of rationality.

This raises the question of whether logic is a proper part of rationality or rationality as a whole.

In the last century several people claimed that logic is rationality as a whole.

For example, this is the case of Frege who, as we have seen in Chap. 17, claims that the laws of logic are the most general laws, prescribing how to think wherever there is thinking at all, thus logic is rationality as a whole.

But, if logic is rationality as a whole, then emotions are opposed to rationality, and human act influenced by emotions must be termed irrational.

However, this conflicts with what was said above about the positive role that emotions play in knowledge, including science and mathematics.

In fact, emotions are not opposed to rationality but complementary to it. They are complementary to it not only because they play a positive part in the analytic method, both in the choice of problems and in the choice of hypotheses. They are complementary to rationality also because they can serve to compensate for the limitations of rationality.

Again, as argued in Chap. 14, while rationality is the capacity to choose appropriate means to given ends, it has nothing to do with the choice of ends.

Emotions can serve to do what rationality cannot, they can help us to choose ends. Our actions are accompanied by an emotional colouring, which has a relevant role in our choice not only of means but also of ends. Such colouring is a result of our emotional side, which never leaves us and silently guides us in each moment of our life.

Of course, the ends that emotion can help to choose cannot be said to be appropriate because, as argued in Chap. 14, one cannot assert of any ends that they are appropriate. Moreover, emotion is a result of natural selection, which has no ends. Nevertheless, emotions can help us to choose ends.

References

Atiyah, Michael. 2010. The art of mathematics. *Notices of the American Mathematical Society* 57:8.

Bacon, Francis. 2004. *The new organon*. Cambridge: Cambridge University Press.

Bechara, Antoine, Hanna Damasio, and Antonio Damasio. 2000. Emotion, decision making and the orbitofrontal cortex. *Cerebral Cortex* 10:295–307.

Bricard, Raoul. 1896. Sur une question de géométrie relative aux polyèdres. *Nouvelles Annales de Mathématiques* 15:331–334.

Crick, Francis. 1988. *What mad pursuit: A personal view of scientific discovery*. New York: Basic Books.

Damasio, Antonio. 2003. *Looking for Spinoza: Joy, sorrow, and the feeling brain*. London: Heinemann.

Dirac, Paul. 1951. A new classical theory of electrons. *Proceedings of the Royal Society of London, Series A* 209:291–296.

Dirac, Paul. 1963. The evolution of the physicist's picture of nature. *Scientific American* 208:43–53.

Dirac, Paul. 1980. The excellence of Einstein's theory of gravitation. In *Einstein: The first hundred years*, ed. Maurice Goldsmith, Alan Mackay, and James Woudhuysen, 41–46. Oxford: Pergamon Press.

Feynman, Richard. 1999. *The pleasure of finding things out: The best short works of Richard P. Feynman*. Helix Books: Cambridge.

Gowers, Timothy. 2007. Mathematics, memory and mental arithmetic. In *Mathematical knowledge*, ed. Mary Leng, Alexander Paseau, and Michael Potter, 33–58. Oxford: Oxford University Press.

Hadamard, Jacques. 1954. *The psychology of invention in the mathematical field*. Mineola: Dover.

Heller, Agnes. 2009. *A theory of feelings*. Lanham: Lexington Books.

Hrbacek, Karel, and Thomas Jech. 1999. *Introduction to set theory*. New York: Marcel Dekker.

Penrose, Roger. 1991. *The emperor's new mind: Concerning computers, minds, and the laws of physics*. London: Penguin.

Pólya, George. 2004. *How to solve it: A new aspect of mathematical method*. Princeton: Princeton University Press.

Rota, Gian-Carlo. 1991. *The end of objectivity: The legacy of phenomenology. Lectures at MIT*. Cambridge: MIT Mathematics Department.

Rota, Gian-Carlo. 1997. *Indiscrete thoughts*. Boston: Birkhäuser.

Rubbia, Carlo. 1997. Asking nature. In *Passionate minds: The inner world of scientists*, ed. Lewis Wolpert and Alison Richards, 195–202. Oxford: Oxford University Press.

Russell, Bertrand. 1949. Answer to questionnaire. In *How to think creatively*, ed. Eliot Hutchinson, 110–112. New York: Abingdon-Cokesbury Press.

Russell, Bertrand. 1995. *An outline of philosophy*. London: Routledge.

Spinoza, Baruch. 1985. *Collected works*. Vol. I. Princeton: Princeton University Press.

Chapter 22
Knowledge and Naturalization

Abstract Developing the view that philosophy is acquisition of knowledge also requires answering the question: What is the relation of knowledge to naturalization? As regards knowledge, 'naturalization' means any position that implies that knowledge is a natural process. There are several views of naturalization. This chapter considers them and argues that they are not valid. Alternatively, the chapter argues that naturalization consists in considering knowledge as a life function that serves to solve problems of biological life and, in the case of human beings, also of cultural life, by the analytic method.

Keywords Naturalization and naturalism · Different views of naturalization · Knowledge as a life function implemented by the analytic method · Arguments against naturalization

22.1 Naturalism and Naturalization

Developing the view that philosophy is acquisition of knowledge also requires answering the question: What is the relation of knowledge to naturalization?

With respect to knowledge, naturalization is any position that involves that knowledge is a natural process.

Naturalization is a special case of naturalism, the view that all reality, including our existence and knowledge, must be explained in terms of nature, because there is nothing beyond nature.

Indeed, naturalism is expressed by Bacon's motto: "Certe ultra naturam nihil [Certainly, nothing beyond nature]" (Bacon 1857–1874, I, 550).

This chapter deals with naturalization. On naturalism in general, see Clark (2016), Sterpetti (2021).

There are several views of naturalization. But this chapter considers them and argues that they are not valid.

C. Cellucci, *A Philosophy for the Future*, Synthese Library 508,
https://doi.org/10.1007/978-3-031-98443-3_22

Alternatively, the chapter argues that naturalization consists in considering knowledge as a life function that serves to solve problems of biological life and, in the case of human beings, also of cultural life, by the analytic method.

22.2 Knowledge as Result of a Natural Process of Inquiry

As already said, there are several views of naturalization.

A view of naturalization is that it consists in considering knowledge as a result of a natural process of inquiry that begins in doubt and terminates in the institution of conditions that remove need for doubt.

Thus, Dewey says that naturalization consists in considering knowledge as a result of a process of inquiry in which "there is no breach of continuity between operations of inquiry and biological operations and physical operations" (Dewey 1938, 19). The process of "inquiry begins in doubt" and "terminates in the institution of conditions that remove need for doubt. The latter state of affairs may be designated" by the word "knowledge" (ibid., 7).

But this view of naturalization is not valid. It is unwarranted to say that knowledge is a result of a natural process of inquiry that begins in doubt and terminates in the institution of conditions that remove need for doubt. For, as argued in Chap. 11, knowledge cannot be certain, so there is always a need for doubt.

22.3 Knowledge as Based on A Posteriori Forms of Intuition

Another view of naturalization is that it consists in considering knowledge as based on forms of intuition that are a priori for the individual but a posteriori for the species, and are what enables living things to survive and orient themselves in the outer world.

Thus, Lorenz says that naturalization consists in considering knowledge as based on forms of intuition that do "not develop anew in every human being as a result of individual experience," and so, "as seen purely from the standpoint of the individual," are "something that exists a priori" (Lorenz 1997, 18). But they have evolved "phylogenetically through confrontation with and adaptation to that form or reality which we experience as phenomenal space" (Lorenz 1977, 9). So, they are a priori for the individual, since they are "fixed prior to individual experience," but are a posteriori for the species, because they "are adapted to the external world" (Lorenz 2009, 233). They are what "enables living things to survive and orient themselves in the outer world" (Lorenz 1977, 9).

But this view of naturalization is not valid. If the forms of intuition evolved phylogenetically, then our earliest human ancestors could not yet have acquired them. But, if the forms of intuition are what enables living things to survive and orient themselves in the outer world, this makes it inexplicable how our earliest human ancestors were able to survive, and therefore how we, their descendants, can be here.

Moreover, if the forms of intuition are a posteriori for the species, this implies that, while we are unable to derive Euclid's fifth postulate, the parallel postulate, from experience, our earliest human ancestors have been able to do it.

In this regard, Poincaré says: "It has often been said that if individual experience could not create geometry the same is not true of ancestral experience. But what does that mean? Is it meant that we could not experimentally demonstrate Euclid's postulate, but that our ancestors have been able to do it?" (Poincaré 2015, 91).

This is absurd, because it would imply that our earliest human ancestors had intellectual powers that we no longer have.

22.4 Knowledge as Natural Phenomenon Explained within Science

Another view of naturalization is that it consists in considering knowledge as a natural phenomenon explained within the present sciences. The only point of view the naturalist philosopher can offer is the point of view of the present sciences. The task of the naturalist philosopher is to improve, clarify, and understand the world theory of the present sciences from within.

Thus, Quine says that naturalization consists in considering knowledge "a natural phenomenon" (Quine 1969, 83). Therefore, "epistemology is best looked upon" as "an enterprise within natural science" (Quine 1975, 68). In fact, "the only point of view" the naturalist philosopher "can offer" is "the point of view of our own science" (Quine 1981, 181). That is, the point of view of the present sciences. The naturalist philosopher "begins his reasoning within the inherited world theory," that is, within the world theory of the present sciences, and "tries to improve, clarify, and understand the system from within" (ibid., 72).

But this view of naturalization is not valid. For, it is limiting to say that the only point of view the naturalist philosopher can offer is the point of view of the present sciences. Indeed, philosophy is concerned with problems that are beyond the present sciences, so the solutions to them cannot be based only on the present sciences. A philosophy dealing with problems that could be solved on the basis only of the present sciences would be reduced to recording the pronouncements of the scientists.

Moreover, it is unrealistic to say that the task of the naturalist philosopher is to improve, clarify, and understand the world theory of the present sciences from within. For, improving, clarifying, and understanding the world theory of the present sciences is an integral part of the scientists' work, and scientists are much more competent to the task than philosophers, who do not have the necessary qualifications.

22.5 Knowledge as Self-Sustaining

Another view of naturalization is that it consists in considering knowledge as self-sustaining, since it is based on a scientific or mathematical practice that is not answerable to the epistemic standards of any external standpoint. Therefore, considerations external to scientific or mathematical practice cannot criticize or justify those practices.

Thus, Maddy says that naturalization consists in considering knowledge as self-sustaining, since it is based on a scientific or mathematical practice that is "not answerable to the epistemic standards of any external" standpoint, "legitimate criticisms of scientific" or mathematical practice "can only come from within scientific" or mathematical "practice itself" (Maddy 1998, 136). Therefore, "considerations external to scientific or mathematical practice cannot justify those practices any more properly than they can criticize them" (ibid.).

But this view of naturalization is not valid. For, in the same way one could say that naturalization consists in considering knowledge as self-sustaining, since it is based on an astrological practice that is not answerable to the epistemic standards of any external standpoint. Therefore, considerations external to astrological practice cannot criticize or justify that practice.

Against this, Maddy argues that, for example, mathematics is essentially different from astrology, because "mathematics is staggeringly useful, seemingly indispensable, to the practice of natural science, while astrology is not," so "there is a strong motivation for the scientific naturalist to give an acceptable account of mathematics with no parallel in the case of astrology" (Maddy 1997, 204–205).

But this argument is based on an evaluation of mathematics from the standpoint of its usefulness to natural science, so from an external standpoint, contradicting the assumption that considerations external to mathematical practice cannot justify that practice.

Maddy also argues: "Suppose mathematicians decided to reject the old maxim against inconsistency—so that both '2 + 2 = 4' and '2 + 2 = 5' could be accepted—on the grounds that this would have a sociological benefit for the self-esteem of school children" (ibid., 198, footnote). If mathematicians insisted that they were pursuing a legitimate goal and "that this goal overrides the various traditional goals, I find nothing in the mathematical naturalism presented here that provides grounds for protest" (ibid.).

But this decision by mathematicians would be based on an evaluation of mathematics from the standpoint of the sociological benefit for the self-esteem of school children, so from an external standpoint, again contradicting the assumption that considerations external to mathematical practice cannot justify that practice.

22.6 Knowledge as Life Function Based on the Analytic Method

Since the above views of naturalization are not valid, an alternative view is needed.

An alternative view is provided by what has been said about knowledge in Chaps. 10 and 18.

In Chap. 10 it has been said that knowledge is a life function that serves to solve problems of biological life and, in the case of human beings, also problems of cultural life.

In Chap. 18 it has been said that knowledge is problem solving by the analytic method.

Then, naturalization consists in considering knowledge as a life function that serves to solve problems of biological life and, in the case of human beings, also problems of cultural life, by the analytic method.

In what follows, 'naturalization' will mean this view of naturalization.

22.7 The Argument of Conflict with Human Essence

Several arguments can be advanced against naturalization.

An argument against naturalization is that it conflicts with human essence, which consists in language.

Thus, Heidegger argues that "the fact that physiology and physiological chemistry can scientifically investigate the human being as an organism is no proof that in this 'organic' thing, that is, in the body scientifically explained, the essence of the human being consists" (Heidegger 1998, 247). Just "as little as the essence of the human being consists in being an animal organism can this insufficient definition of the essence of the human being be overcome or offset by outfitting the human being with an immortal soul, the power of reason, or the character of a person" (ibid.). In fact, there is an essential difference between human beings and animals and plants, that is, animals and plants "lack language" (ibid., 248). And "in its essence, language is not the utterance of an organism; nor is it the expression of a living thing. Nor can it ever be thought in an essentially correct way in terms of its symbolic character" (ibid., 248–249). The "human being is not only a living creature who possesses language along with other capacities. Rather, language is the house of being in which the human being ek-sists by dwelling, in that he belongs to the truth of being, guarding it" (ibid., 254). Thus, language is "the home of the human being's essence" (ibid.).

But this argument is not valid. As Pinker says, "language is no more a cultural invention than is upright posture," in fact, "it is a distinct piece of the biological makeup of our brains," human beings "know how to talk in more or less the sense that spiders know how to spin webs" (Pinker 1995, 18). Language reflects the biological makeup of human beings, just like the spider web reflects the biological makeup of spiders, and both language and the spider web serve biological purposes. Thus, language is not "the ineffable essence of human uniqueness," but "a biological adaptation to communicate information" (ibid., 19). It is "an evolutionary adaptation, like the eye, its major parts designed to carry out important functions" (ibid., 24).

Therefore, it is not valid to say that human essence consists in language.

22.8 The Argument of Conflict with Rational Thought

Another argument against naturalization is that naturalization cannot account for the emergence of rational thought.

Thus, Lewis argues that, according to naturalization, "the type of mental behaviour we now call rational thinking" must "have been 'evolved' by natural selection, by the gradual weeding out of types less fitted to survive. Once, then, our thoughts were not rational," those "which had a cause external to ourselves at all were (like our pains) responses to stimuli" (Lewis 1947, 28). Reason and rational thinking cannot have emerged by natural selection. For, "natural selection could operate only by eliminating responses that were biologically hurtful and multiplying those which tended to survival," and "it is not conceivable that any improvement of responses could ever turn them into acts of insight, or even remotely tend to do so" (ibid., 29). For, "the relation between response and stimulus is utterly different from that between knowledge and the truth known," so an improvement of responses could not lead "to their conversion into valid inferences" (ibid.). Thus, naturalization cannot account for the emergence of rational thought. Therefore, naturalization "discredits our processes of reasoning or at least reduces their credit to such a humble level that it can no longer support" naturalization "itself" (ibid., 40)

But this argument is not valid. According to it, reason cannot have emerged by natural selection. On the contrary, as Norman says, "reason is one of the most powerful and interesting traits natural selection has yet devised" (Norman 2016, 703). In particular, "reason emerged and persisted primarily because it helped our ancestors" to "resolve conflicts rooted in discrepant states of mind. We use reason" mostly to align "other people's mental states "with our own, and sometimes to align them with circumstances in the world" (ibid., 703–704). But although "we sometimes argue so as" to "advantage ourselves at the expense of those we argue with," more often "we reason in ways that are mutually advantageous," and "this must be so" (ibid.).

Therefore, it is not valid to say that naturalization cannot account for the emergence of rational thought.

22.9 The Argument of Conflict with Science

Another argument against naturalization is that, according to it and evolution, the principal function or purpose of our cognitive faculties is not that of producing true beliefs, but instead that of contributing to survival. So, there is no assurance that they produce true knowledge of the world. Therefore, there is a conflict between naturalization and science. There is assurance that our cognitive faculties produce true knowledge of the world only if we grant that God creates human beings in his image.

Thus, Plantinga argues that, according to naturalization and evolution, the "principal function or purpose" of "our cognitive faculties is not that of producing true" beliefs, "but instead that of "contributing to survival" (Plantinga 2011, 315–316). So, "if I believe both" naturalization "and evolution, I have a defeater for my intuitive assumption that my cognitive faculties are reliable" (ibid., 314). Then, "I have a defeater for any belief I take to be produced by my cognitive faculties. That means that I have a defeater for my belief that" naturalization "and evolution are true" (ibid.). Therefore, "I cannot rationally accept it. And if one can't accept both" naturalization "and evolution, that pillar of current science, then there is a serious conflict between" naturalization "and science" (ibid.). There is assurance that our cognitive faculties produce true knowledge of the world only "from the perspective of theism," according to which "God creates human beings in his image, a crucial component of which is the ability to know worthwhile and important things about our world" (ibid., 285).

But this argument is not valid. For, the argument assumes that the aim of science is truth. However, this assumption is unjustified because, as argued in Chap. 11, the aim of science is not truth but plausibility. Our cognitive faculties contribute to survival by producing hypotheses about the world, and are reliable because the hypotheses they produce are plausible. So, there is no conflict between naturalization and science. A conflict arises only if one mistakenly assumes that the aim of science is truth instead of plausibility.

The argument also assumes that there is assurance that our cognitive faculties produce true knowledge of the world only if we grant that God creates human beings in his image. But this assumption is unjustified. Even if we grant that God creates human beings, God certainly does not create them in his image. For, God is supposed to be immortal, omnipotent and omniscient, while human beings are neither immortal, omnipotent, nor omniscient. Therefore, even if we grant that God creates human beings, this in no way implies that human beings should have the ability to know worthwhile and important things about our world.

22.10 The Argument of Conflict with the Method of Science

Another argument against naturalization is that, according to it, a naturalistic methodology is an inductive theory of science, but this gives rise to inconsistencies. Therefore, naturalization conflicts with the method of science. Scientific theories are not the result of inductions, they are not discovered by rational processes, therefore we must admit a certain priority of irrationalism.

Thus, Popper argues that naturalization implies that "a naturalistic methodology" is an "inductive theory of science" (Popper 2005, 31). We "may consider and compare two different systems of methodological rules; one with, and one without, a principle of induction. And we may then examine whether such a principle, once introduced, can be applied without giving rise to inconsistencies; whether it helps us" (ibid.). From this examination it is clear that the principle of induction "does not help us; and that it even gives rise to inconsistencies. Thus I reject the naturalistic view. It is uncritical" (ibid.). In fact, "scientific theories" are not the result of inductions, they are "inventions" (Popper 2002, 61). They are not discovered by rational processes, we must admit "a certain priority of irrationalism" (Popper 2013, 437).

But this argument is not valid. Scientific theories are discovered by the analytic method, in which induction is one of the non-deductive rules by which hypotheses are obtained. Then hypotheses are subjected to the process of establishing whether they are plausible and are accepted only if they are plausible. Therefore induction, and non-deductive rules in general, do not give rise to inconsistencies. Moreover, the analytic method is a rational method, so there is no need to admit a certain priority of irrationalism.

22.11 The Argument of Conflict with Mathematical Truth

Another argument against naturalization is that, according to it, mathematical truths are not truths about abstract entities but truths about material entities, so truths about the world. However, truths about negative numbers or complex numbers are not truths about the world. So, there is a conflict between naturalization and mathematical truth.

Thus, Brown argues that, according to naturalization, "there is no other avenue to learn anything about reality except through the natural sciences," but "this makes mathematics a real problem" (Brown 2021, 79). For, then "mathematical facts, like two plus three equals five, are actually not abstract truths" but "truths about the world. It just means that combining any two objects with three objects gives you five material objects" (ibid., 80). But "even if you can make that work for very low-level arithmetic of small numbers, you run into serious problems as soon as you start talking about negative numbers or complex numbers or something like that" (ibid.).

But this argument is not valid. For, it assumes that naturalization implies that mathematical objects are material entities, and mathematical truths are truths about material entities. But this is unwarranted. As argued in Chap. 18, mathematical objects are hypotheses made to solve mathematical problems by the analytic method, and mathematical results are solutions to problems obtained by the analytic method, so they are not true but plausible.

22.12 The Argument of Conflict with the Method of Mathematics

Another argument against naturalization is that, according to it, the method of science is the hypothetico-deductive method, which consists in formulating hypotheses, deducing predictions from them and testing their predictions against experience. On the contrary, mathematics does not involve experience in any way. Therefore, there is a conflict between naturalization and mathematics.

Thus, Williamson argues that, according to naturalization, the scientific method consists in "formulating theoretical hypotheses and testing their predictions against systematic observation and controlled experiment. This is the hypothetico-deductive method" (Williamson 2014, 30). Then one challenge to naturalization "is to find a place for mathematics," because, "natural sciences rely on it," but this raises the problem of whether we should count mathematics "a science in its own right" (ibid.). Now, on the one hand, if we count pure mathematics as science, then the view of naturalization that the scientific method is the hypothetico-deductive method "is wrong, for it does not fit the science of mathematics, which proves its results by pure reasoning, rather than the hypothetico-deductive method" (ibid.). Indeed, "when mathematicians assess a proposed new axiom, they look at its consequences within mathematics, not outside" (ibid.). Thus, they do not test its consequences against experience. On the other hand, "if we do not count pure mathematics as science, we thereby exclude mathematical proof by itself from the scientific method" (ibid.). But this discredits naturalization, because naturalization "privileges the scientific method over all others, and mathematics is one of the most spectacular success stories in the history of human knowledge" (ibid.).

But this argument is not valid. For, it assumes that mathematics proves its results by pure reasoning, thus mathematics does not involve experience in any way. This assumption is not valid, because most mathematical problems arise from the external world and hence from experience, and are solved using hypotheses that are accepted because they are plausible, that is, such that the arguments for them are stronger than the arguments against them, on the basis of experience (see Cellucci 2025).

22.13 The Argument of Conflict with Incompleteness Theorems

Another argument against naturalization is that, according to it, the working of the human mind can be reduced to the working of the brain. But, by the incompleteness theorems, the working of the human mind cannot be reduced to the working of the brain. Therefore, there is a conflict between naturalization and the incompleteness theorems.

Thus, Gödel argues that, by the incompleteness theorems, "either the human mind surpasses all machines," or "else there exist number-theoretical questions undecidable for the human mind" (Wang 1996, 185). But the second alternative is unacceptable because "if it were true it would mean that human reason is utterly irrational in asking questions it cannot answer, while asserting emphatically that only reason can answer them. Human reason would then be very imperfect and, in some sense, even inconsistent" (ibid., 317). Therefore, we must conclude that the human mind surpasses all machines. This implies that "the working of the human mind cannot be reduced to the working of the brain, which to all appearance is a finite machine with a finite number of parts, namely, the neurons and their connections" (ibid., 186). This "is opposed to the materialist philosophy" (ibid., 187). Thus, it is opposed to naturalization.

But this argument is not valid. It depends on the assumption that, according to naturalization, the working of the human mind can be reduced to the working of the brain. But this assumption is not valid because, as argued in Chap. 13, the mind consists of certain faculties of the body which do not reduce to the brain, and are implemented not only by processes internal to the body, but also by processes external to the body. Therefore, the working of the human mind cannot be reduced to the working of the brain. Only the Neanderthals and archaic Homo sapiens had a mental activity that, compared to ours, appears to have been substantially more reliant on their brains alone.

That the working of the human mind cannot be reduced to the working of the brain is not in conflict with naturalization. For, the faculties of which the human mind consists are faculties of the body.

22.14 The Argument of Conflict with Applicability of Mathematics

Another argument against naturalization is that it implies that the human race is in no way privileged. On the contrary, the applicability of mathematics to the world can be accounted for only by recognizing that the human race is in some way privileged, as there is a correspondence between mind and the physical world as a whole. Therefore, there is a conflict between naturalization and the applicability of mathematics.

Thus, Steiner argues that naturalization is in "opposition to anthropocentrism—the teaching that the human race is in some way privileged, central to the scheme of things" (Steiner 1998, 55). But the applicability of mathematics to the world can be accounted for only by recognizing that the human race is in some way privileged, as there is a "correspondence" between mind "and the physical world as a whole" (ibid., 176). It is this that makes the universe "an intellectually 'user friendly' universe, a universe which allows our species to discover things about it" (ibid., 8). The applicability of mathematics to the world depends on this "relation between mind and the cosmos" (ibid., 45).

But this argument is not valid. It is unwarranted to say that the applicability of mathematics to the world can be accounted for only by recognizing that the human race is in some way privileged, as there is a correspondence between mind and the physical world as a whole. Mathematics is not applicable to the world because there is such correspondence, it is applicable to it because of Galileo's revolution according to which, the object of science is not to know the essence of things, but only to know some phenomenal properties of things, mathematical in kind. (For more on the applicability of mathematics, see Cellucci 2022, Chapter 16).

22.15 The Argument of Conflict with Teleological Principles

Another argument against naturalization is that, according to it, everything that exists and everything that happens must be explained by physical laws. But this conflicts with the fact that teleological principles are at work in the history of nature, and these principles are essential for the intelligibility of the world. In particular, a theory of teleology must show how our lives are part of the process of the universe gradually waking up and becoming aware of itself.

Thus, Nagel argues that, according to naturalization, "there is ultimately one way that the natural order is intelligible, namely, through physical law—everything that exists and everything that happens can in principle be explained by the laws that govern the physical universe" (Nagel 2012, 18). But this assumption is contradicted by the fact that "principles of a different kind are also at work in the history of nature," namely principles that are "teleological rather than mechanistic" (ibid., 7). These principles are essential for "the intelligibility of the world" (ibid., 17). In particular, a theory of teleology must show how "each of our lives is part of the lengthy process of the universe gradually waking up and becoming aware of itself" (ibid., 85).

But this argument is not valid. For, there is no evidence that teleological principles are at work in the history of nature, let alone that our lives are part of the process of the universe gradually waking up and becoming aware of itself.

22.16 Confirming the Proposed View of Naturalization

Since the arguments against naturalization are not valid, it seems fair to conclude that naturalization consists in considering knowledge as a life function that serves to solve problems of biological life and, in the case of human beings, also of cultural life, by the analytic method.

This gives substance to the concept of naturalization. At the beginning of this chapter it has been said that naturalization is any position that involves that knowledge is a natural process. But, if saying this must not remain a generic statement of principle, it is necessary to specify in what sense knowledge is a natural process.

This is obtained by saying that knowledge is a life function that serves to solve problems of biological life and, in the case of human beings, also of cultural life, by the analytic method.

This concept of naturalization encompasses the main characters of knowledge that have been presented in Part II of this book.

References

Bacon, Francis. 1857–1874. *Works*. London: Longmans.

Brown, James. 2021. Plato's heaven: A user's guide. In *Conversations about philosophy*, ed. Howard Burton, vol. 1, 63–117. Toronto: Open Agenda Publishing.

Cellucci, Carlo. 2022. *The making of mathematics: Heuristic philosophy of mathematics*. Cham: Springer.

Cellucci, Carlo. 2025. Mathematics and experience. *Foundations of Science* 30:473–487.

Clark, Kelly. 2016. *The Blackwell companion to naturalism*. Hoboken: Wiley-Blackwell.

Dewey, John. 1938. *Logic: The theory of inquiry*. New York: Holt, Rinehart and Winston.

Heidegger, Martin. 1998. *Pathmarks*. Cambridge: Cambridge University Press.

Lewis, Clive. 1947. *Miracles: A preliminary study*. London: Geoffrey Bles.

Lorenz, Konrad. 1977. *Behind the mirror*. London: Methuen.

Lorenz, Konrad. 1997. *The natural science of the human species: An introduction to comparative behavioral research: The 'Russian manuscript' (1944–1948)*. Cambridge: The MIT Press.

Lorenz, Konrad. 2009. *Kant's doctrine* of the a priori in the light of contemporary biology. In *Philosophy after Darwin: Classic and contemporary readings*, ed. Michael Ruse, 231–247. Princeton: Princeton University Press.

Maddy, Penelope. 1997. *Naturalism in mathematics*. Oxford: Oxford University Press.

Maddy, Penelope. 1998. V=L and maximize. In *Logic colloquium 1995*, ed. Johann Makowsky and Elena Ravve, 134–152. Berlin: Springer.

Nagel, Thomas. 2012. *Mind and cosmos: Why the materialist neo-darwinian conception of nature is almost certainly false*. Oxford: Oxford University Press.

Norman, Andy. 2016. Why we reason: Intention-alignment and the genesis of human rationality. *Biology and Philosophy* 31:685–704.

Pinker, Steven. 1995. *The language instinct*. New York: Harper-Collins.

Plantinga, Alvin. 2011. *Where the conflict really lies: Science, religion, and naturalism*. Oxford: Oxford University Press.

Poincaré, Henri. 2015. *The foundations of science: Science and hypothesis—The value of science—Science and method*. Cambridge: Cambridge University Press.

Popper, Karl. 2002. *Conjectures and refutations: The growth of scientific knowledge*. London: Routledge.
Popper, Karl. 2005. *The logic of scientific discovery*. London: Routledge.
Popper, Karl. 2013. *The open society and its enemies*. Princeton: Princeton University Press.
Quine, Willard. 1969. *Ontological relativity and other essays*. New York: Columbia University Press.
Quine, Willard. 1975. The nature of natural knowledge. In *Mind and language*, ed. Samuel Guttenplan, 67–81. Oxford: Oxford University Press.
Quine, Willard. 1981. *Theories and things*. Cambridge: Harvard University Press.
Steiner, Mark. 1998. *The applicability of mathematics as a philosophical problem*. Cambridge: Harvard University Press.
Sterpetti, Fabio. 2021. *La logica del naturalismo: Metafilosofia e filosofia della scienza alla luce della sfida naturalista*. Milan: Mimesis.
Wang, Hao. 1996. *A logical journey: From Gödel to philosophy*. Cambridge: The MIT Press.
Williamson, Timothy. 2014. What is naturalism? In *The armchair or the laboratory?* ed. Matthew Haug, 29–31. London: Routledge.

Part III
Conclusion

Chapter 23
Concluding Remarks

Abstract This chapter recapitulates and makes some concluding remarks on the main thesis of the book, that philosophy can be useful again only if it is acquisition of knowledge, that is, an inquiry aimed at acquiring knowledge of the world. Abandoning the view that philosophy is acquisition of knowledge has led to the present irrelevance of philosophy.

Keywords Philosophy as acquisition of knowledge · Questions beyond the present sciences · The heuristic view of philosophy · Perspectives for the future of philosophy

23.1 Main Thesis of the Book

The great developments in the sciences since the nineteenth century have created the need to find a new role for philosophy. However, the attempts in this direction have failed.

This raises the question: Can philosophy be useful again, and what type of philosophy can be such?

The main thesis of the book is that philosophy can be useful again if and only if it is acquisition of knowledge, that is, an inquiry aimed at acquiring knowledge of the world.

This thesis may seem peculiar because the great developments in the sciences since the nineteenth century have led to the belief that only the present sciences can be acquisition of knowledge.

But the thesis is not peculiar because, although the present sciences answer many questions about the world, they do not answer many other questions about the world, questions that are beyond the present sciences, in the sense that they are not open questions of any of the present sciences.

Only philosophy can seek and address these questions. For, scientists today are busy tackling major challenges in their own fields, so they cannot be expected to seek and address questions beyond their fields.

C. Cellucci, *A Philosophy for the Future*, Synthese Library 508,
https://doi.org/10.1007/978-3-031-98443-3_23

This requires a fundamental change with respect to the dominant philosophy today, analytic philosophy, which assumes that philosophy is not and should not be acquisition of knowledge, that is, an inquiry aimed at acquiring knowledge of the world. Acquiring knowledge of the world is the aim of science, not of philosophy, philosophy aims only to understand what we already know.

This assumption has had a negative effect on philosophy. Philosophy has become more and more inward-looking and self-referential, of no interest to people working in other fields and to cultured people at large. As a result, philosophy has never been considered so irrelevant.

Only if philosophy is acquisition of knowledge, philosophy can escape from irrelevance.

23.2 The Heuristic View of Philosophy

Developing the view that philosophy is acquisition of knowledge requires answering several questions about knowledge.

A particularly important question is whether there is a method for acquiring knowledge, that is, a method of discovery.

The book argues that there is such a method, that is, the analytic method. It consists of non-deductive rules for finding hypotheses, deductive rules for deducing solutions to problems from hypotheses, and a procedure for establishing that the hypotheses are plausible, that is, the arguments for them are stronger than the arguments against them, on the basis of experience.

The analytic method is the method of mathematics, science, and knowledge in general. So, if philosophy is acquisition of knowledge, then the analytic method is the method of philosophy.

Since the analytic method is a method of discovery, the view that philosophy is acquisition of knowledge can also be called the heuristic view of philosophy.

23.3 Perspectives for the Future of Philosophy

That the analytic method is a method of discovery is the basis for asserting that, if philosophy is acquisition of knowledge, then it can seek and address questions about the world that are beyond the present sciences, and, if successful, it can open new ways to knowledge and lead to the birth of new sciences.

The view that philosophy is acquisition of knowledge is in keeping with the developments of philosophy since the seventeenth century.

In fact, philosophy has played an essential role in the birth of modern science, and subsequently has played a role in the birth of new sciences. It has also done so in the past century, when it has led to the birth of, for example, computer science, cognitive science, artificial intelligence, and linguistics.

There is no reason to suppose that new sciences will not arise in the future and that philosophy will not lead to their birth. Let philosophy try.

Index

C. Cellucci, *A Philosophy for the Future*, Synthese Library 508,
https://doi.org/10.1007/978-3-031-98443-3

MIX
Papier aus verantwortungsvollen Quellen
Paper from responsible sources
FSC® C105338
FSC
www.fsc.org

If you have any concerns about our products,
you can contact us on
ProductSafety@springernature.com

In case Publisher is established outside the EU,
the EU authorized representative is:
Springer Nature Customer Service Center GmbH
Europaplatz 3, 69115 Heidelberg, Germany

Printed by Libri Plureos GmbH
in Hamburg, Germany